# A Professional G̶

# Room Divisio̶ ̶ ̶ ̶ ̶ ̶ ̶ ̶ ̶ ̶ ̶ ̶ ̶S

**Manoj Kumar Yadav**

*Assistant Lecturer*
Institute of Hotel Management,
Catering Technology & Applied Nutrition
Gwalior (MP)

## I.K. International Pvt. Ltd.

NEW DELHI

*Published by*

I.K. International Pvt. Ltd.
4435-36/7, Ansari Road, Daryaganj
New Delhi–110 002 (India)
E-mail: info@ikinternational.com
Website: www.techsarworld.com

ISBN: 978-93-85909-93-1

Published/Printed by Krishan Makhijani for I.K. International Pvt. Ltd., 4435-36/7, Ansari Road, Daryaganj, New Delhi–110 002.

# Preface

The origin of this book lay in the lack of any one convenient source to which hospitality students and practitioners could refer in seeking basic, management knowledge and information on industry. This professional guide is the compilation of material from various sources – electronic, books, magazines, electronic journals and notes, open library, literature and so forth.

A comprehensive detail of hospitality industry operating in India as well as in abroad has been provided, along with a brief description of their definitions, origins, nature and functions. Additionally, most importantly, it identifies and defines many of the most commonly used terms and concepts in hospitality field, especially in room division.

Hospitality education has undergone a drastic series of changes over the past years. It demands new approaches to teaching and new directions for learning. "A Professional Guide to Room Division Operations" is a practical-cum- theoretical hand on text to help you become more efficient and effective professional. Students will embark on a journey to learn more innovative hospitality methods and their application.

This professional guide is written with the needs of the following broad groups in mind:

- Students undertaking various hospitality programmes at different hospitality educational institutions.
- Academicians and teachers involved in the field of hospitality industry.
- Consultants to hospitality industry.
- Industry observers or other interested parties.

The professional guide is divided into two parts:

- Front Office Operations (Part 1)
- Housekeeping Operations (Part 2)

In addition to the above two sections, almost all chapters are powered with a range of chapter activities and quick scroll boxes for the purpose of exercise and instant glance in reference to always tinted texts.

**Manoj Kumar Yadav**

# Acknowledgements

I would like to give sincere thanks to all those professionals and individuals who have given me assistance and support in the preparation of this guide. The completion of this book has drawn upon a variety of experience and literature. I also want to owe my sincere gratitude to the following people who gave their precious time for my book and assisted in its completion through their knowledge, experience and guidance:

- First, I give sincere gratitude to my institute management that it always support and inspire me for self-developmental activities.
- Thanks to my all students who always raise excellent queries which force me to study as more as I can. My students will always be motivator for me.

After all, special thanks to my god (Shri Hanuman Ji), my parents (Father – R. S. Yadav – always inspire me to move forward, think bigger, do right things in life and always help others) and my wife Poonam. Last but not the least, my good luck my two cute daughters – Aastha Yadav and Yashika Yadav because I have taken out time from their schedule for developing this book.

**Manoj Kumar Yadav**

## Dear Readers

Welcome to the world of Hospitality Industry

- An industry which always gives you opportunity to grow.
- An industry which gives you opportunity to see the world.
- An industry which always supports nation's wealth.
- An industry which always gives you chance to learn culture of others.
- An industry which gives two basic necessity of life – food and boarding as well as money.
- An industry which always gives you chance to meet with others and show your skill and talent.
- An industry which gives you career building opportunity in almost every public relation organizations.
- An industry in which you can meet renowned personalities of the world.
- An industry which shows you what is luxury and what could be luxurious in life.
- An industry which supports tourism and transport sector.

# Contents

*Preface* .................................................................................................... *v*

*Acknowledgements* ................................................................................ *vii*

## PART A   Front Office Operations

### 1. Introduction ................................................................................. 3

| | | |
|---|---|---|
| 1.1 | Hotel or Inn | 3 |
| 1.2 | Role and Importance of Tourism Industry | 4 |
| 1.3 | Classification of Hotel | 6 |
| 1.4 | Standard Organization Structure of Hotel | 16 |

### 2. Room Division ............................................................................ 22

| | | |
|---|---|---|
| 2.1 | Room Division | 22 |
| 2.2 | Sections of Room Division | 23 |
| 2.3 | Other Major Departments of a Hotel | 28 |
| 2.4 | Supportive Departments of Hotel | 29 |

### 3. Front Office ................................................................................. 33

| | | |
|---|---|---|
| 3.1 | Front Office | 33 |
| 3.2 | Organizational Structure | 33 |
| 3.3 | Functions of Front Office | 34 |
| 3.4 | Layout of Front Office/Hotel Entrance | 35 |
| 3.5 | Front Office Equipment | 36 |
| 3.6 | Front Office Sections | 37 |
| 3.7 | Duties and Responsibilities of Front Office Staff | 42 |

## 4. Hotel Revenue Sources and Guest Segmentation ........52

4.1  Revenue Source    52
4.2  Guest    52
4.3  Guest (or Market) Segmentation    53
4.4  Revenue Sources    53
4.5  Segmentation of Guest    54

## 5. Prologue with Rooms ........62

5.1  Room    62
5.2  Types of Rooms    63
5.3  Additional Room Types    67

## 6. Room Tariff ........73

6.1  Room Tariff    73
6.2  Basis for Room Tariff Fixation    74
6.3  Methods of Determining Room Rates    78
6.4  Room Rate Types    80

## 7. Reception Section ........88

7.1  Reception Section    88
7.2  Functions of Reception Section    89
7.3  Role of Reception Section in Different Stages of Guest Cycle    90
7.4  Preparation for Guest Check-in    91
7.5  Guest Check-in Procedure    93
7.6  Guest Check-out Procedure (At Reception Desk)    96
7.7  Other Tasks    98

## 8. Cash Section ........104

8.1  Cash/Cashier Section    104
8.2  Functions of Cash Section    106
8.3  Role of Cashier Section in Different Stages of Guest Cycle    107
8.4  Opening and Settlement of Account/Bill    108
8.5  Account/Bill Settlement Modes    111
8.6  Guest Check-out Procedure (at Cashier Desk)    118
8.7  Modes of Check-out    123
8.8  Processes of Guest Check-out    124
8.9  Standard Conversation between Checking-out Guest and Cashier    125

## 9. Telephone Section .........................................................131

9.1 Telephone Section 131

9.2 Functions of Telephone Section 132

9.3 Role of Telephone Section in Different Stages of Guest Cycle 133

9.4 Wake-up Call Procedure 134

9.5 Telephone Protocol 135

9.6 Standard Phrases for Conversation between Telephone
Operator and Guest 139

## 10. Reservation Section.......................................................141

10.1 Reservation Section 141

10.2 Functions of Reservation Section 142

10.3 Role of Reservation Section in Different Stages of Guest Cycle 143

10.4 Process of Room Reservation 143

10.5 Preparation for Guest Reservation 145

10.6 Types of Room Reservation 148

10.7 Tools for Determining Room Availability 150

10.8 Systems of Reservation 153

10.9 Sources of Reservation 157

10.10 Overbooking 158

10.11 Standard Conversation between Prospective Guest and
Reservation Desk/Agent 159

## 11. Bell Desk .......................................................................165

11.1 Bell Desk 165

11.2 Functions of Bell Desk 166

11.3 Role of Bell Desk Section in Different Stages of Guest Cycle 167

11.4 Task Force of Bell Desk 167

11.5 Procedure for Guest Luggage Handling 168

11.6 Guest Check-out Procedure at Bell Desk 171

11.7 Standard Phrases for Conversation between the Bell Hop and Guest 174

## 12. Supportive Centres .......................................................177

12.1 Supportive Centre 177

12.2 Uniformed Service 178

12.3 Guest Relation Desk 179

12.4 Concierge Desk 181

12.5   Business Centre                                                              183
12.6   Spa                                                                         184
12.7   Recreational Activities                                                      185
12.8   Game Zone                                                                   185

## 13. Mail and Message Section ....................................................189

13.1   Mail and Message Section                                                     189
13.2   Functions of Mail and Message Section                                        190
13.3   Types of Mails and Messages                                                  190

## 14. Handling Situations ...........................................................195

14.1   Situation Handling                                                           195
14.2   Usual Situation Handling                                                     196
14.3   Unusual Situation Handling                                                   210
14.4   Guest Complaints and Their Types                                             215

## 15. Guest Registration..............................................................220

15.1   Guest Registration                                                           220
15.2   Concept of Registration                                                      222
15.3   Methods of Guest Registration                                                227
15.4   Procedure for Check-in                                                       229

## 16. Guest Cycle........................................................................240

16.1   Guest Cycle                                                                  240
16.2   Stages of Guest Cycle                                                        241
16.3   Stages of Guest Cycle (Along with Various Activities that
       Associate with Each Stage)                                                   242
16.4   Guest Cycle Activities Under Different Operating Modes                       250
16.5   Task Force in Different Stages of Guest Cycle at Front Desk                   254

## PART B   Housekeeping Operations

## 17. Role of Housekeeping .........................................................263

17.1   Housekeeping                                                                 263
17.2   Functions of Housekeeping Department                                         264
17.3   Housekeeping Equipment                                                       264
17.4   Guest Room Amenities and Supplies                                            266

17.5  Layout of Housekeeping Department     269
17.6  Structural Foundation of Housekeeping Department     269

## 18. Housekeeping Department: Organizational Structure .......... 278

18.1  Organizational Structure     278
18.2  Organizational Structure of Housekeeping Department     279
18.3  Duties and Responsibilities of Housekeeping Staff     282

## 19. Linen and Uniform Room ......................................... 290

19.1  Linen     290
19.2  Uniform     291
19.3  Functions of Linen Room     292
19.4  Layout of Linen Room and Linen Exchange Cycle     296
19.5  Bed-Linen or Bedroom Linen     297
19.6  Bathroom Linen     301
19.7  Restaurant Linen     303
19.8  Uniform     307

## 20. Cleaning and Polishing ........................................... 311

20.1  Cleaning and Polishing     311
20.2  Cleaning Process     311
20.3  Cleaning Schedule/Frequency     312
20.4  Cleaning Agents     316
20.5  Cleaning Methods     326
20.6  Cleaning Areas     327
20.7  Polishing     329

## 21. Guest Room Cleaning, Inspection and In-Room Services .......... 336

21.1  Guest Room Cleaning     336
21.2  Room Cleaning Procedure     337
21.3  Bed Making     340
21.4  Bathroom Cleaning     342
21.5  Holistic Cleaning     343
21.6  Turndown or Evening/Night Service     347
21.7  Offer Complementary Items and Supplies in VIP Rooms     348

## 22. Public Area Cleaning and Inspection........................................352

22.1   Public Area Cleaning                                              352
22.2   Care and Cleaning of Housekeeping Equipment                       360
22.3   Cleaning of Fixtures and Fittings                                 362
22.4   Cleaning of Metal Surface                                         362
22.5   Care and Cleaning of Sundry Things                                364

## 23. Laundry Services ....................................................................368

23.1   Laundry Service                                                   368
23.2   Laundry Options                                                   369
23.3   Laundry Supporters                                                372
23.4   Laundry Operation Cycle                                           376
23.5   Laundry Process/Laundry Cycle                                     380
23.6   Dry Cleaning                                                      383

## 24. Fabrics and Fibres....................................................................386

24.1   Fabric                                                            386
24.2   Fibre                                                             386
24.3   Natural Fibre                                                     387
24.4   Artificial Fibre                                                  388
24.5   Fabric                                                            393
24.6   Stain Removal                                                     394

## 25. Interior Decoration..................................................................402

25.1   Interior Decoration                                               402
25.2   Flower Arrangement                                                403
25.3   Floor Decoration (Rangoli)                                        405
25.4   Hotel Lighting                                                    406
25.5   Colour                                                            410
25.6   Paint and Painting                                               412

## 26. Flooring and Wall Covering .......................................................417

26.1   Flooring                                                          417
26.2   Wall Covering                                                     417
26.3   Types of Flooring                                                 417
26.4   Types of Wall Covering                                            420

## 27. Pest Control and Waste Management ..................................428

27.1 Pest and Pest Control 428
27.2 Control of Rodents 428
27.3 Control of Insects 429
27.4 Control of Wood Rot 432
27.5 Waste Management 433

## 28. Safety and Security ...............................................437

28.1 Safety and Security 437
28.2 Internal Threats 438
28.3 Fire and Bomb Threat 441
28.4 External Threats 443
28.5 Basic First Aid Guidelines 445
28.6 Key and Key Control 447

## 29. Room Selling Techniques ......................................453

29.1 Selling/Room Selling 453
29.2 Seven Ps 454
29.3 Room Selling Strategies 457
29.4 Hotel's Most Effective Selling Tools 459

## 30. Night Auditing ....................................................463

30.1 Night Audit 463
30.2 Functions of Night Auditor 464
30.3 Step-by-Step Process for Night Auditing 465

## 31. Coordination of Front Office with Other Departments ...........470

31.1 Coordination and Cooperation 470
31.2 Front Office Coordination with Interrelated Departments 471
31.3 Coordination with Supportive Departments 471
31.4 Coordination with Intrarelated Sections 473

## Appendices ...............................................................477

## Bibliography ............................................................495

## Index ......................................................................499

# PART A

## Front Office Operations

# Introduction

## OBJECTIVES

*After reading this chapter, students will be able to...*

- define a hotel or inn and its classification;
- discuss types of hotel on various bases;
- understand supplementary accommodation; and
- identify the importance of hotel rating system.

## 1.1 HOTEL OR INN

British law defines hotel or inn as a *"place where a bona fide traveller can receive food and shelter, provided he is in a position to pay for it and is in a fit condition to be received"*. The word *hotel* derives from the French term hôtel (coming from hôte meaning host and represent the old French hostel, from Latin Hospes, Hospitis, a stranger, foreigner, thus a guest), which refers to a French version of a townhouse or any other building that provides care, rather than a place offering accommodation.

The history of the hospitality industry dates all the way back to the Colonial Period in the late 1700s. The modern concept of a hotel derives from 1794, when the *City Hotel* opened in New York City. The building was quite large and had 73 rooms. Hotels took a distinct step in style and class when the *Tremont House* (with 170 rooms) opened in Boston in the year 1829. This hotel was considered by many to be the beginning of what was regarded as the first class service. Many hoteliers also consider Tremont House as a first five-star hotel of the world.

In 1908, the Buffalo Statler was opened which marked the beginning of the modern commercial hotel era. In the 1920s, many famous hotels were opened, including the Waldorf Astoria, New York's Hotel Pennsylvania and the Chicago Hilton & Towers, which was originally named as Stevens.

In the 1950s and 1960s, the practice of franchising appeared within the industry. During the same decade, the American concept of *motel* also began to replace roadside cabins (the first motel was opened in year 1925) as use of the automobile that spread throughout the society in America. The industry as we know it today began to take form in the early 1950s and 1960s, leading the way for growth into the dynamic industry that we know today.

............................................................................

**Activity**

✓ Browse the website and find out the origin of English Inn, American Inn, and Japanese Inn and also trace the features that were first time introduced by the Buffalo Statler.

✓ State five reasons that created base to consider the Tremont House of Boston as a beginning of first class hotel or why many hoteliers consider it the first five star hotel of the world.

............................................................................

## 1.2   ROLE AND IMPORTANCE OF TOURISM INDUSTRY

The hotel (or hospitality) industry is a fraction of one of the oldest and largest industry which is known as the *travel and tourism industry*. The United Nation's World Tourism Organization (UNWTO) defines tourism as *"the activities of people travelling to and staying in places outside their usual environment for not more than one consecutive year for leisure, business or other purposes."* Broadly, tourism can be classified into three groups: inbound, outbound and international tourism. The term *inbound tourism* is concerned with the travel or tourism into a country by foreign nationals whereas *outbound tourism* indicates travel or tourism by the residents of any country into alien country. Tourism is the business of providing and marketing services and facilities for leisure travellers. Thus, the concept of tourism is of direct concern to governments, carriers, lodging, restaurant and entertainment industries, and of indirect concern to virtually every industry and business in the world.

Therefore, the term "tourism industry" relates to all those industries that are involved to enable these activities. The offered products are not only tangible goods but also consist of different services. Typical services are transport and accommodation along with lot of other services that are needed to fulfil the wishes of the travellers and tourists. The term *traveller* is used for every person who travels but the word *tourist* is particularly used for that traveller who starts his journey and returns back to the same point after 24 hours. Remember, every tourist is a traveller but every traveller is not a tourist.

**TOURISM IMPORTANCE**

- Generate foreign currency
- Reduce balance of payments
- Generate employment
- Open business opportunities
- Open career opportunities
- Share cultural heritage
- Support related industries
- Give multiplier income

According to American researcher S. Plog (1974), in widely cited tourist typologies tourist fall along three continua based on their travel preferences, i.e., allo-centric, psycho-centric and mid-centric. The term *allo-centric tourist* indicates those holiday-makers who every time actively seeks out new destinations to travel whereas *psycho-centric* is used to describe those holiday-makers who seek out familiar destinations to travel. Lastly, *mid-centric* tourist falls between allo-centric and psycho-centric tourist who seeks out some familiar and few new destinations.

## Activity

In table below, write down the features of each type of tourist:

| S. No. | Base/Criteria | Allo-centric tourist | Psycho-centric tourist | Mid-centric tourist |
|--------|---------------|----------------------|------------------------|---------------------|
| 1 | | | | |
| 2 | | | | |
| 3 | | | | |
| 4 | | | | |
| 5 | | | | |

There are three important fractions or components of tourism industry; these are accessibility, accommodation and attraction (or AAA). Each fraction reflects different set of industry.

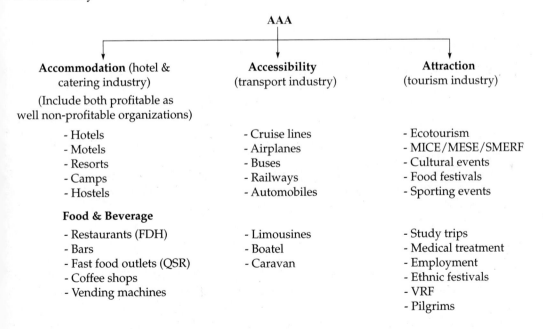

**AAA**

**Accommodation** (hotel & catering industry)
(Include both profitable as well non-profitable organizations)

- Hotels
- Motels
- Resorts
- Camps
- Hostels

**Food & Beverage**

- Restaurants (FDH)
- Bars
- Fast food outlets (QSR)
- Coffee shops
- Vending machines

**Accessibility** (transport industry)

- Cruise lines
- Airplanes
- Buses
- Railways
- Automobiles

- Limousines
- Boatel
- Caravan

**Attraction** (tourism industry)

- Ecotourism
- MICE/MESE/SMERF
- Cultural events
- Food festivals
- Sporting events

- Study trips
- Medical treatment
- Employment
- Ethnic festivals
- VRF
- Pilgrims

## 1.3   CLASSIFICATION OF HOTEL

### Resort Hotels

✓ Hotel sited in exotic location and usually targeted towards family clientele and vacationers. It mainly offers leisure/recreational services to their guests. The food service is predominantly offered through buffet counter.

✓ Resort hotels can be classified into three broad categories, i.e. *sea side resorts* (it mainly offers water sports), *hill resorts* (it mainly offers mountain adventure) and *mega resorts* (it mainly offers indoor/outdoor game zones).

### Motel and Suburban Hotels

✓ *Motels* are situated on interstate highways, away from the rumpus of the city and usually targeted towards travellers. It mainly offers parking facility with limited/ overnight accommodation. Due to its location peculiarity, motels are also known as *highway hotels*.

✓ There are trivial differences between motel and suburban hotel. *Suburban hotels* are located away from city (or at some remote location) but it is not obligatory for these hotels to have parking facility. Consequently, it is also not necessary that suburban hotel should be on interstate highways which is necessary in motel concept.

### Floatel and Boatel Hotels

✓ *Floatel hotels* surf on the surface of the water in international oceans and usually target pleasure travellers. Cruise lines are excellent example of floatel that offer short as well as long duration of journey.

✓ *Boatel hotels* also surf on the surface of the water but in local water bodies like canals. In India, *shikara* is a perfect example of boatel.

## Roatel and Metal Hotels

✓ *Roatel hotel* means hotel moving on wheels and often also known as *mobile hotel*. Straightforwardly, hotels which provide accommodation and other hotel facilities during the journey are called "roatel". Broadly, roatel can be classified as railways, roadways and airways roatel.

✓ *Metal hotels* indicate to totally mechanized hotels where preponderance of services is offered via vending machines like tea, coffee, hot and cold snacks and so forth.

## No-frill Hotels (hotel concept without luxury)

✓ These are the hotels which offer simple rooms, without any provision of meals or drinks during the guests stay. *No-frill* is a term used to describe any service or product for which the non-essential features have been removed to keep the price low.

✓ There are also *no frill airlines* which offer simple cabins without offering any meals or drinks during the journey.

## Garni Hotels (a component of no-frill concept)

✓ It is a European term for a small hotel without dining facilities. These are budget hotels that provide only the facility of boarding and lodging but not (or limited) food & beverage provisions.

## Hostel

✓ Hostel style accommodation is predominantly found in educational bodies like schools, colleges, institutions and universities. It is usually targeted towards the budget travellers (especially students, apprentices), thus generally attracting a *backpacker* or younger clientele.

✓ Bedrooms are usually offered in a dormitory style (or on sharing basis) and bathrooms are also shared, although private bedrooms with attached bathroom are also be available.

## Transient Hotels

✓ The term *transient hotel* indicates the accommodation facility where majority of guests are not permanent residents, like airport hotels which primarily depend on transitory guests/general passengers.

✓ Often, these hotels are also called *airport* or *momentary hotels* in which guest stays for a day or even less. These hotels are situated nearby the airports, therefore usually targeted towards the airlines passengers plus crew members.

| Target market | Primary classification | Duration of stay | Ownership & affiliation | Theme/ environment | Size of property | Geographic location | Standard and level of services | Rating (star/diamond/ crown, etc.) |
|---|---|---|---|---|---|---|---|---|
| • Families<br>• Businessmen<br>• Students<br>• Transients<br>• Pilgrimages<br>• Gamblers<br>• Travellers | • Resorts<br>• City hotels<br>• Motels<br>• Airports<br>• Floatels<br>• Boatels<br>• Rotels | • Residential<br>• Semi-residential<br>• Non-residential | • Independent<br>• Chain<br>• Franchise<br>• Management contract<br>• Joint venture | • Heritage<br>• Ecotel<br>• Boutique<br>• Spa | • Small<br>• Medium<br>• Large<br>• Grand | • Exotic location<br>• Highways<br>• Airport<br>• Water surface<br>• Heart of city<br>• Remote location | • Economy<br>• Mid-market<br>• Luxury<br>• All suite<br>• Limited service<br>• Full service | • Five<br>• Four<br>• Three<br>• Two<br>• One |

Basis for classifying hotels

**Note:** In India, minimum 10 rooms are required for getting the five-star classification.

## Residential Hotels

✓ These hotels are also known as *apartment hotels or houses* that provide long-term accommodation usually for one month to one year. Rooms may or may not be fully furnished.

✓ Residential hotels range from the luxurious to the moderately priced. Luxury residential hotels offer housekeeping services, a dining room and sometimes room service provision too.

## Semi-residential Hotels

✓ These hotels incorporate the features of both, transient and residential hotels. Because semi-residential hotels offer accommodation for a week or month like residential hotels as well as on per day basis like transient hotels offer.

✓ The offered room rates are usually lower for long staying guests while for transient guest, the room rate is usually higher.

## Independent Hotels

✓ These hotels are autonomous bodies which mean a hotel with no chain or franchise affiliation. It may be owned by an individual proprietor or by a group of investors (meaning it is governed through sole proprietor style ownership).

## Franchise

✓ It is a business operating system that is offered for sale and which generally provides such benefits as the use of a company name, participation in promotional activities and access to management expertise and systems.

✓ Purchaser (or franchisee hotel) commonly pay an initial fee plus an ongoing amount to cover promotional and other costs &/or a percentage of gross sale.

## Management Contract

✓ It is a contractual arrangement, commonly established between the owners of hotel buildings and hotel management firms, for the provision of operational/ organizational services and expertise in return to a set fee &/or percentage of profit earned.

## Chain Hotels

✓ A hotel company that operates several properties, such as *Holiday Inn Worldwide* or *Hilton Hotels Corporation*. In simple terms, chain hotels are defined as all hotels under the ensign of a hotel group.

✓ Such chain properties provide both a trademark and a reservation system as an integral part of the management of their managed properties.

## Activity

Fill the below given blanks with appropriate answers (i.e., hotel chain/name, parent company or product line) in both Indian and international chain hotels.

### Indian Chain Hotels

| Hotel Chain (or Hotel Name) | Parent Company | Product Line |
|---|---|---|
| Taj Group | ? | Gateway, Residency, Palace & Resorts, Vivanta, Cultural Centre, Garden Retreat, Ginger hotels. |
| ? | East India Hotel Limited (EIHL) | ? |
| ITC | ? | Heritage, Fortune, ITC (Luxury Collection), Welcome/Sheraton hotels |
| The Ashok Group | India Tourism Development Corporation (ITDC) | ? |
| ? | ? | Chef Air Mumbai and Chef Air Delhi (these are flight catering units) |

### International Chain Hotels

| Hotel Chain (or Hotel Name) | Parent Company | Product Line |
|---|---|---|
| Le Meridian | Starwood Group of Hotels | ? |
| ? | ? | Comfort Inn, Comfort Suites, Sleep Inn, Clarion, Quality Inn, Cambria Suites. |
| ? | Intercontinental Group of Hotels | ? |
| Hyatt | ? | Grand, Park, Regency, Place, Summerfield Suites, Andaz, Resorts and Vacation Club. |
| Radisson | Carlson Group of Hotels | ? |

## Timeshare Hotels

✓ The concept of timeshare hotels started from Europe. Time share is an *exchange program* which offers membership (with multiple ownerships on same unit) based accommodation and generates business for long period of duration. Timeshare concept is mainly offered by resort hotels (of chain group).

✓ Timeshare is available in four major forms, i.e., fixed, floating, rotating and points club. Interval International II (a product line of Interval Leisure Group, Inc.) is the world's largest chain of timeshare hotel with more than 2800 properties/resorts in over 75 countries around the world.

**Activity**

Give an example of a timeshare hotel that fulfils the description provided.

| Timeshare type | Description | Give example |
|---|---|---|
| Fixed | Hotel unit/room is only allotted for a specified/ particular week or days of the year and rest of the year, others can buy and utilize it. | ? |
| Floating | Hotel unit/room is again allotted for a particular period of time (usually a week) but without specification of dates. | ? |
| Rotating | It combines the benefit of both—fixed and floating timeshare. Rotation of holiday can go either backward or forward on the season. This gives an opportunity to all owners on a rotational basis. | ? |
| Points club | Also known as *vacation club*. It enables buyers to enjoy various properties of same chain, depending upon the number of points they have accumulated. It works on first come first serve basis. | ? |

## Condominium Hotels

✓ The concept of condominium hotels started in US. Condominium is again a type of membership programmes (but with single ownership) offered by resort hotels (of chain group).

✓ It also generates business for long period of time. RCI (a product line of Wyndam Worldwide) is the world's largest chain of condominium hotels with more than 3800 properties in over 80 countries worldwide.

## Heritage Hotels

✓ These hotels are also called "historic, fort, haveli, or palace hotels". In it, hotel guest is graciously welcomed, offered room that have their own history, serve traditional cuisine and are entertained by folk artists.

✓ These hotels put their best efforts to give the glimpse of their region and an opportunity to experience royal pleasure in traditional ambience. Heritage hotel concept is buoyant in three categories.

### Activity

Write down the description regarding the hotels, along with the year in which they were built.

| Heritage hotels | Description | Built around (or between the years) |
|---|---|---|
| Heritage | Hotels should have a minimum of 5 rooms (with 2 beds in each guest room) in traditional architectural design and offer traditional cuisine. | ? |
| Heritage Classic | ? | 1920-1935 |
| Heritage Grand | ? | ? |

## Ecotel Hotels (nature-friendly hotel)

✓ These hotels are also called "eco-hotels, eco-friendly hotels or green hotels". These are environment-friendly hotels.

✓ The ecotel hotels use eco-friendly items in the guest rooms as well as in operational areas. For example, *Orchid Mumbai* which is considered **Asia's** first and most popular five-star ecotel hotel.

## Boutique Hotels (architecture of rooms is based on theme)

✓ The concept of boutique hotel first surfaced in North America in the year 1984. Boutique hotels are furnished in a themed, stylish and/or inspirational manner and offer an exceptional and personalized level of services and facilities. In many boutique hotels, every guest room is based on a different theme.

✓ These are primarily targeted towards corporate travellers because these guests give high importance on privacy, luxury and service delivery. Its example include - in **India** the *Park Bangalore* is a perfect example of boutique hotel. Boutique hotels may also offer the provision of *honesty bar*.

## Mid-Market Hotels (a component of limited service hotel)

✓ It is a "standard, tourist class, suite or mid range hotel" that offers small living room with appropriate furniture, small bedroom with king sized bed (some suite hotels also offer a compact kitchenette with fridge and mini bar) and private bath.

## All-suite Hotels (a luxury hotel with all suite rooms)

✓ These hotels create a hard-hitting competition for mid-market and luxury hotels, as the offered services & facilities are above average but at mid-market rates.

✓ Condominium hotels are considered a variation of all suite hotels but a typical condominium hotel is located in a resort area and marketed as an alternative destination to full service hotels.

# Luxury Hotels (usually a full service hotel)

✓ These hotels are also called "deluxe, superior, executive or high class hotels" that offer luxury/world class services. Generally a luxury hotel provides *executive floor*, *EVA floor*, fine dine restaurant, lounges, concierge service, meeting rooms, dining facilities, etc. Bath linen is provided to each guest and is replaced accordingly.

✓ The guest rooms (luxury suites, two or more dining rooms and cocktail lounge) contain high level furnishing, artwork, etc. Prime market for these hotels are celebrities, wealthy people, business executives and high ranking political figures.

# Limited Service Hotels

✓ It can be an independent as well as part of a chain hotel. Chain hotels belong to chain/good brand hotels with franchise memberships but built within limited areas, especially without restaurant and banquet hall.

✓ These hotels are located near business areas such as industrial parks, cities, and airport terminals. Limited service hotels are usually small properties with 150 or fewer rooms or suites.

# Budget Hotels

✓ These hotels are also known as *economy hotels* and meet the basic needs of the guest by providing comfortable and clean room.

✓ These hotels provide limited services and are known for their low prices, as these hotels meet just the basic needs of travellers. In India – *Red Fox group of hotels* and *Lemon Tree group of hotels* are excellent examples of budget hotels.

## Activity

The following chart shows the name of a chain hotel. You need to write down the name of any 5 hotels, their parent company, type of hotel/geographic location, target market, number of rooms, standard/ level of service offered.

| S. No. | Hotel Name | Parent Company | Type of hotel/ geographic location | Target market | Number of rooms | Standard/ level of service |
|---|---|---|---|---|---|---|
| 1 | Usha Kiran Palace, Gwalior | Taj Group | Heritage | Leisure Travellers | 40 rooms | Mid-level |
| 2 | | | | | | |
| 3 | | | | | | |
| 4 | | | | | | |
| 5 | | | | | | |

## SPAS

- ✓ The ellipsis SPA stands for "Sanum Per Aqua" and it is related with health treatment through water (or *balneotherapy*). This concept originated in Belgium.
- ✓ Spas are the resorts which provide therapeutic bath and massage facilities along with other features of luxury hotels. The excellent example of spa hotel in India is *Ananda Spas* in the Himalayas.

## Bed & Breakfast (B&B hotel)

- ✓ These are small inns or lodges that provide a room and breakfast. In simple terms, a bed and breakfast (B&B) is a private home in which guests can be accommodated at night in private bedrooms (which may or may not have attached bathrooms). Breakfast is included in the price and other meals may be available by arrangement.

| Activity |
|---|

The following chart shows that how a particular hotel can be defined/classified in terms of various parameters. In this chart, some columns have been filled whereas remaining columns are kept empty for your practice. So try it:

| S. No. | Hotels/ Type | Example | Location | Target market | Ownership | Level of service | Standard |
|---|---|---|---|---|---|---|---|
| 1 | | Star cruise | | | | | |
| 2 | | | Highway | | Sole Prop. | | |
| 3 | Starwood | | | | | | |
| 4 | | | | Family | | | |
| 5 | | | | | | Full service | |
| 6 | Business | | | | | | Luxury |
| 7 | | ANA hotels | | | | | |
| 8 | | | Airport | | | | |
| 9 | Garni | | | | | Limited | |
| 10 | Heritage | | | | | | |
| 11 | | Ginger | | | | | Economy |
| 12 | ICH | | | | Chain | | |
| 13 | | | | Students | | | |
| 14 | | Four points | | | | | |
| 15 | Ecotel | | | | | | |

# Supplementary Accommodation

✓ It includes Youth hostel, Tourist house, Circuit house, Paying guest accommodation, Holiday camp, Gastoft & Auberge, Conventional hotel, Conference centre,
Country hotel, Caravan, Parador, Garden hotel, Tavern, Chalets, Ryokan, etc.

## Activity

Fill in the blanks with correct description (or supplementary accommodation type).

| Supplementary Accommodation | Description |
|---|---|
| Caravan | ? |
| Tavern | ? |
| Chalets | It is a Swiss hotel concept which is mainly found in Alpine areas. These hotels commonly have a steeply pitched roof and wide eaves. |
| Ryokan | ? |
| Auberge | ? |
| Gastoft | ? |
| ? | Budget accommodation which is mainly targeted towards students. Here, washroom/bathroom facilities are often shared and only the basic services are provided. |
| Circuit house | ? |
| Tourist house | ? |
| Country lodge | ? |
| ? | A rehabilitation centre or hospice which mainly offers temporary accommodation during any natural or human generated calamities. It mainly offers basic food, drinking water and shelter facilities to the people free or at nominal price. |

| Capsule | ? |
|---|---|
| RHV park | ? |
| Dormitories | ? |
| Garni hotel | It refers to a European hotel concept. These are small hotels which do not have dining provisions or may have limited dining provisions. |
| Rest house | ? |
| Farm house | ? |
| ? | It refers to a French hotel concept which is a kind of holiday home that is usually fully furnished plus equipped with self-catering provisions. In other words, it is a simple accommodation which encompasses most form of holiday cottages/apartments. |
| Losmen | ? |
| Bure | ? |
| Cabarets | In general, it is a small club or restaurant that offers singing and/or comedy sketches to their guests. |
| Guest house | ? |
| Boutique hotel | ? |
| Apartotel | It is an apartment house or building which is used as a residential establishment. |
| Chop house | ? |

## 1.4   STANDARD ORGANIZATIONAL STRUCTURE OF HOTEL

The term *organizational structure* indicates to a diagram of chain of command of people working in various departments, showing how a company or office is organized. It facilitates in developing *job description* and *job specification* in terms of duties, responsibilities, position title, who reports to whom, and so forth of every directly as well as indirectly involved department and its *job positions*.

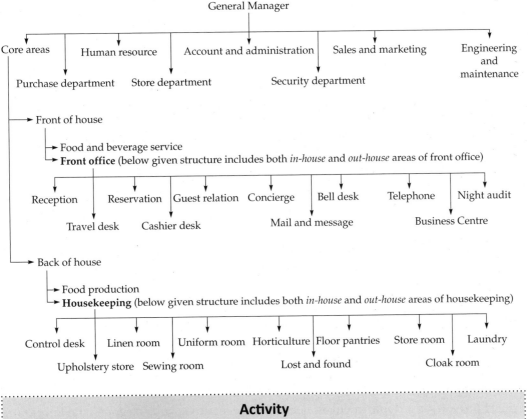

In below equation, write down the correct division from the given-hotel room division or food & beverage division

| Front of house | | Back of house | | Indicate to which division |
|---|---|---|---|---|
| Front office | + | Housekeeping | = | ? |
| Food & beverage | + | Food production | = | ? |

A brief description of major departments is given below:

# Front Office

✓ Front office is responsible for making room reservations, registering guests, collecting payments and settling guests account, providing money exchange (or forex) facilities and providing all information required by the guest in relation to property, sometimes related to locality also.

✓ Front office department is headed by front office manager under the direct supervision of room division manager (RDM). The department has some supervisory positions (such as lobby manager, duty manager) along with various intersections' supervisors (like reservation supervisor) and operational level in-house (intersections' agents like reservation, reception, guest relation agent, etc.) and out-house staff (airport representative, chauffeur, etc.).

## Food & Beverage

✓ Food and beverage department is responsible for offering meals and beverages to resident as well as non-resident patrons into all available food & beverage outlets. It is mainly held responsible for service quality, simultaneously also ensuring food quality, food costing, service standard and implementation of "standard operating procedure". Many hotels have prepared *service blueprints* in order to properly implement their standard operating procedures.

✓ This department is headed by food and beverage manager under the direct supervision of food and beverage director. The supervisory work of different food and beverage outlets is performed by respective outlet's supervisors and operational level work is performed by captain, steward and their assistants.

## Housekeeping

✓ Housekeeping department is responsible for keeping the hotel neat and tidy (both in-house as well as out-house areas). In addition, it is also responsible for offering pleasant environment, ensuring safety and security, interior decoration, laundering guest clothes and hotel linens, placing amenities and supplies into guest rooms and so forth.

✓ Housekeeping department is headed by executive housekeeper who reports to room division manager. Nowadays, public area cleaning (like parking space) is out-sourced but guest rooms and several in-house areas are under the direct control and supervision of housekeeping department. The administrative work of all housekeeping sections is performed by their respective supervisors whereas operational work is performed by attendants.

## Food Production

✓ Food production or kitchen is mainly responsible for preparing and cooking all available range of food choices for guests which serves in different food and beverage outlets of the establishment. In large establishments, there are separate kitchens (or *decentralized kitchen* system) for separate food and beverage outlets.

✓ In smaller establishments, there is a common or *centralized kitchen* for the preparation of all types of food for all food and beverage outlets. This department is headed by executive chef under the direct supervision of food and beverage director. Additionally, it offers the job of sous chef, chef-de-parti, commis, etc.

## Accounts and Administration

- ✓ The accounts and administration department is responsible for financial management, budget correction and approval, calculating salaries, wages and fringe benefits of employees, passing office orders and circulars, collect cash, cheques, drafts and other negotiable payment options and deposit it with bank/ send them for payment.

- ✓ The department is headed by accounts officer (or administrative officer). It depends on the organization hierarchy of hotel. All operational works are performed by appointed clerks and other office staff.

## Sales and Marketing

- ✓ Sales and marketing department is responsible for marketing and promoting hotel products and services via different modes of advertisement. In small hotels, this department is usually found in back area whereas in large hotels a range of sales and marketing offices are built in different regions and cities which accept business for all involved sister hotels.

- ✓ Due to multifaceted work profile, sales and marketing departments should always be in contact with banquet sales executive, hotel reservation section, travel agents and tour operators. This department is headed by sales and marketing manager/ director under whom sales team or sales agents work.

## Engineering and Maintenance

- ✓ Engineering and maintenance department is responsible for all kinds of repair and maintenance work of entire hotel property. For example, if any problem occurs such as bulb, fuses, AC, etc., are not functioning properly, then housekeeping needs to contact the hotel's engineering & maintenance department; they are available 24 hours.

- ✓ The department is headed by the chief engineer under whom chief officer and other maintenance staff works. This department performs preventive, breakdown and scheduled maintenance. Apart from this, it also performs repair works, tracks guarantee and warranty policy/procedure of purchased and fitted products.

## Purchase Department

- ✓ Purchase department is responsible for purchasing all equipment and commodities requisitioned by the different departments of hotel. This department is headed by purchase manager; he must have sound knowledge of prevailing market conditions and available suppliers.

## Store Department

✓ Store department is responsible for placing orders then receiving purchased commodities/equipment/appliances then marking, afterwards securely transferring them into storeroom (or to respective sections/departments). Store room is headed by store manager who is responsible for proper stocking, issuing (FIFO or LIFO), counting and reordering of commodities.

## Security Department

✓ Security department is responsible for keeping safe the hotel employees, hotel property, guests and their belongings. Security mainly deals with guest room security, loss prevention, alarm system, closed circuit television and in case of police interference, it aids to them.

✓ Due to such legal responsibilities, security agency must have all required appropriate licenses. This department is headed by security officer/security manager. Employees who are working in security department must pass through police verification process.

## CONCLUSION

Hotels provide boarding and lodging facility against payment to the selected market segment/s. Major hotel types involve motel hotel, city hotel, resort hotel, airport hotel, luxury hotel, membership hotel and so forth. Apart from these given hotel types, hostel, rest house, guest house and so forth are also considered supportive/supplementary hotel types. In order to smoothly run any hotel organization, *departmentalization* of hotel is most important. Departmentalization is concerned with division of hotel establishment into *responsibility centres*. Some of these responsibility centres are front-of-house whereas others are back-of-house division. Along with this, some departments are working as supportive centres like security, purchase, maintenance and so forth.

| Terms (with Chapter Exercise) | |
|---|---|
| Room division | ? |
| FIFO/LIFO | It is an inventory rotation system. FIFO stands for *first-in-first-out* whereas LIFO stands for *last-in-first-out*. |
| Reordering point | Pre-determined minimum stock point whereby placed inventory items are required to be ordered. |
| Chef-di-parti | Kitchen position which is also called 'station chef' or in short CDP. These chefs are in-charge of their own particular areas (or sub-kitchen units) of production. |
| Departmentalization | ? |

| | |
|---|---|
| Sous chef | Kitchen position who works under the direct supervision of executive chef and is mainly responsible for supervising daily kitchen activities. |
| Commis | ? |
| In-house | Various internal hotel areas which are located inside the main hotel building. It involves both guest rooms and in-house public areas like corridors and lobbies. |
| Out-house | Various external hotel areas which are located outside the main hotel building. It mainly includes public areas like parking space, garden, etc. |
| Service blue prints | ? |
| ? | Originally, it is related with porter but nowadays the term is used for information desk/agent. |
| Front-of-house | It is mainly concerned with direct revenue generating department such as front office, food and beverage, gift shop and so forth. Often it also includes places where guest and staff interactions directly take place. |
| EVA floor | A floor on which all rooms are designated to women only and in-room service and room cleaning is also done by women staff only. Often it is also known as *pink floor*. |
| Back-of-house | Indirect revenue generating departments such as housekeeping, food production department, etc. Often it also includes places where guest and staff interactions indirectly take place. |
| ? | Hotel units. |
| Vacation club | ? |
| Executive floor | It is also known as *club floor* that is mainly found in the five star hotels. It provides superior quality rooms along with additional services & facilities like business centre, private lounge, priority check-in and check-out. Often *honesty* bar provision is also found on executive floor. |
| ? | An exchange programme and upto some extent it is similar to condominium programme. |
| Limousine | Hotel's vehicle that provides pick-up and drop facility to its guests. |
| Mini bar | Guest room bar which contains limited stock of alcoholic and non-alcoholic drinks. |
| Honesty bar | It is an unattended beverage bar, typically in the hotel lobby or lounge area where payment is left to the guest. It is different from in-room mini bar, where any consumption automatically transfers in the respective room accounts. |

# 2
## CHAPTER

# Room Division

**OBJECTIVES**

*After reading this chapter, students will be able to...*

- define room division and its components;
- identify and describe the functions of departments that come under room division; and
- differentiate between the job-profile of room division and front office manager.

## 2.1   ROOM DIVISION

Nowadays, majority of hotels use *room division* word instead of front office. But in reality, the term refers to those departments and their sections that particularly generate (or support in generating) revenue through room sales. Broadly speaking, room division consists of two major departments: front office (including *uniformed service department*) and housekeeping. A brief description of both is given later in this chapter. Room division department is headed by room division manager (RDM) who is predominantly responsible for front office manager and executive housekeeper. Few hotels may also include security department in their room division.

The organizational structure of room division varies from hotel to hotel. These variations are due to differences in the size of hotels, the type & level of services offered and finally the preference of management. Staff working in room division must possess optimistic attitude with plenty of desirable attributes. The term *attribute* indicates the qualities and etiquettes of an employee which reflects the excellence of the organization too. If the staff is well trained and efficient, they can, to a certain extent, make up for other deficiencies in the services provided. These attributes are classified into two broad groups, i.e., work oriented and corporeal attributes.

The term *work oriented* attribute is concerned with the virtues which all room division staff must own while they perform their duties and responsibilities. For instance, good communication, punctuality, loyalty, sincerity, sense of responsibility and so forth. On the other side, the term *corporeal attribute* is concerned with physical qualities like how the staffs presents itself and the first impression they create on others. All room division employees should be aware of the personal attributes such as take bath daily, shave daily, wear clean uniform, gargle before coming on duty and so forth. It is the sole responsibility of every employee to ensure that these physical and personal attributes are put into practice.

| **Activity** | | | |
|---|---|---|---|
| Fill in the below cited blanks with appropriate attributes/personality traits (of *room division staff*) or its description. | | | |

| Work oriented | Description | Corporeal attributes | Description |
|---|---|---|---|
| Diplomatic | ? | Smiling face | ? |
| Confident | ? | Physically fit | ? |
| Loyal | ? | ? | Take proper care of body cuts/burns |
| ? | Feeling of sense of responsibility | ? | Does not have foul mouth odour |
| ? | Language & communication power | ? | Use light make-up |
| Sharp memory | ? | Bathe & brush teeth | ? |
| Reference point | ? | ? | Pleasant body odour |
| ? | Decision making ability | Jewellery | ? |
| Sincere | ? | ? | Cufflink and button |
| Punctual | ? | Nails and their cleanliness | ? |
| Salesmanship | ? | ? | Short hair and side locks |
| ? | A successful forward planner | Shoes & socks | ? |

## 2.2  SECTIONS OF ROOM DIVISION

"Room Division Department = Front Office Department + Uniformed Service + Housekeeping Department"

The term *room division department* is a collective word for all those departments/sections which directly as well as indirectly create a mammoth impact on room sales. *Room division department* is headed by room division manager who is responsible for the supervision of front desk, reservation, housekeeping, concierge, guest services, security, and communication/telephone. So, major departments in the room division are *front office, housekeeping* and *uniformed service departments*. Room division manager (RDM)/

Director of Room Sale (DRS) directly supervises these departments and their functions. A brief description of major departments that work under the direct supervision of room division department is given below.

## Front Office Department

✓ The front office is the nerve centre of a hotel and acts as a *public face* of the hotel. It interacts with all hotel guests in different stages of communiqué phases (collectively it is known as *guest cycle*) i.e., during pre-arrival, arrival, stay, departure and after departure stage. In fact in most of the cases, the only direct contact most of the guests have with hotel employees, other than in the restaurants, is with members of the front-office department.

✓ Mostly, the first interaction between front office employees and guests takes place in reservation stage/section (as it deals with all guest queries) then at reception desk (it receives, registers and handovers room keys to guests). During the stay, guests often interact with concierge (to obtain information) and guest relation desk (for benevolence and complaints). Thereafter, during departure stage when guests settle their bills with cashier and bellboys handle guests' luggage. Finally, after departure, when hotel's sales and marketing division, reservation section or guest relation desk keep themselves continuously in touch with previous guests with an inherent attempt to promote sales.

| Activity | |
|---|---|
| Ascertain the pecuniary role of front office department in generating revenue and creating benevolent environment of the hotel. Some of them are cited below. You are required to fill in the given blank spaces with appropriate description: | |

| Role | Description |
|---|---|
| Create goodwill | Because front office firstly interacts with guest and provides all hotel/ locality related information, consequently forms a good positive/negative impression of hotel in front of guest. |
| Public face | ? |
| Nerve centre | ? |
| Command hub | ? |
| Revenue source | Because usually more than 50% revenue is generated from room sales, concurrently front office also works as a marketing hub (or generates business for other departments/outlets). Thus, affects RevPAC, RevPAR & ADR. |
| Business point | ? |

| Complaint core | Because in most of the cases guest doesn't know where they have to complain, thus they come at front desk/guest relation desk for their product/service/ mechanical complaints. |
| Support centre | ? |
| Ingress/egress window | ? |

# Housekeeping Department

✓ Housekeeping is one of the crucial parts/departments of room division which is headed by executive housekeeper who is responsible for the integral work of the department, for example, cleanliness, maintenance, laundering and maintaining the aesthetic appeal of the hotel.

✓ In unison, housekeeping also maintains *guest privacy* and ensures security for the room assets and guest's belongings. Secondly, it also provides all required *room amenities* and *supplies*. Apart from it, housekeeping also provides *guest loan items* on rental fee, prepare and maintain its records and finally retract to given items before guest departure.

✓ Eventually, it is not wrong to conclude that housekeeping department plays a vital role in accomplishing *guest expectations* and *their satisfaction*.

| Activity |
|---|

Ascertain the contributory role of housekeeping department in creating aesthetic appeal and upkeep of the hotel. Some of them are cited below. You are required to fill in the blank spaces with appropriate description.

| Role | Description |
|---|---|
| Create goodwill | ? |
| Energy management | ? |
| Hygiene & sanitation | Of public areas (like restaurants, lobby, corridors) and guest rooms |
| Guest privacy | ? |
| Guest satisfaction | ? |
| Offer guest loan items, supplies and amenities | ? |
| Safety & security | Of hotel employees, hotel property, guests and guest belongings/ valuables |
| Guest expectation | ? |

| Hotel decor | ? |
|---|---|
| Laundering | Of staff uniforms, hotel linens (like bed linens and bath linens) and guest clothes |
| Pest control | ? |
| Offer provisions | ? |
| Support maintenance | ? |

## Uniformed Service

✓ Uniformed service department (or USD) is also known as *guest service department* because it is usually responsible to take care of guests and their needs. The department primarily provides most personalized services to the hotel guests. Uniformed service department comprises all those members of the hotel staff who wear uniforms. This uniform is of a distinctive colour but the basic prototype is same among all hotels.

✓ Uniformed service department is a division/subdivision of room division department (and often a constituent of front office department) and mainly comprises bell desk, concierge desk, valet parking attendants and doorman. This division/subdivision is headed by *bell captain* who gives direction and instruction to all staff members who work under uniformed service. If there is no separate uniformed service department then front office manager will distribute its work to other front office members and supervise all of them.

Uniformed service

Concierge    Bell desk    Travel desk    Airport representative    Lift operator    Doorman    Valet parking

| Activity |
|---|
| Carefully read the description (or may go through reference materials like books, magazines, etc.) then fill in the below given spaces with appropriate subdivision names. |

| Sub-Divisions | Description |
|---|---|
| ? | Originally, the position is known as *porter* but nowadays mainly responsible for providing information to guests. |
| ? | This section is responsible for circulating the *movement list* (thereafter ANS in manual front office) so that respective section/s can prepare themselves for the new arrivals. |
| ? | This section never deals with *chance guest* but includes the name of these chance guests at the time of preparing *movement list* (in context to expected departure) for very next day. |

| Guest Relation desk | • It deals with welcoming of guests by offering flower bouquet or garland to *check-in* guests (especially to VIPs). In India, almost all hotels welcome their guests in Indian traditional style, i.e., *aarti & tilak*. Thus, the basic function of this section is to take care of guests' needs. Guest relation desk particularly focuses on guests with the specific needs such as non-English speaking guests, or perhaps large conference groups. |
|---|---|
| Bell desk | • It deals with the handling of guests' luggage to and from the guest room, as well as undertaking minor tasks such as running *errand* cards during guest arrival and departure stage. It also provides left luggage facility to the departing guest, deliver amenities and supplies (like newspaper delivery) into guest rooms. In many hotels, bell desk also works as an information desk (or perform all tasks of concierge desk), depositing C forms into FRRO, deal with guest paging, handling mails and messages, etc. |
| Concierge desk | • Concierge is a French word which means porter. But nowadays, concierge is considered an information desk whereas the porter is only responsible for dealing with guest luggage. Concierge desk is responsible for making reservations for dining, obtaining tickets for theatre & sporting events, arranging for transportation by limousine, airplane or train and provide information on babysitting, amusement and on cultural events plus local points of interest. |
| Travel desk | • An intermediary desk which derives financial gain (in the form of mediating expenses) by linking supplier of travel, tourism and recreational services (such as sightseeing, museum, theatre, etc.) with hotel guests through the provision of reservation, ticketing and other services. |
| Airport Representative | • Airport representative is a person who is principally responsible for representing a hotel on airport terminal.<br>• He should also ensure that all staff maintains a good relationship with airport authority for the smooth arrival/departure of all prospective guests. |
| Valet Parking Attendant | • Valet parking attendants are usually held responsible for assisting in opening/closing car doors during guest arrival (in front of hotel portico). Afterwards park the car on behalf of guest then raise *car docket* and handover it to guests. Finally, on guest departure, bring the car back from parking zone and hand it over to the guest. |
| Doorman | • He is principally responsible for opening and closing of hotel's main door for guest's ingress and egress. After valet parking attendant, he is the first person who creates first impression (by greeting & welcoming) of hotel in front of the guest. In India, ideally a giant person in traditional uniform (or themed uniform) is preferred for this work. |

## 2.3   OTHER MAJOR DEPARTMENTS OF A HOTEL

Besides front office, there are three core departments in a hotel, i.e., housekeeping, food production and food & beverage. Housekeeping and food production department works as indirect revenue generating departments whereas food & beverage is a direct revenue generating department. A brief profile of these three departments is given below.

| Housekeeping Department | • A brief description of housekeeping department has already been given in introductory phase of this chapter. It is primarily responsible for the cleaning of all guest rooms and public areas, maintaining the laundry, gardens & supplying flowers to all departments.<br>• The housekeeping department is further divided into various sections like-control desk (housekeeping supervisor's desk), floor linen room pantry (in decentralized housekeeping operation), meeting point/room for public area cleaning staff, linen room & uniform room, laundry, horticulture room, etc. |
|---|---|
| Food Production Department | • It (or *centralized kitchen*) is mainly responsible for the preparation of food that is served in all food and beverage serving outlets of the hotel. *Decentralized kitchen system* is used in large hotels. Nowadays, *satellite kitchen system* is also common.<br>• Broadly, food production department can further be classified into two, i.e. hot kitchen and cold kitchen. *Hot kitchen* comprises roast/grill, vegetable, confectionery, hot soup, tandoor and so forth whereas *cold kitchen* comprises pastry/bakery, garde manger, butchery, cold store and so forth. |
| Food & Beverage Department | • It is mainly responsible for serving food & beverage items to guests in public dining outlets and in guest rooms. *Permit room* and other outlets that offer alcoholic beverages must have required licences for storing, selling and serving beverages.<br>• Food & beverage department consists of various outlets like room service, restaurant, coffee shop, banquet, bar, etc. All these outlets play a pecuniary role in generating revenue for hotel plus in providing outstanding services. |

## 2.4  SUPPORTIVE DEPARTMENTS OF HOTEL

Apart from core areas, there are a range of departments in a large hotel. Some generate business for the hotel whereas others support in generating revenue and providing efficient and effective services to the guest. A brief introduction of some supportive departments of hotel is listed below.

| Sales & Marketing | • It is mainly responsible for selling products and services of the entire establishment. The director of sales, who employs several sales executives and clerical staff, usually heads this department. Their job is to sell the hotel's services and facilities to wider, often international clientele.<br>• Its major role involves placing *promotional calls*, selling products & services of the hotel and to do the advertisement of the property. |
|---|---|
| Engineering & Maintenance | • This department deals with the repair and maintenance of all kinds of appliances used in a hotel. It performs routine, preventive, breakdown and scheduled maintenance.<br>• Its major duties involve general maintenance and repair in the hotel like ground & swimming pool, cabinet making; repair & maintenance of tools and equipment used by different departments; maintenance of HVAC (heating, ventilation, air conditioning, etc); fire safety & protection and energy management. |
| Accounts & Administration | • It deals with the collection and payment of bills which are settled either via cash or credit mode. The account office has the responsibility of overall financial management, producing financial statements and other financial documents.<br>• Accounts section also deals with expenses related to advertising; expenses related to sales & promotion and account related to employees' salaries, wages, employees' meals prices, fringe benefits, etc. |
| Human Resource Department | • It is responsible for managing the work force for the entire hotel. It deals with employees benefit programme and monitor compliance with laws related to equal opportunity in hiring and promotion.<br>• Its major responsibility includes deciding employees remuneration and other benefits; selecting, recruiting & improving the staff; job evaluation; determining salaries & wages for the staff; providing training programmes to operational as well as middle level employees. |
| Purchase & Store | • Purchase department is responsible for purchasing every item on behalf of entire hotel which comes through purchase order. Therefore, purchase manager must have thorough knowledge of market and a keen eye for the best product at best price.<br>• Once the goods have been purchased & received carefully, they must be stored correctly, to avoid waste by deterioration, damage or theft. The main objectives are to keep supplies secure and in good condition, to make them easy to find & count and to ensure that property does not run out of them when they are needed immediately. |

| Security | • It is principally responsible for the protection of guests, their property, hotel employees and hotel property. Owner/s and operators are charged under law to take all reasonable precautions to protect guests from robbery, arson, rape and other kind of offences. During serious incidents it coordinates with police department like during death of guest in room. <br>• Security department has to deal with guest room security; loss prevention/ locks, alarm system, closed circuit camera and safe deposit boxes. |
|---|---|

## Activity

Fill in the right side column with the correct hotel department(s)/section/staff involved so as to meet guest needs in the following scenarios.

| S. No. | Questions | Department/ Section/ Staff Involved |
|---|---|---|
| 1 | A resident guest wants a flower bouquet and chocolate tray (in his room) for his wife, on the occasion of his 25th anniversary. | |
| 2 | A resident guest wants to order early morning breakfast (through Door Knob Menu Card) as he needs to catch a flight. | |
| 3 | Room air conditioner is not working properly, therefore guest wants to request for room change. | |
| 4 | Guest wants to place order for regular house water/filtered drinking water. | |
| 5 | A business traveller discovers that his personal computer left in the room has been stolen. | |

## CONCLUSION

Core areas of any hotel establishment can be broadly classified into two groups, i.e., room division and food and beverage division. *Room division* relates to the combination of housekeeping, uniformed service and front office department. It is headed by room division manager (RDM). On the other side, food and beverage division relates to the food production and food & beverage outlets. Food and beverage division is headed by food and beverage director (FBD). Both these divisions comprise various subsections/ subdivisions and play a crucial role in the functioning of a hotel. Apart from core areas, hotel establishments also have additional or supportive departments which assist in day-to-day function and shore up guest satisfaction.

| | Terms<br>(with Chapter Exercise) |
|---|---|
| Amenities | Items which give the guest an extra convenience or allow them to get something done, like a toothbrush, shoeshine cloth or extra pen. |
| RDM | RDM stands for Room Division Manager who is principally responsible for managing housekeeping, uniformed service and front office operation. |
| FBD | ? |
| Door knob menu card | Disposable (or single use) in-room menu card which is mainly used for serving morning breakfast into guest rooms. |
| Guest cycle | Gyratory series of events/activities which take place between guest and hotel during different stages of interaction. |
| Movement list | Expected arrival-cum-expected departure list that is prepared by reservation section and circulated one day in advance among different relevant sections. |
| FRRO | Foreigners Regional Registration Office. |
| ? | Car parking voucher or ticket or receipt that is issued after parking car in parking lot. |
| Errand card | Luggage card which contains information about the guests and their luggage. The bellboy takes charge of this card. It is used in manual front office and prepared during both arrival (arrival errand card) as well as departure stage (departure errand card). |
| Guest expectation | ? |
| Guest supplies | All those items that are conducive to the guest's material comfort and convenience. It includes guest room amenities, expendables and loan items. |
| Concierge | It is a French term which indicates the porter but nowadays it is totally related to a person who is primarily responsible for providing hotel related and locality related information to guests. |
| Bell captain | ? |
| Cold kitchen | All those kitchen sections which are primarily responsible for cold preparations such as garde manger, pastry section and so forth. |
| Hot kitchen | All those kitchen sections which are responsible for preparing hot dishes such as soup kitchen, Chinese kitchen and tandoor section and so forth. |

| | |
|---|---|
| Satellite kitchen | Kitchen management system whereby satellite kitchen caters to a particular outlet because central or regional kitchen cannot cater due to its location disadvantage. It is also known as *receiving kitchen*. |
| Centralized kitchen | Kitchen management system whereby all food orders are processed from single kitchen to different food and beverage outlets. |
| Guest satisfaction | ? |
| Uniformed service | Section of room division which includes all those front office employees who wear uniform like bell desk staff and doorman. |

# 3

# Front Office

<hr>

## OBJECTIVES

*After reading this chapter, students will be able to...*

- elucidate the importance of organization structure;
- define the duties and responsibilities of different front office employees;
- draw the layout and discuss the functions of front office department;
- identify the various sections of front office department; and
- identify the different manual, semi-manual and automatic front office equipment.

## 3.1 FRONT OFFICE

*Front office* department is the first place/office where hotel and guests interact with each other. It is the department which holds the prime responsibility for the sale of *room inventory*. Hotel room is a perishable product because on every night, every unsold room leads the hotel towards a loss of revenue, which can never be recuperated in future. Therefore, many hoteliers also consider that a hotel room has a shelf-life of 24 hours.

So front office department should make every effort to sell all available lettable rooms on every night. Reception should not only focus towards the room sales but also strive to sell the available services and facilities of all *revenue centres* within the hotel like coffee shop, bar, restaurant, spa, etc. This kind of sales approach, in reference to point of sale outlets, is known as *in-house sales* and is a means of encouraging guests to make use of the hotel's facilities & services.

## 3.2 ORGANIZATIONAL STRUCTURE

It is a chain of command of people working in various departments, showing how a company or office is organized. It describes the responsibility, duty, position, who reports to whom, who supervises whom, etc., for every department and staff involved in it.

In a small hotel, front office functions may be carried out by a single person at a reception area or front desk like answering the telephone calls, taking bookings, welcoming and registering guests, billing and processing payments and so forth. But in large hotels, there may be separate/subdivisions of front office department and each subdivision performs their assigned work.

### Activity

Below you have been provided with certain parameters. You need to write down the differences between front office department and front desk/reception section in accordance with the given parameter.

| Parameters | Front office department | Front desk/Reception section |
|---|---|---|
| Nature | | |
| Departmentalization | | |
| Span of control | | |
| Manpower | | |
| Work pressure | | |
| Found where? | | |

## 3.3 FUNCTIONS OF FRONT OFFICE

Front office also has a complimentary role of image building because it is the first as well as the last point of contact of every in-house guest. The fundamental work profile of front office department includes reservation, reception, registration, room assignment to guests and acts as a continuous source of information for guests. Major functions of any hotel's front office department may include:

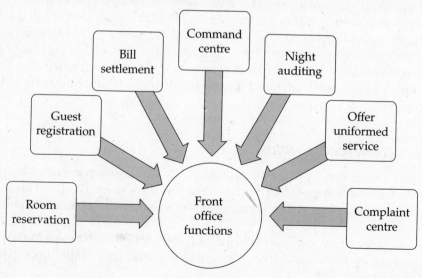

## 3.4   LAYOUT OF FRONT OFFICE / HOTEL ENTRANCE

In general, all sections of front office department are found in lobby area (or in *foyer*). Nowadays, *atrium concept* is quite famous in designing the hotel lobbies. The *atrium concept* is an architectural design in which guest rooms overlook the lobby from the first floor to the roof. It was first used in the 1960s by Hyatt Hotels. The location of different front office sections should always be well planned in accordance with their work role, stage of interaction with guests, stream of work, working relation with each other and so forth.

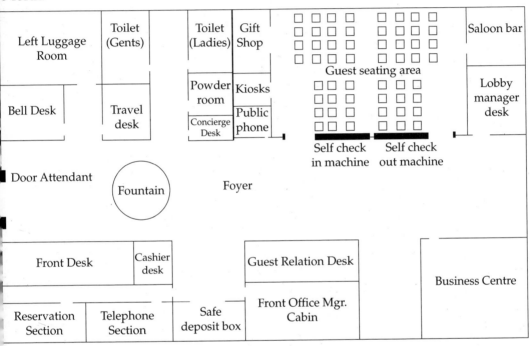

---

### Activity

✓  Arrange 3 trips for your students to visit small, medium and large/chain hotels in your locality. Instruct students to observe the work environment and job profile of various front office employees. Thereafter, students need to submit the project in which they state the vision, mission, goal and objective statement of respective visited hotel.

✓  Students may also be assigned to draw the layout of visited hotel's front office department in which they can display the location of various front office sections along with reasons why these sections are located at their respective places.

✓  Students may also justify the statement, "Many hoteliers believe that employees of small or medium hotels are multi-skilled in comparison to large/chain hotels, as they have to deal with each and every aspect of front office". Support this statement with the help of your observation.

---

## 3.5 FRONT OFFICE EQUIPMENT

The term *front office equipment* is concerned with all those devices/tools/appliances which support while performing different front office activities. It also includes wooden/metal made furniture such as desk/counter, chairs, tables; daily miscellaneous stationery such as voucher, pen, pencil, stapler and other similar items used in the front office department. A list of manual, semi-manual and mechanical equipment used in front office is given below.

| Manual (non-automated front office equipment) | Semi-manual (semi-automated front office equipment) | Mechanical (fully automated front office equipment) |
|---|---|---|
| • Rack (room, key, mail, information, reservation)<br>• Folio bucket<br>• Room compendium<br>• Stationery items<br>• Supportive documents (forms, voucher, folio, brochure, etc.)<br>• Reader board | • Rotatory rack<br>• Bell cart<br>• Typewriter<br>• National cash register<br>• Postal weighing machine<br>• Currency authenticator<br>• Stamps & punching device<br>• Hotel clock | • Self check-in terminal<br>• Video check-out TV<br>• Property management system<br>• Call accounting system<br>• Electronic cash register<br>• Wake-up device<br>• Franking machine |

### Activity

Match the equipment with its operating mode, thereafter browse website and collect pictures of each equipment. Finally paste/draw that picture in given photo column.

| S. No. | Name of Equipment | Operating Mode | | | Photos |
|---|---|---|---|---|---|
| | | NA | SA | FA | |
| 1 | Call accounting system | | | | |
| 2 | Room compendium | | | | |
| 3 | Self check-in machine | | | | |

| 4 | Bell cart | | | | |
|---|---|---|---|---|---|
| 5 | Rotatory rack | | | | |
| 6 | Typewriter | | | | |
| 7 | Franking machine | | | | |
| 8 | Folio well | | | | |
| 9 | Room status board | | | | |
| 10 | Rack (like room rack/mail & message/information, etc.) | | | | |
| 11 | Property management software | | | | |
| 12 | Room occupancy sensor | | | | |
| 13 | National cash register | | | | |
| 14 | Wake-up device | | | | |
| 15 | Point of sale | | | | |
| 16 | Account posting machine | | | | |
| 17 | Fax machine | | | | |
| 18 | Bulletin board | | | | |
| 19 | Luggage net | | | | |
| 20 | Bird cage | | | | |

**Note:** The above given acronym/symbol specifies as: NA (non-automatic), SA (semi-automatic) and FA (fully automatic)

## 3.6  FRONT OFFICE SECTIONS

The *front office* is often confused with reception. The difference between both comes in size, sections, expenses and number of employees. The front office department comprises various sections (or sub-departments) like reservation, reception, concierge, etc., whereas reception is a part of front office department. The term *front office* is used in large organizations whereas small establishments term it *reception*. The functional division of front office department can be categorized into the following general areas:

| | |
|---|---|
| **Reservation Desk** | • Responsible for receiving and processing reservation requests for future overnight accommodations. It also works closely with the hotel's sales & marketing division, especially when large group reservations are being solicited or processed.<br>• Desk must maintain accurate records and closely track room availabilities to properly handle overbooking as well as underbooking. It also deals with reservation modification and cancellation requests. |
| **Registration Desk** | • It welcomes, registers and allots rooms to guest/s. Lastly, gives farewell to guest. The desk usually deals with guest's check-in and check-out. During check-in, desk confirms guest's reservation status. If guest holds reservation, it executes registration formalities like updating GRC, issuing room keys and lastly rooming the guest.<br>• If guest is walk-in then it checks whether room is available or not. If available then provides room information & complete registration otherwise advises alternate hotel. |
| **Cashier Desk** | • It deals with opening of account, collection of payments (either via cash, plastic money or deferred mode) and giving clearance to bell desk to release guest's luggage after settlement of account (during check-out). Straightforwardly, it also processes with guest check-outs.<br>• In many hotels, cashier also involves in dealing with forex whereas other hotels have separate position for forex work. |
| **Forex Desk** | • It deals with exchange/collection of foreign currency, thus desk requires to update itself with daily currency exchange rate. Desk agent must possess knowledge about different licences that requires dealing with foreign currency, i.e., RLM series of licences.<br>• Desk agent should also be aware of other legal formalities like issuing of *encashment certificate* to guest during every foreign exchange transaction. |
| **Mail & Message Desk** | • It deals with collection and dissemination of mails, parcels and messages on behalf of hotel guests and employees. In large hotels, the desk is known as *mini post office.*<br>• Mails of resident guests are directly sent to their respective rooms whereas future guests' mails are sent to reservation office. Mails of past guests are diverted to their *mail forwarding address.* Mails of hotel's operational level employees are sent to time office and managers' mails sent to their respective offices. |
| **Telephone Desk** | • Telephone desk deals with incoming & outgoing calls, locates registered guests and management staff, deals with emergency communication, and assists the front desk clerk and cashier when necessary.<br>• Monitor guests' telephone charges and prepare *traffic sheet.* It contains call details like telephone number, call duration, call timing, call charges and so forth. |
| **Night Auditor** | • Night auditor (works at front desk during the night shift i.e. 11 p.m. to 7 a.m.) posts room rates & pending charges, balances the daily financial transactions of guests & non-guests accounts, house accounts, departmental accounts & open new date (or *date roll*).<br>• It also prepares *early bird report,* room discrepancy report, report of guest with high balance, night report, etc., and submits to concerned authorities. During the night shift, he assists night front desk staff in handling check-in and check-out of guests. |

**Front Office Sections**

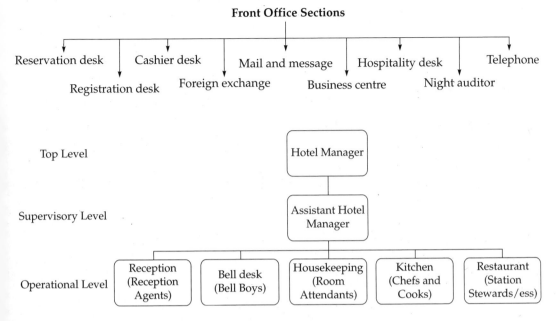

**Organizational Structure of a Small Hotel**

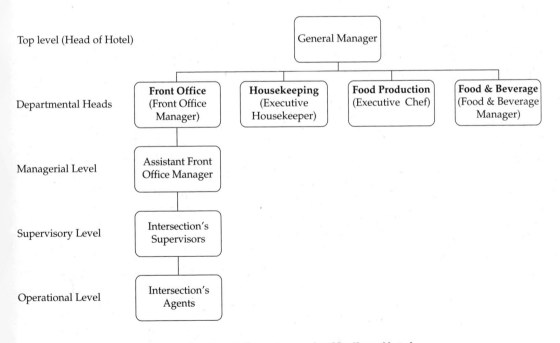

**Organizational Structure of a Medium Hotel**
(With special reference to front office department)

## Organizational Structure of a Large Hotel
### (With special reference to front office)

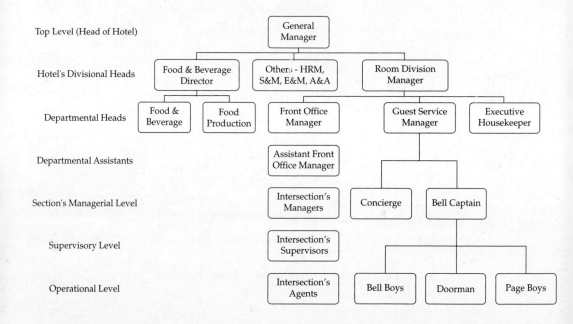

## Activity

Match the levels of skill required with their role the below given positions need to play in any hotel organization.

| Position | Level of skill | Role |
|---|---|---|
| Front office inter-section's agents (like reservation agents) | Technical | Interpersonal |
| Departmental heads (like front office manager) | Conceptual | Decisional |
| Front office inter-section's supervisors (like front desk supervisor) | Human relation | Informational |

# Organizational Structure of Front Office Department of a 5-Star Hotel

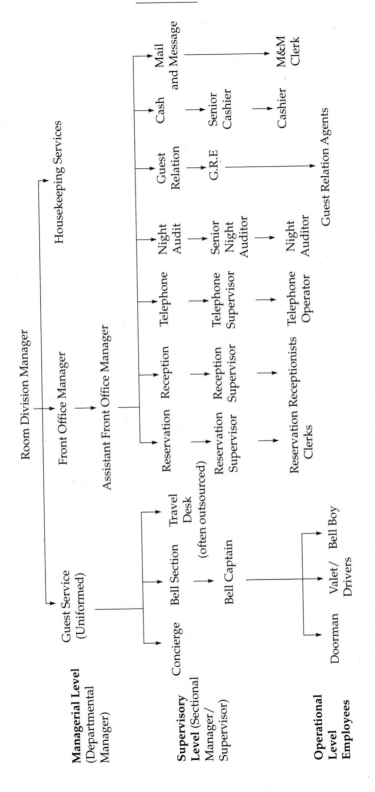

## 3.7　DUTIES AND RESPONSIBILITIES OF FRONT OFFICE STAFF

### Room Division Manager (RDM)

He is accountable for planning, organizing, staffing, directing, controlling, purchasing, budgeting and supervising the front office, uniformed service and housekeeping department. He creates plans to maximize the average daily rate, percentage of occupancy, approve *master budget*, assists in developing room rate strategies/special room rates/discounts/allowances and go through suggestions. He also maintains coordination and open communication with other departmental heads. He requires to stay up-to-date with events taking place in the market and in the competition.

| JOB SPECIFICATION OF RDM | | |
|---|---|---|
| Reports to | - | General Manager |
| Qualification | - | HM Graduate |
| Experience | - | 10-12 years |
| Expertise | - | Management |
| Proficiency | - | Conceptual skills |
| Supervise | - | FO, US and HK |

### Front Office Manager (FOM)

He is accountable for preparing monthly reports, reviews and approvals all room movement and room rate changes to ensure that they were necessary. He handles the guests' complaints and follows them in order to reduce future complaints. If there is separate uniformed service department then front office manager will also supervise it. He also forecasts future room availability to ensure that the optimal level of occupancy is attained. He assists in the training and cross-training of front office employees and in the formulation and implementation of front office policies and procedures. He also fulfils manpower requirement and cost control in front office area.

| JOB SPECIFICATION OF FOM | | |
|---|---|---|
| Reports to | - | RDM |
| Qualification | - | HM Graduate |
| Experience | - | 8-10 years |
| Expertise | - | Management |
| Proficiency | - | Conceptual skills |
| Supervise | - | All front office sections |

### Front Desk Supervisor (FDS)

He/she is also called "reception supervisor" and is responsible for smooth check-in and check-out of guests. He trains and cross-trains front desk personnel in the tasks of registration, mail handling, information services, and check-in and check-out procedures. He also resolves the room discrepancy/*room variance* and prepares the weekly/monthly schedule (duty

| JOB SPECIFICATION OF FDS | | |
|---|---|---|
| Reports to | - | FOM |
| Qualification | - | HM Graduate |
| Experience | - | 6-8 years |
| Expertise | - | Human relations |
| Proficiency | - | Supervisory skills |
| Supervise | - | All front desk activities |

rota and determine weekly offs) of front desk employees. He checks and ensures that daily and hourly computer reports are run properly and distributed to the required personnel. He assists the group coordinator with all group arrivals. He works as liaison between the guests and management and is responsible for assigning VIP rooms and ensure guest satisfaction.

## Front Desk Agent (FDA)

Usually, women are preferred for this post thus it is also known as "receptionist". She mainly attends to guest enquiries and handles check-in and check-out process and sometimes settles guests' hotel bills (in the absence of front desk cashier). She registers the check-in guests and allots room and hand overs room keys to them. She attends the reception phone calls in the absence of staff (switchboard operator). Due to her multifaceted work profile, she requires to come at right time on duty and in the absence of reception supervisor, she also has to manage all his/her work.

| Job Specification of FDA | |
|---|---|
| Reports to | - FDS |
| Qualification | - HM Graduate |
| Experience | - 3-4 years |
| Expertise | - Communication |
| Proficiency | - Operational skills |
| Supervise | - FO Industrial Trainees |

## Reservation Supervisor/Manager (RM)

He prepares daily and monthly room occupancy forecasts. He reviews and gives feedback on the effectiveness of room revenue strategies & tactics. He is mainly responsible for processing group reservation requests, preparing & distributing weekly/monthly reservations & revenue forecast report to the concerned departments. He must ensure that personal service is stressed and that sales techniques being used should be appropriate. He leads, manages and trains reservation agents as well as helps guests to modify or cancel their reservations. He also develops and maintains a good working relationship with the central reservations office and travel agents. He must assure that all reservations, both group and individual, are recorded and followed up.

| Job Specification of RM | |
|---|---|
| Reports to | - FOM |
| Qualification | - HM Graduate |
| Experience | - 6-8 years |
| Expertise | - Human relations |
| Proficiency | - Supervisory skills |
| Supervise | - Reservation Section |

## Reservations Agent/Clerk (RA)

He is mainly responsible for taking reservations and providing future guests with information about the facilities of the hotel. He should give friendly and courteous service to future guests while being involved in telephone sales. He is mainly accountable for answering all reservation phone calls, taking reservations, and dealing with reservations correspondence. He also deals with group bookings, cancellations, changes and rooming lists. He also inspects to perceive that all equipment are working properly and that the needed amount of supplies is in hand or not. He must conduct telemarketing under the direction of the director of sales and marketing.

| JOB SPECIFICATION OF RA | | |
|---|---|---|
| Reports to | - | RM |
| Qualification | - | HM graduate |
| Experience | - | 3-4 years |
| Expertise | - | Telephonic sales |
| Proficiency | - | Operational skills |
| Supervise | - | Industrial trainees |

## Bell Captain (BC)

He is principally responsible for organizing the duties & responsibilities of bellboys who are directly working under him & ensure that every bell staff must follow the prescribed SOPs of the bell desk. He is also directly held responsible for managing and controlling the operation of

| JOB SPECIFICATION OF BC | | |
|---|---|---|
| Reports to | - | FOM or LM |
| Qualification | - | HM graduate |
| Experience | - | 6-8 years |
| Expertise | - | Staff handling |
| Proficiency | - | Supervisory skills |
| Supervise | - | Bellboys or bell desk activities |

left luggage room. He prepares the records of bell staff for the purpose of controlling & promotion. In certain hotels, he maintains a departure sheet/register and informs the housekeeping department about the guest departures. Preparation of staff duty roster and maintaining bulletin board in the lobby section and overseeing left luggage room operations are also his duties. He looks after newspaper/service required by the guest and handles crew & group baggage at the time of arrival & departure. He also ensures the timely & proper recording of *log book* and maintaining the bellboy's worksheet.

## Bellboy/Bell Hop/Bell Attendant/Porter

Bellboy is principally responsible for transporting the guest luggage to and from the hotel room. During rooming the guest, he also gives the information about various services, amenities and facilities provided by the hotel. He presumes all duties and responsibilities of bell desk operation in the absence of bell captain. He supervises the doorman, pageboys and lift operators. He also supervises

| JOB SPECIFICATION OF BB | | |
|---|---|---|
| Reports to | - | Bell captain |
| Qualification | - | Short-term course |
| Experience | - | 1-2 years |
| Expertise | - | Staff handling |
| Proficiency | - | Supervisory skills |
| Supervise | - | Industrial trainees (if any) |

the deposit, receipt and handling of luggage on arrival & departure of guests. In small hotels, bellboy is also responsible to run *errand card*, deliver newspaper in guest rooms, mail handling, key services, telephone operation and placing wake-up calls to guests. In certain circumstances, he also submits the daily reports to the hotel management.

## Concierge/Le Clefs (LC)

*Concierge* is mainly responsible for arrangement of reservations and tickets to shows, events, tours, and theatres. Sometimes he also runs *errands* and assists with airline bookings and reconfirmations. His major responsibility is to work as a guest's liaison with both hotel and non-hotel services and for arranging secretarial services for guests on demand. He directly provides

| JOB SPECIFICATION OF LC | |
|---|---|
| Reports to | - LM or FOM |
| Qualification | - HM graduate |
| Experience | - 4-6 years |
| Expertise | - Communication |
| Proficiency | - Multilingual |
| Supervise | - Industrial trainees (if any) |

courteous, prompt, and tactful service to guests. He is also accountable for reserving space for dining in restaurant and other food & beverage outlets. He assists in arranging sightseeing tours and transportation for the guests in coordination with travel desk. At times, he must also ensure the prompt delivery of messages, mails, telexes, and faxes.

## Telephone Operator (TO)

This area is commonly referred to as *switchboard* and is staffed by switchboard/ telephone operator. Many large size hotels offer room-to-room dialing by which guests of one room can dial directly to another room. The utmost duty of a telephone operator is to transfer/connect calls from outside the hotel to the appropriate guest room. For security measurement, operators must do this without giving out the room number of a hotel guest. The telephone operator may rarely have face-to-face interaction with

| JOB SPECIFICATION OF TO | |
|---|---|
| Reports to | - MOD or FOM |
| Qualification | - Graduate |
| Experience | - 2-4 years |
| Expertise | - Communication |
| Proficiency | - Call handling |
| Supervise | - Industrial trainees/ New appointees |

guests of the hotel, but plays an important role in representing the hotel to the guest. For this position, a friendly and courteous tone of voice is very important.

## Front Office Cashier (FOC)

Front office cashier prepares and settles off guest as well as non-guest accounts. Cashier should fully be aware of company's credit policies, rules regarding acceptance of *credit card*, *traveller cheque* and *foreign currency*. All cashiers should be fully conversant with the use of cash register, credit card approval systems and other machines in relation to their job. Cashier should follow proper procedure & policy

| JOB SPECIFICATION OF FOC | |
| --- | --- |
| Reports to | - CA or FOM |
| Qualification | - HM Graduate |
| Experience | - 4-6 years |
| Expertise | - Accounting |
| Proficiency | - Accountancy |
| Supervise | - New appointees (if any) |

usually concerning with *bill cancellation, rebate, paid out, discount* and so forth. He/she also administers the *safe deposit boxes*, maintains initial *cash bank*, deposits collected cash and other negotiable items and provides a *foreign currency exchange* service.

## Night Auditor (NA)

The basic duties and responsibilities of a night auditor is to perform *audit trail*; prepare daily transcript, supplemental transcript & recapitulation sheet; post-room charges; check and complete guest, non-guest, departmental and house accounts. Night auditors are also responsible for the front desk operation during the overnight shift. His primary responsibilities include

| JOB SPECIFICATION OF NA | |
| --- | --- |
| Reports to | - FOM/AO |
| Qualification | - Commerce Stream |
| Experience | - 2-3 years |
| Expertise | - Bookkeeping |
| Proficiency | - Accounting |
| Supervise | - Industrial Trainees (if any) |

preparing daily reports (like high balance report, night/flash report, guest ledger balance report, room discrepancy report, etc.), balancing transactions, *date roll* and conducting security walks. Due to multifaceted work profile, a night auditor must be able to work independently and with minimal supervision. They must also be able to solve problems and troubleshoot in order to resolve guest issues that may arise and respond to emergency situations.

## Duty Manager (DM)

This position title is also known as "Manager on Duty (MOD)". As the captain of the hotel's front office (especially for hotel lobby), the duty manager has the responsibility of ensuring that every traveller arrives as a guest and leaves as a friend. He/she manages the daily operations of the hotel and contributes to helping all guests have a pleasant and memorable check-in and check-out experience. His/her job specifications include

- Training all front office personnel that how to handle guest requests, and check-in and check-out process.
- Assisting in maintaining the established credit policies.
- Training how to handle guests with scanty baggage, probable skipper and for so forth situations.
- Counsel, guide, and instruct personnel in the proper performance of their duties.

| JOB SPECIFICATION OF DM | |
|---|---|
| Reports to | - FOM or LM |
| Qualification | - HM graduate |
| Experience | - 4-6 years |
| Expertise | - Guest dealing |
| Proficiency | - Multilingual |
| Supervise | - Lobby activities |

## Business Centre Personnel (BCP)

These employees are generally responsible for fax, photocopying and other secretarial services. They also assist hotel as well as guests who are unable to speak English by translating the language of any foreigner into English or local language. They sometimes hire equipment (e.g., laptop, computer & mobile phone) and arrange meeting room for the hotel guests (when guests demand) in hotel's business centre.

| JOB SPECIFICATION OF BCP | |
|---|---|
| Reports to | - LM or FOM |
| Qualification | - HM graduate |
| Experience | - 2-4 years |
| Expertise | - Multilingual |
| Proficiency | - Secretarial service |
| Supervise | - Business Centre Activities |

## Airport Representatives (AR)

Airport representatives are usually responsible for greeting the hotel guests at airport while they are going to receive the guests. If hotels have their own facilities for pick-up and drop then he escorts guest to the hotel's vehicle, meanwhile assisting guest in carrying baggage. Otherwise arrange taxis from airport to hotel for

| JOB SPECIFICATION OF AR | |
|---|---|
| Reports to | - LM or FOM |
| Qualification | - HM graduate |
| Experience | - 2-4 years |
| Expertise | - Multilingual |
| Proficiency | - Custom formalities |
| Supervise | - Pageboys & bellboys at airport |

the guest. At times, he is also responsible for taking the transient bookings (at airport terminal) and assist in departing guests at airport. Often he may also assist guests in operating *self-check in machine* that is installed at airport terminals. Airport representative must maintain good relationship with airline personnel, immigration and custom officers.

## Doorman/Linkman/Door Attendant (DA)/Commissionaire

His place of duty is outside or inside (or nearby) the main entrance of the hotel and he is the first person who meets the guest at the hotel gateway. He is principally responsible for opening and closing the front door during ingress and egress. He oversees the arrival and departure of guest's or hotel's vehicles. During the arrival & departure, he also opens

| Job Specification of DA | |
|---|---|
| Reports to | - Bell captain/LM |
| Qualification | - 12th/craft course |
| Experience | - 1-2 years |
| Expertise | - Communication |
| Proficiency | - Guest welcoming |
| Supervise | - In & out-house activities |

the doors of the car or taxi and escorts the guest to/from the hotel (if *chauffeur* is no appointed). In small hotels, along with door operation, he also performs the job o bellboy and summons car or taxi on guest request. He is also responsible to ensure tha only authorized hotel members enter from the main entrance while other hotel staff use the staff entrance or back entrance.

## Page Boys (PB)/Groom

He is principally responsible for paging which means searching & finding the strange guest at given locations like - in hotel itself (at restaurant, lobby, lounge or any other public areas), at airport terminals, at railway stations & so forth. In certain hotels, he also delivers the messages, letters, telegrams,

| Job Specification of PB | |
|---|---|
| Reports to | - Bell captain |
| Qualification | - 12th/craft course |
| Experience | - 1-2 years |
| Expertise | - Location facts |
| Proficiency | - General aptitude |
| Supervise | - Trainees/new employees |

etc., into guest rooms. In the absence of bellboys, he performs all his duties like—ru errands, escort guests to their allotted rooms, during rooming he can broadcast th various hotel facilities, amenities & services to guests. During rush hours (especiall during the check-in & check-out hours), he may also do the job of carrying luggag and valet attendant.

## Valet Attendant (VA)/Chauffeur

In simple terms, the position title is also known as "driver" who is principally responsible for driving the hotel's vehicle– it may be a car, van or mini bus. Due to his driving oriented work profile, he must have a *driving licence* (DL) and should also have knowledge of routes to different destinations, localities & nearby cities or states. It is also

| Job Specification of VA | |
|---|---|
| Reports to | - LM or DM |
| Qualification | - 12th/craft course |
| Experience | - 4-6 years |
| Expertise | - Driving & routes |
| Proficiency | - Mechanical skills |
| Supervise | - Parking activities |

preferable to have mechanical ability to keep the vehicle in good condition and should be a reliable person with a sense of responsibility. When parking guest's vehicle, he may also need to raise *car docket* and handover it to guest.

| Activity |
| --- |
| Suppose you are a front office manager of a large-size chain hotel. Due to the financial crises, your hotel decided to cut cost by streamlining the organizational structure. As a front office manager, how would you redesign the organizational structure of your department, referring to the previously given organizational structure of large/chain hotel, to suit the need of your hotel? |

## CONCLUSION

Front office department can be called public face of any hotel establishment. It must be equipped with a range of manual, semi-manual and mechanical equipment. Front office department comprises various sections in which different job positions are available from top level to bottom level. The term *organizational structure* is a collective term for establishing formal relations among all these job positions. Resultant, the authority and responsibility flows from top level to bottom level. Remember, the organizational structure may vary with type of establishment, i.e., small, medium, large or chain establishment. The layout of any front office department is mainly responsible for forming first impression as well as also affects the efficiency, effectiveness and work flow of the department.

| Terms (with Chapter Exercise) | |
| --- | --- |
| Divisional head | Head of hotel's major division, i.e., director of food and beverage division and room division manager. |
| Early bird report | *Early bird* refers to cheaper room rate which is offered to those guests who book their rooms well in advance. A report related to all such room rates is collectively known as early bird report. |
| Hotel clock | Clock that shows the official time of the hotel, so the time of this clock must be checked on regular basis. Nowadays, hotel clock also assists in framing time for placing wake-up calls, thus it is usually kept near the switchboard area. |
| Conceptual skills | Skills related with an ability to see beyond the technical aspects of a job. It include skills like recognizing the interdependence of various departments & functional areas within the organization & understanding how the organization fits into the structure of the industry, the community & the wider world at large. |

| ? | A discount allocation or may also be used for refund of money. |
|---|---|
| Safe deposit box | Facility like *bank vault* (or bank locker) which is offered to hotel's resident guests in order to keep their valuables at secured place. This safe-keeping facility is known as *safe deposit box*. |
| Rebate | ? |
| Paid out | Situation whereby hotel employee pays on behalf of resident guest in lieu of purchasing any additional facility or thing from outside the organization like movie tickets. Afterwards, hotel raises a visitor paid out voucher (or guest disbursement voucher) in order to charge the guest. |
| Readers board | Display board on which hotel provides a printed schedule of daily in-house events or/and other relevant in-house activities or information; this display board is also called as "Bulletin board". |
| Operational skills | Functional skills (or work related skills). For instance, skill of a cook, waiter, room attendant, bellboy and so forth. |
| ? | In terms of structural design, it is a large open air or skylight covered space surrounded by a building. |
| Concierge | A French term used for a porter (or porter desk) who is responsible for offering luggage assistance and also provides hotel and locality information to guest during *rooming* process. But nowadays, it is particularly known as information agent. |
| Supervisory skills | Directorial skill of any employee. The term supervision is the combination of two words i.e. super + vision. Here, *super* indicates to excellence whereas *vision* indicates to observatory visualization. |
| Check-in | It is a collective term for all those activities (especially registration related activities) which take place between a hotel and guest during the arrival stage. |
| ? | Systematic process of guest departure from the hotel, thus also known as *departure*. |
| Interpersonal skills | Human relations skills. It is the ability of a manager to work efficiently and effectively with employees at every level in the organization. For this, managers need interpersonal skills that enable them to relate to guests, bosses, peers and employees. |
| Custom formalities | All those activities which take place at airport or seaport during the arrival stage. For example, luggage declaration, currency declaration (in CDF), filling of disembarkation card, etc. Remember, there are two channels for customs clearance: red channel and green channel. |

| Car docket | ? |
|---|---|
| Mechanical skills | Technical skill, especially related with mechanical devices/equipment, for instance, motor vehicle/auto-part handling related skills. |
| Traffic sheet | Telephone department's document (or telephonic call details related sheet) that shows the detail of all calls, especially long distance calls like STD or ISD. It shows information like time of calling, duration of call, place where guest has called, rate of call and final amount of call, etc. |
| Commissionaire | ? |
| Self check-in machine | Automatic system of self-registration for guest. A walk-in guest as well as guest-with-reservation can opt this facility. |
| Log book | Bound book that is used to pass instructions, information or any other notifications to the next shift. It is also used for inter-departmental communication. |
| Encashment certificate | Certificate issued by authorized currency exchange dealer (or licence holder) to *alien guest* (or foreign guest) in lieu of foreign currency exchange. |
| Room variance report | Room discrepancy report. |
| Chauffeur | Post of driver. |
| ? | Hotel entrance, nowadays it is very spacious, having salon bar, business centre and various other front office sections. |
| Errand | Card that contains guests' information in relation to their arrival/departure (time & date), number of pax, number of luggage, etc. It is mainly used by bellboys during arrival & departure. |

# 4 CHAPTER

# Hotel Revenue Sources and Guest Segmentation

$\bigcirc$ **OBJECTIVES**

*After reading this chapter, students will be able to...*

- .identify different revenue sources and their margin of contribution in total hotel revenue;
- discuss different market/guest segment/s and their needs, wants and demands;
- describe the purpose of travel behind targeted market segment/s;
- identify the geographic and demographic pattern of targeted market segment/s; and
- express different group business: MICE, MESE and SMERF.

## 4.1 REVENUE SOURCE

The term *revenue source* is concerned with all those possible revenue points/gateways from where a hotel can generate revenue. The two fundamental revenue sources of a hotel are products and services. *Product* is concerned with tangible things whereas *service* is related with intangible things. Remember hotel industry offers both. It is necessary that the developed product/service(s) should meet/exceed customers' need and want; therefore, it is vital to accurately identify the perceived value of the product/service. In hotel industry, revenue is generated via three major sources, i.e., guest rooms, food and beverage outlets and events space.

## 4.2 GUEST

Guest can also be called a "bona fide traveller". It means, a person who is in fit condition (like not drunken), not carrying any bad record (or not blacklisted), plus able and willing to pay for the demanded services & facilities. A guest may be:

- Anyone who buys (or wants to purchase) a room, meeting space, food & beverages or other services from the hotel.
- Someone who has certain needs and wants (in relation to hotel and its services & products) to be fulfilled. But he/she must be able and willing to pay for the required needs & wants.

- A valued customer who is willing to pay a fair price for a quality product/service and does not want to be overcharged.

## 4.3 GUEST (OR MARKET) SEGMENTATION

It is concerned with division of market (or guest) and selection of few/all/particular market as a *target segment/s*. In order to select right market segment, a researcher requires to develop a profile of each segmented market, thereafter analyze and perform comparative study among them. It is not necessary that a hotel can select only one market segment; selection depends upon hotel's market coverage strategies.

On the basis of selected market segment/s, hotel organization determines its rest of the policies and steps. But before this, it is necessary to understand selected segment's needs and wants, simultaneously marketplace where hotel operation will be carried out (for example, market potential, degree of market competition, average spending power of guests/selected market segment/s, governing laws and so forth).

## 4.4 REVENUE SOURCES

### Guest Rooms (Component of Room Division)

- Revenue is generated through room sales; these rooms can be sold at premium, rack or special rates. Hotel rooms are mostly available in three configurations, i.e., standard, suite and promoted/enhanced. A hotel can sell more number of rooms than available room inventory by offering day use room provision.
- Room rate differs with level of configuration; like suite room will always be costlier than standard room. Remember room sale is a major source of revenue for any residential hotel as it generates more than 50% of total hotel revenue.

### Food and Beverage Outlets (Component of Food and Beverage Division)

- Revenue is generated through the sale of meals and beverages. These can be sold in different food and beverage outlets such as restaurants, coffee shop, through room service, bar and in banquets and so forth.
- Remember food and beverage outlets are ancillary revenue generating sources, thus always count its effect on revenue while accepting or denying any room booking.

### Event Space (Component of Food and Beverage Division)

- It generates revenue through banqueting functions which mainly take place in banquet halls or in open spaces (like ODCs). Banqueting function can be broadly classified into three segments i.e. formal, semi-formal and informal banqueting functions.

- Remember, banqueting functions are bulk revenue generating points. Therefore, booking for banqueting function should always analyze, compared and then only accepted.

## Supportive Services (Component of Hotel's Additional Services)

- It indicates various additional services, provisions and facilities provided by hotels to their guests against extra charges. These supportive centres generate small amount of money but add value to the core product. For example, revenue generation from spa, gym, game zone and so forth definitely add value to property standard.
- Specially, resort hotels are famous for their supportive recreation facilities. For instance, *sea side resorts* offer different kinds of water sports such as scuba diving, boating, rafting and so forth whereas *hill resorts* offer mountain tracking, bungee jumping, ice skiing and so forth.

## 4.5 SEGMENTATION OF GUEST

It is concerned with classification/segmentation of guests on various bases. The classification of guests provides significant information of market/guest turn-in which support in designing marketing strategies of hotel as well as how to efficiently and effectively formulate these marketing strategies. Guest classification also offers opportunities to determine which guest segment is more profitable and vice versa. In addition, also support in developing proper room-sales-mix-ratio/guest-sales-mix-ratio. Guests can be classified in various ways based on their geographic, demographic, psychographic and behavioural factors. But in hotel industry we generally classify the entire market into two broad segments, i.e., group and transient market segment.

## Group Market Segment

It indicates to volume business which generates volume revenue for the hotel. Group market segment can be further divided into several fractions which will vary with group size, purpose of visit and so forth like MESE, MICE and SMERF. A brief description of each is given below. The room reservation pattern of group market segment differs from transient market segment, as in group market segment a hotel can forecast well in advance about the group room demand. And on the basis of historical data, hotel can pre-determine potential *wash factor*, *lead time* and *discount margin* of group market segment. The reservation avenue of group market segment is almost same as of transient market segment. Alike to transient market segment, hotels can also pre-determine their group room sales-mix-ratio; consequently block rooms for targeted group market segments. Rooms kept aside for group market segment is known as *group rooms*.

## *MESE* (It stands for **Meetings, Exhibitions, and Special Events**)

- Generally, the meeting venue has specialized facilities designed to meet the demands of group of people convening for the purpose of a meeting, exhibition or any other special function. The term *special event* is associated with one time or in-frequently occurring event.

- To the attendee, a special event represents an opportunity for leisure, social or cultural experience beyond which is normally available to them. The special event facet of the hospitality industry is the business of conceiving, designing, developing and producing ideas.

## *MICE* (It stands for **Meetings, Incentives, Conventions and Exhibitions**)

- Nowadays, it is a general term for meeting industry. The players involved in MICE business purposely seek to provide services & products to corporations, associations, institutions and government organizations, to facilitate the conduct of their meetings. This segment caters to various forms of business related activities like meetings, international conferences and conventions, events and exhibitions, etc.

- The MICE industry converts the annual business meetings and conferences into a glamorous and enjoyable event for the delegates and attendants. It may also be called "business tourism".

## *SMERF* (It stands for **Social, Military, Educational, Religious and Fraternal Groups**)

- ✓ It indicates the market segment that seeks banquets and meeting facilities. Meetings are mostly organized by SMERF (social, military, educational, religious & fraternal groups), corporations, associations and so forth. In Asian countries, large organizations are using MIDAS (Meeting & Incentives Direct Access System) for generating meeting & incentive oriented business.

- ✓ This database system contains information about the associations, corporations and travel companies that specialize in the meeting related services. Meeting room arrangement is set in accordance to the wishes of the meeting organizer.

### Activity

Enlist several examples to show how a chain hotel could customize a frequent group guests' hotel experience by using the frequent group guests' programme.

# Transient Market Segment

These are non-group market segment/s which are mainly classified into two broad groups, i.e., *business transient* and *pleasure transient* groups. The room reservation pattern of transient market segment differs from group market segment, as there is no way to forecast when transient guests will arrive. Generally, front office manager uses historical data to determine the most likely demand prototype of transient guests. On the other side, the reservation avenue of transient market segment includes CRS, GDS, internet, fax, mail, through travel agency and so forth. Hotels also predetermine their transient room sales-mix-ratio, consequently block rooms for business and pleasure transient market segment. Rooms kept aside for transient market segment are known as *transient rooms*.

## *Pleasure Market Segment*

This specific market segment includes vacationers or family travellers, travellers who travel for the purpose of recreation like pleasure travel by elderly and travel by singles or couples. Some common features of pleasure guest are:

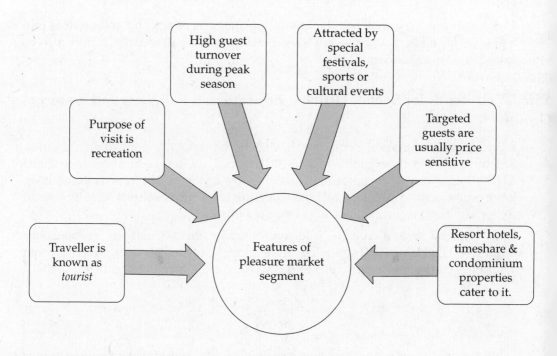

| Activity | | | | | | | |
|---|---|---|---|---|---|---|---|

In below given table, fill in the blank spaces with correct descriptions/differences.

| Differences | | | | | | Guest Types | |
|---|---|---|---|---|---|---|---|
| 1 | 2 | 3 | 4 | 5 | 6 | | |
| ? | ? | ? | ? | ? | ? | Regular | Walk-in |
| ? | ? | ? | ? | ? | ? | Pleasure | Leisure |
| ? | ? | ? | ? | ? | ? | Individual | Group |
| ? | ? | ? | ? | ? | ? | FITs | GITs |
| ? | ? | ? | ? | ? | ? | Domestic | Alien |
| ? | ? | ? | ? | ? | ? | Skippers | Blacklisted |
| ? | ? | ? | ? | ? | ? | VIP | CIP |
| ? | ? | ? | ? | ? | ? | SPATT | Lady |
| ? | ? | ? | ? | ? | ? | MICE | MESE |

## *Business Market Segment*

Business travellers mainly include sales representatives and travelling professionals. Such guests almost invariably occupy single rooms and stay for one or two nights only. Regular business traveller is an important source of business for many lodging properties. Some common features of business travellers are:

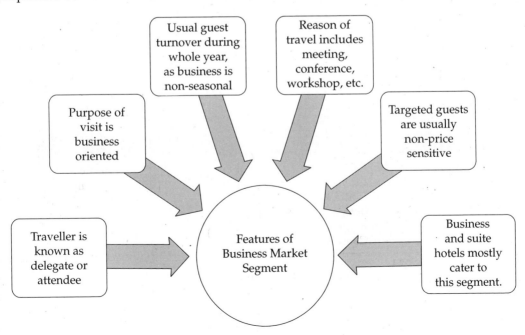

## Activity
### (Brainstorming)

Referring to the information you learnt in this chapter, imagine yourself as a pleasure or business traveller. What kind of a hotel would you like to stay in? In five minutes, brainstorm with your partner a list of the qualities of an idyllic hotel. Use as many adjectives or phrases as possible to describe your idyllic hotel. Here are examples of statements to use to start the brainstorming process:

| As a pleasure traveller | As a business traveller |
|---|---|
| ✓   An ideal hotel should be.................. | ✓   An ideal hotel should have.................. |
| ✓ | ✓ |
| ✓ | ✓ |
| ✓ | ✓ |
| ✓ | ✓ |

| Number of pax | Geographic | Behavioural | Importance | Past history | Reservation status |
|---|---|---|---|---|---|
| • Individual<br>• En block | • Domestic<br>• Foreigner | • Timid<br>• Fussy<br>• Irritating<br>• Socializing | • VIP<br>• CIP<br>• MIP | • Blacklisted<br>• Hotlisted<br>• Skipper<br>• Regular | • Chance guest<br>• Guest-with-guaranteed booking<br>• Guest-with-confirmed booking |

| Nature of travel | Business source | Recognition | Purpose of visit |
|---|---|---|---|
| • Touring<br>• Airline<br>• Transit | • Family<br>• Incentive<br>• Company delegates<br>• Employees<br>• Consortia | • Incognito<br>• HWC<br>• SPATT | • Business<br>• Pleasure |

Guest segmentation parameters

## Activity

Match the descriptive sentences with their correct guest types.

| S. No. | Description | Guest |
|:---:|---|---|
| 1 | Guest travel to attend workshop | Individual |
| 2 | Team achieved their sales target, thus awarded with a foreign tour | Domestic |
| 3 | Physically disabled guest | Fussy |
| 4 | Nervous or shy guest | Group |
| 5 | Guest skip without settling account | Skipper |
| 6 | Guest doesn't want to disclose his identity | Transit |
| 7 | Guest travelling with his family for recreation | Timid |
| 8 | Guest travelling alone | Chance |
| 9 | En-block guests | Incentive |
| 10 | Guest with awful previous record | VIP |
| 11 | Delegates of summit | Blacklisted |
| 12 | Guest travel within the region | Foreigner |
| 13 | Airline passenger | Business |
| 14 | Picky or choosy guest | Incognito |
| 15 | FFIT | Pleasure |
| 16 | Guest without advanced booking | SPATT |

## CONCLUSION

There are varies sources of revenue for a hotel organization. The principal product of hotel to produce revenue is *room inventory*, simultaneously room sales also contribute around 50% of total hotel revenue. The remaining 50% revenue is generated by other points of sale outlets and supportive departments which mostly include food and beverage sale. Like any other organization, hotels also segment to market then select suitable segment/s for the purpose of selling products and services. Generally, hotels are classified according to market segment/s into two broad groups i.e. *transient segment* and *group segment*. The transient segment is classified on the basis of purpose of travel (leisure or business) whereas group segment classifies on the basis of source of origin

(MICE, MESE or SMERF). For effective selling, hotels profoundly study the geographic, demographic and behavioural pattern of selected market segment/s.

| Terms (with Chapter Exercise) | |
|---|---|
| Wash factor | Estimation/prediction of rooms in a group that may be cancelled in future. It is totally based on historical data, market trend and demand. It is also known as *wash down* or simply *wash*. |
| Lead time | Time gap between room reservation and its confirmation/guarantee date. |
| MICE | Meetings, Incentives, Conventions and Exhibitions. It indicates to a group business segment. |
| MESE | Meetings, Exhibitions and Special Events. It also indicates to a group business segment. |
| SMERF | ? |
| Bonafide traveller | All those travellers who do not have any blacklisted (or unwanted) kind of blameworthiness. In simple terms, it indicates to a guest with legitimate background. |
| Consortia rate | Room rate that can be negotiated between a hotel and a travel agency group. A consortium is a conglomerate of travel agencies, for example, American Express, Navigant, and Carlson Wagon Lit. |
| En-block | ? |
| Alien | Foreign clientele. |
| Blacklisted | A list of undesirable people who are not allowed to be entertained into the hotel as they may have a bad police record or previously skipper or any other reason. |
| ? | All those guests who display unnecessary excitement, activity or intent in all situations. These guests will always show unnecessary concerns about the products and services provided by the hotel to them. |
| Timid | All those guests who lack confidence thus often found that these guests are very shy in nature and avoid interacting not only with hotel staff but even with other guests of the hotel too. |
| Incognito | Famous personalities like celebrities and sportsman who want to stay undercover. |

| | |
|---|---|
| Condominium | Concept or type of hotel establishment where membership programme, in lieu of renting hotel rooms for a particular period of time in the long run, are offered to guests. It offers *full ownership* and RCI (Resort Condominium International) is the world's largest condominium resort hotel chain. |
| Timeshare | ? |
| GDS | Global Distribution System and indicates to a reservation network (or software) in which different sectors/industries participated together. For example, SABRE, Amadeus, Galileo, Quantas, etc. Conventionally, it is known as ADS {Airline Distribution System as it is a reservation network (or a kind of CRS) for consortium of airlines} |
| CRS | Central Reservation System which is a reservation network of hotel industry such as Holidex 2000, MARSHA III, etc. CRS of chain organization is known as affiliated CRS whereas CRS of referral groups (i.e. group of independent hotels) is known as non-affiliated CRS. |
| Revenue source | All those channels/points from where revenue is generated for the organization. |
| Business tourism | A particular tourism type (or tourism segment) which is commerce oriented. |
| Discount margin | Differential profit margin (DPM) between expected (or pre-determined) selling price of a product/service and actual (or charged) selling price of the product/service. |
| ? | Non-group business segment like chance guest, incentive guest, family guest, corporate client, company guest. |
| Transient rooms | Rooms which are booked at either rack, corporate, incentive, package, government, or foreign traveller rates. It also includes occupied rooms booked via a third party websites (except group booking). |

# 5
**CHAPTER**

# Prologue with Rooms

**OBJECTIVES**

*After reading this chapter, students will be able to...*

- define a room and how it can be classified;
- discuss types of hotel rooms and their configuration;
- diagrammatically understand the layout of hotel rooms; and
- identify the importance of these types of hotel rooms.

## 5.1  ROOM

*Room* is a basic product of every hotel and it has limited shelf-life of 24 hours, therefore room is also featured as a perishable product of hotel industry. Every hotel room generates different level of revenue even when their room size, layout, configuration, amenities & services are similar. It is due to difference in booking *lead time*, type of clientele, number of pax, required number of rooms, season/market demand; pre-determined room *sales-mix-ratio*, stage of strategy implementation and ancillary revenue generation by guest. As all hotels have explicit room inventory therefore it is always advantageous that every room must be lettable at all times.

The term *room inventory* indicates the maximum number of rooms available with hotel for sale on per day basis but when it is calculated for a week (or more); in this case the total number of room nights for a week will be known as *room supply*. Remember room inventory is different from lettable rooms. *Lettable rooms* indicate total number of saleable rooms. It excludes house use rooms, out-of-order rooms (OOO) and sometimes out-of-service rooms (OOS) too. Therefore, it is always desirable to minimize OOO and OOS rooms as much as possible but it doesn't mean that hotel should compromise with the situation. Instead hotel should undertake renovation (or repair) work of OOO rooms during shoulder/off-season. Remember, there is significant difference between OOO and OOS rooms. OOO rooms are usually kept aside for long duration whereas OOS rooms are blocked for short periods.

At the time of accepting room reservation request, reservation agent must consider the guest's room preference because room preference will lead towards guest satisfaction which create base for future repeat business. *Room preference* is an individual guest's choice in terms of room type and its configuration, designation and location. The term *room configuration* is related with the physical characteristics (or makeup) of the guest room. On the other side, the term *room designation* indicates whether the room is a smoking or non-smoking room and *room location* specifies the room position on a particular floor/place. Remember, the term *room number* is different from all these terms; room number stands for a specific room. All these terms are searching parameters for *room identification*.

## Activity

Match the below given standard rooms with their nature of expense, number of beds placed in and pre-determined occupancy level (or standard number of pax can accommodate).

| S. No. | Description | Terms |
|--------|-------------|-------|
| 1 | Perishable product of any hotel | Room preference |
| 2 | Room composition (or room make-up) | Room inventory |
| 3 | Guest choice for room | Room location |
| 4 | Particular room only | Rooms |
| 5 | Physical position of rooms | Room configuration |
| 6 | Whether room is a smoking or non-smoking | Room number |
| 7 | Total number of rooms | Room designation |
| 8 | Room searching parameters | Room identification |

## 5.2  TYPES OF ROOMS

Hotel room is usually categorized by its layout that includes bathroom, sleeping room, *parlour room* (or convertible parlour room) and other features such as room size, configuration, amenities, services & facilities. Generally, it is the room configuration that support in determining the rack rate and margin of discount for different targeted market segment/s. This room configuration can be classified as *standard, suite* and *promoted*. The size of room and bathroom also plays an important role in determining room rates and classifying hotel. For example, dimension of a single room will always smaller than double room, thus double room will be always costlier than a single room. But size of both rooms as well as bathrooms varies with category of hotel. This classification criterion has been given in Indian government hotel classification system. But remember, there is no room and bathroom size as prescribed in classification system for any heritage hotel.

| Category | Single Room | Double Room | Bathroom | Remarks |
|---|---|---|---|---|
| 5 stars deluxe & 5 stars | 180 sq. ft. | 200 sq. ft. | 45 sq. ft. | Fully AC |
| 4 stars & 3 stars | 120 sq. ft. | 140 sq. ft. | 36 sq. ft. | AC / Non-AC |
| 2 stars & 1 star | 100 sq. ft. | 120 sq. ft. | 30 sq. ft. | AC / Non-AC |

**Standard configuration or standard rooms** (offer single room for single plus multiple occupancy)

The basic category of hotel room which comprises maximum number of rooms in total room inventory is of *standard rooms*. It is the standard room group on which hotel management offers different margins of discount, consequently attract targeted market segment/s. Once the market demand flourishes for the hotel rooms, front desk agent or reservation agent may perform suggestive selling and upselling in order to generate more room revenue. Standard room types are based on the intended number of occupants. Hotel can further categorize to its standard room types, based on their intended number of occupants like, single, double, triple and quad rooms.

| Single room | • It is designed for single occupancy, thus contains one standard size single bed. |
|---|---|

| Double room | • It is designed for double occupancy, thus contains one double bed or king size bed. |
|---|---|

| Twin room | • It is also designed for double occupancy but contains two single beds. It is also known as twin bedded room or queen double. |
|---|---|

| Triple room | • It is designed for triple occupancy thus contains three single beds or one double bed with one extra bed. It is also known as triplet room. |
|---|---|

| Quad room | • It is designed for quad occupancy thus contains four single beds or two double beds. |
|---|---|

## Activity

Match the below given standard rooms with their nature of expenses, number of beds placed in and pre-determined occupancy level (or standard number of pax can accommodate).

| S. No. | Rooms | Description | Beds | Occupancy |
|---|---|---|---|---|
| 1 | Single | More expensive than double room | 2 single beds | 2 pax |
| 2 | Double | More expensive than triple room | 1 double + 1 single bed | 3 pax |

| 3 | Twin | More expensive than single room | 1 single bed | 4 pax |
| 4 | Triple | Less expensive than twin room | 2 double beds | 1 pax |
| 5 | Quad | Less expensive than double room | 1 king size bed | 2 pax |

**Suite configuration or suite rooms** (set of two interconnected rooms = living room + sleeping room)

It is the most expensive type of room (or room configuration) because suite rooms are bigger in size (in terms of floor area) and also offers greater level of services & facilities along with extra amenities in comparison to standard rooms. These rooms are usually sold to elite clientele. In suite rooms, there are two interconnected rooms, i.e., *living room* (also known as parlour room or welcome room) and *bedroom* (or sleeping room), thus expensive in nature. Suites are named according to the décor or its view. Generally, suite rooms are located on the higher floors (or on bi-level floors). Suite rooms can be further subdivided into several types/grades; some of them are as follows:

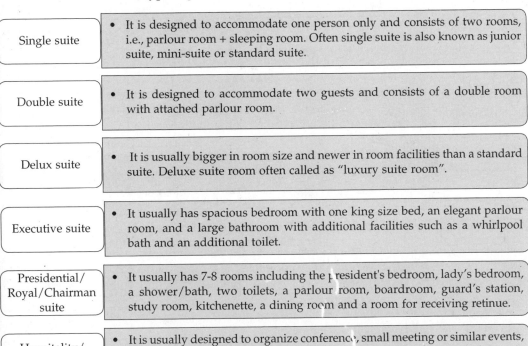

| Single suite | • It is designed to accommodate one person only and consists of two rooms, i.e., parlour room + sleeping room. Often single suite is also known as junior suite, mini-suite or standard suite. |
| --- | --- |
| Double suite | • It is designed to accommodate two guests and consists of a double room with attached parlour room. |
| Delux suite | • It is usually bigger in room size and newer in room facilities than a standard suite. Deluxe suite room often called as "luxury suite room". |
| Executive suite | • It usually has spacious bedroom with one king size bed, an elegant parlour room, and a large bathroom with additional facilities such as a whirlpool bath and an additional toilet. |
| Presidential/ Royal/Chairman suite | • It usually has 7-8 rooms including the president's bedroom, lady's bedroom, a shower/bath, two toilets, a parlour room, boardroom, guard's station, study room, kitchenette, a dining room and a room for receiving retinue. |
| Hospitality/ Hospos suite | • It is usually designed to organize conference, small meeting or similar events, for the purpose of entertaining selected meeting delegates. Therefore, it is often known as function room. |

**Promoted configuration or promoted rooms** (rooms with promoted goods = soft goods + hard goods)

It is concerned with additional goods, amenities, facilities and services that are offered to guests to meet their expectations. A room with promoted arrangement may include anything from nice view, balcony to any amenity which add value and give base to hotel to charge more. This room configuration falls between 'standard and suite' configuration (or next level after standard configuration) and often also known as *upgraded/enhanced configuration* and *concierge* or *business level rooms*. Ideally, the promoted room configuration includes standard room configuration with additional amenities and/or service. The promoted configuration carries a higher room rate than the standard room but lower than suite rooms. These rooms are usually targeted towards VIPs, regular and loyal guests and often to unhappy guests also. It include:

| | |
|---|---|
| Penthouse | • It is a room that usually opens to a private rooftop courtyard area. These are generally located on the top floor of the building. The room configuration is also very much similar to suite room, therefore it is expensive too. |
| Efficiency room | • It is a room that has kitchen facilities like cooking range, microwave, wash basin, etc., provided as part of the room. Often these rooms are also known as cottage rooms in resort hotels. Nowadays, many properties also offer whiteware in these rooms. |
| Lanai room | • It is a Hawaiian term for a hotel room with a balcony or patio with an overlooking of landscaped area or scenic view. Thus, these rooms are usually found in hill resorts. |
| Sico room | • It is a multi-purpose room that can be used as a meeting room (for the gathering of 10-12 people) as well as sleeping room. |
| Hospitality room | • It is a promoted or suite room that is arranged for the convenience, comfort and socialization of guests, often with drinks and snacks. |

## Activity

Browse the website of Indian hotels then complete the table by filling the columns with five appropriate soft goods and hard goods. The first one has been done as an example for you.

| Soft Goods | Hard Goods |
|---|---|
| Drapery | King size bed |
| ? | ? |
| ? | ? |
| ? | ? |
| ? | ? |

## 5.3 ADDITIONAL ROOM TYPES

Room types which are usually not found in every hotel organization. It depends on individual property whether to construct these additional room types or not. Therefore, the below given additional room types may also go beyond it.

| | |
|---|---|
| Duplex | • It is a double storied suite room that is connected by staircases internally. Duplex rooms are usually the parlour rooms which are on the lower levels/ground floor and the sleeping room on the higher levels/second floor. |
| Adjoining room | • Two or more rooms immediately adjacent to each other but without a connecting door. In simple terms, rooms which have a common wall but no interconnecting door. |
| Adjacent room | • Rooms which are close to each other but may not have a common wall. In simple terms, rooms close by or across the corridor, but are not side by side. |
| Interconnecting room | • Rooms having an individual entrance from outside with a common connecting door between them. The common door is locked when not in use. One basic advantage of interconnecting rooms is that, both these rooms can be sold collectively as well as separately. |
| Cabana room | • A room on a beach or next to a swimming pool which usually opens for 24 hours and may have been designed as an accommodation unit, especially as a changing room, or as a structure to provide shade and a place to sit. But never as a sleeping room. |
| Alcove room | • The concept of Alcove room was borrowed from Spain. It is a small room attached to a bigger room and often designed to accommodate a bed, pianos, etc. |
| Siberia | • It is an inferior kind of room that is mainly situated nearby or under the staircases. If this kind of room is sold, then it must be informed to the guest in advance. |
| Hollywood twin room | • It has two single beds with a common head board and can accommodate two people only. Sometimes, it is also known as hollywood living room. |

| | Activity | |
|---|---|---|
| **S. No.** | **Questions** | **Room Type** |
| 1 | A room for single occupancy but may be fitted with a single, double or queen-size bed. | ? |
| 2 | Rooms that are side by side, but do not have a connecting door between them. | ? |
| 3 | Rooms close by or across the corridor, but are not side by side. | ? |
| 4 | Two rooms that are side by side and have a connecting door between them. | ? |
| 5 | A room for double occupancy and has two twin beds. | ? |
| 6 | Room with one or more bedroom/s and a living space. The bedrooms might be single, double or twin double. | ? |
| 7 | A room for double occupancy and has two twin-beds but joined together by a common headboard. | ? |
| 8 | A room for triple occupancy and has been fitted with three twin beds or one double bed with one twin bed. | ? |
| 9 | A room for double occupancy and has been fitted with a double or queen size beds. | ? |
| 10 | A room for multiple occupancy and has been fitted with two twin, double or queen size beds. | ? |
| 11 | A room usually targeted for family guest and fitted with kitchenette facility. | ? |
| 12 | A room fitted with all electronic switches at lower height. | ? |
| 13 | An inferior room which is mainly situated under the staircases. | ? |
| 14 | A room that is constructed nearby swimming pool and mainly used for changing clothes. | ? |
| 15 | Composition of two rooms that split floor-wise, thus interconnected by staircases. | ? |
| 16 | Room on top of the floor with swimming pool facility. | ? |

## Room Status Terminology

| | | | |
|---|---|---|---|
| Complimentary | A room offered free of charge (FOC). | **Room Status** | **Code** |
| Did not check out | A room which is still occupied. | No need to serve | NNS |
| | | Extra bed | XB |
| Sold out | It designates full occupancy. | Light baggage | LB |
| | | Check-out | CO |
| Day use room | A room that is offered for day use only. Day rate is charged to guest. | Out of service | OOS |
| | | Free of charge | FOC |
| Do not disturb | It means that guest has requested not to be disturbed, by hanging DND or switched on red light. | Sleeper | SLP |
| | | On change | O. Ch. |
| Vacancy rate | It denotes the average number of empty rooms in a hotel over a period of time. | Due out | DO |
| | | Do not disturb | DND |
| | | Out of order | OOO |
| Due out | It denotes that the guest is expected to leave the hotel room by today itself. | Out of town | OOT |
| | | No baggage | NB |
| | | Sleep out | SO |
| Late check-out | It denotes that the guest is allowed to check-out after the standard check out time. | Stayover/stay on | Sty. O |
| | | Skipper | SKP |
| Lock-out | It denotes that the hotel locked the guest room due to some legitimate reason like non-settlement of bill. | Lock out | LO |
| | | Did not check-out | DNCO |
| Occupied | It denotes that the room is currently occupied by the resident guest. | Occupied & clean | O/C |
| | | Occupied & dirty | O/D |
| On change | It denotes that the guest room is under the cleaning process for new arrival. | Vacant & clean | V/C |
| | | Vacant & dirty | V/D |
| | | Expected departure | ED |
| Out of Order | It denotes that the room is not ready for sale (for long duration), as it may be under working/renovation. | Vacant & Ready | V/R |
| | | Occupied | Ocup. |
| | | Late check-out | LCO |
| Out of Service | It denotes the same status/situation as above; the room is not ready for sale but for short duration. | Day use room | DUR |
| | | Complimentary | Compl. |
| | | Did not arrive | DNA |
| Skipper | It denotes that the guest has left the hotel room without settling his/her bill/account. | Arrival notification slip | ANS |
| | | Departure notification slip | DNS |

| Walk out | It denotes the same situation as above guest has left without settling his/her bills/account. |
|---|---|
| Makeup room | It denotes the process of changing linen and cleaning a guest's room (or room under change). |
| Sleep out | It denotes that the guest has been registered into the room but does not use the hotel's room bed. |
| Sleeper | It denotes that the guest has left the room (after proper settlement of account) but status shown by front desk indicates that he/she is in the room or room is still occupied. |
| Stayover | It denotes that the resident guest is not expected to check out today (as stated earlier) and will remain at least one more night. |
| Overstay | It denotes that guest remains (with prior permission from hotel) in the hotel room beyond the scheduled date of departure. |
| Vacant & ready | It denotes that the room has been vacated then cleaned and is fully ready for sale. |

## Activity

Differentiate between the following:

| Sold out vs walk out | Due-in vs due-out | Room make up vs make up room |
|---|---|---|
| Sleep out vs walk out | Sleeper vs skipper | Out of order vs out of service |
| Stay-over vs overstay | Lock out vs late check out | Vacancy rate vs rack rate |
| Scanty baggage vs left luggage | | |

## CONCLUSION

As *room inventory* is responsible for generating maximum revenue for hotel. Similarly, it is also a fact that it is highly perishable in nature. Hotel room has 24 hours of self-life within which room can also sell at multiple times, particularly twice a day like in case of day use room. Every hotel has maximum number of standard rooms in comparison to suite and promoted rooms. Each room generates different level of revenue at every sale because room rate varies with guest type, *reservation horizon* (or *lead time*), meal plans, day-use provision, *occupancy level* and so forth. Sometimes even the same room when sold to same clientele, it may generate different amount of revenue. But every hotel always tries to achieve as much revenue as possible on all occasions. In lieu to generate maximum revenue, hotel offers additional room types along with most available room types like interconnected rooms, *disable friendly rooms*, rooms on EVA floor, efficiency room and seco/sico room. Nowadays, at every sale (or reservation call) hotel enlightens to guest that they have rooms in accordance to their needs and wants whereas most of the hotels achieve this target by adopting overbooking strategies.

| Chapter Terms (with Chapter Exercise) | |
|---|---|
| Parlour room | It is also known as *welcome room* or *living room* which usually contains sofa-cum-bed (or couch), television, etc. |
| Soft goods | Amenities like drapery, bedspreads, bathrobes, etc., which increased room rate. |
| Room configuration | It is concerned with in-room facilities, amenities, supplies, whiteware and other arrangements. |
| Hard goods | ? |
| Couch | It usually indicates a sofa-cum-bed. |
| Room inventory | Total number of available rooms on particular revenue day in the property including OOO and OOS. |
| Room preference | Room choice of guest like on ground floor or nearby elevator, etc. |
| Room location | Physical location of any room like room is located on the second floor. |
| Room designation | ? |
| White ware | Freezer, washing machine and sometimes, a dryer also. It is mainly offered in cottage rooms. |
| ? | Rooms that are used by hotel for its own purpose, for example, rooms used by hotel employees for executive offices, hotel storage, etc. |
| Room supply | Number of rooms in a hotel or set of hotels (or room inventory, i.e., number of rooms on per day basis) multiplied by the number of days in a specified time period. For instance, room inventory of 100 rooms in subject hotel × 31 days in the month represents room supply of 3100 for the month. |

| Room status | It is a daily report or a term that gives information on the condition of rooms available in a hotel. |
|---|---|
| Room code | Acronyms made of letters or number written on lists to mark room status. |
| Occupancy level | ? |
| Reservation horizon | Reservation booking duration or time horizon. Simply, it is related with time period of booking and for its confirmation. |
| Convertible parlour | Room with couch. Simply, sitting room (or parlour room) with sofa-cum-bed. |
| Whirlpool bath | Heated pool (or bath tub) in which hot aerated water is continuously circulated. |
| OOO | Rooms sited as out-of-order due to some major faults like heavy repair & maintenance, rooms going under renovation and so forth reasons. It affects occupancy level and forecast. |
| ? | Rooms temporarily blocked off due to minor repair works like AC, general cleaning or due to low occupancy level. It doesn't affect occupancy forecast. |
| Function room | Meeting room or business centre but often found to be used as a banqueting function room also. |
| Vacancy rate | Average number of empty rooms in a hotel over a period of time. |
| Sales-mix-ratio | Fragmented composition of room inventory for different market segment/s or vice versa (i.e., fragmented composition of market segment/s for available room inventory). |
| Disable friendly room | ? |
| Due-in | Expected number of rooms/guests going to check-in into the hotel. |
| ? | Expected number of rooms/guests going to check-out from the hotel. |
| ? | Room with balcony and from where scenic view appears. |
| Lead time | Latency (or gap period/duration) between guest's non-guaranteed (or room enquiry) room reservation and its confirmation date. |

# 6
CHAPTER

# Room Tariff

*After reading this chapter, students will be able to...*

- define room tariff and prepare room tariff card;
- discuss different types of room rates and different basis for determining room rates;
- elucidate how discounted rates negatively affect DPM (differential profit margin);
- explain inclusive and exclusive room rate plans; and
- find out different approaches of room rates, i.e., rule of thumb, Hubbart, market condition & others.

## 6.1 ROOM TARIFF

The rates charged by the various hotel establishments for their rooms are called "room rates or room tariff". There are different types of room rates like premium rate, rack rate, BAR, ARR and special room rates (or discount rates) which erratically affect the profitability (or DPM) of the hotel establishment. Hotel room is a perishable product and its cost will never be recovered in future if not sold on any particular day or night. Therefore, hotel needs to offer different room tariffs to different targeted market segment/s during all seasons. For example, transient market segment offers higher room tariff than group market segment. Similarly, off-season rates are always lower than shoulder season and peak season rates.

Every hotel uses *room tariff card* to dictate its planned room tariff to selected market segment/s. This tariff card mostly indicates to rack rates (in accordance with different *meal plans*), rates for every extra/additional bed (if guest demands) and applicable taxes & service charges. The size, design, structure, length and quality of paper used to make room tariff card vary with hotel to hotel.

Hotel room rates are similar in many ways to airline fares. For example, airlines have different rates for different classes (like $1^{st}$ class is expensive than economy class). Similarly, hotels also offer different rates for different room configurations. They are both *quantifiable* (as they can be measured & structured to meet certain criteria) and *qualifiable* (as large amount of discretion is allowed in which rates are implemented).

| Activity |
|---|
| • Can a *room tariff card* be used as a marketing tool? Collect at least two room tariff cards from a nearby hotel. And list down three changes you would like to make on the tariff card to improve its marketing potential. |
| • Suppose you are a personal assistant to the Managing Director of Company Hot Lemon in Indonesia. Your boss needs to travel to Australia, thus he asks you to book a hotel having rooms with all the facilities & amenities required for a comfortable stay but within a budget of around INR 20, 000 per night. Now you have to compare the room configuration of different hotels by browsing their websites and recommend to the Managing Director on the choice of hotel within the assigned budget. |

## 6.2   BASIS FOR ROOM TARIFF FIXATION

### Market Based Rates (or Competitors Based Room Rates)

✓ It is also known as *competitors based* or *market condition based pricing policy* because many hotels set their room rates on the basis of competitors' prices or marginally lower/higher than the competing hotels. This pricing policy is usually followed by small hotels. There are two types of market competition in hotel industry, i.e., price competition and product/service competition.

✓ *Price competition* is concerned with impact of change in price on hotel's room demand. The general principle here is the higher the price level the less responsive the demand and *vice versa*. On the other side, *product/service competition* is concerned with the impact of quality of product/service on hotel's room demand. If you revert to the problem of the pricing situation then put great emphasis on the excellence of products & services of the hotel establishment.

### Customer Based Rates (or Market Segment Based Room Rates)

✓ It means a specific market segment/s to which a hotel establishment has chosen to sale its products (rooms) and services. Therefore, hotel management has to direct its marketing efforts towards that specific targeted market/s. It is also a vital room rate determinant, as management needs to consider in advance that offered rate should be within the *average spending power* of selected/targeted market segment/s. Here, hotel can classify its market segment into two broad groups, i.e., transient market segment and group market segment.

✓ *Transient market segment* (or non-group market segment) includes room sale to any individual guest like walk-in, FIT, FFIT, family, incentive guest, company representative, corporate client, incognito guest, regular individual guest, crew members and so forth. *Group market segment* includes room sales to group or its members like GIT, MICE, MESE and SMERF, association, corporations, tour groups and so forth.

✓ Remember, at the time of determining group (or giving *group designation*) it is vital to consider the size or number of pax in any group. In hotel industry, usually the term *group designation* is given to a cluster of twelve or more people. Lastly, group rate will vary as group size varies.

| Activity |
|---|
| In the given table, fill in the blank spaces with correct description thereafter match with suitable room rate. |

| Guest Type | Description | Rate Offered |
|---|---|---|
| Incognito | Famous personalities like celebrities who want to stay undercover | Rack rate |
| Transient | ? | |
| Regular | Hotel's permanent guests who predominantly book the same hotel | |
| Group | ? | Special rate |
| Reservation | ? | |
| Chance | Walk-in guests who do not hold any advance booking with hotel | |
| Incentive | ? | Premium rate |
| Business | Targeted guest segment/s which belongs to business sector. | |
| Corporate | ? | |

# Capital & Cost Based Rates (Based on Capital Invested & Operating Cost)

✓ Capital represents the amount originally contributed by the proprietor/partners that are increased by profit and decreased by losses and drawings. At the time of determining the room rates for different room configurations, owner/management always keeps in mind as to what will be the rate of return (and profit) on capital invested (or ROI). Management splits this invested capital into number of years then determine profit margin for each financial year and finally conclude the room rates.

✓ Cost measures the relationship between total room revenue for break-even and required profit margin and number of room sales. Total cost comprises *operating cost* which is concerned with fixed (cost remains same whether production is elastic or inelastic), variable cost (cost varies with elasticity of production) and marginal cost (cost varies with every additional production/unit). In it, mainly two methods are used, i.e., Hubbart formula and rule of thumb approach.

## Meal Based Rates (Meal Inclusive or Exclusive Room Rates)

✓ This term refers to the meal/s inclusive or exclusive room rates, i.e., whether meal/s are included or not within the room tariff design by accommodation establishment. Obviously, meal inclusive room rates will always be higher than meal exclusive room rate.

✓ There are different *meal plan*s EP (European Plan), CP (Continental Plan), AP (American Plan), MAP (Modified American Plan), BP (Bermuda Plan) and ASP (All-Suite Plan).

| Quick Scroll on Meal Plans | | |
|---|---|---|
| EP | Non-inclusive room plan | Only room tariff |
| CP | Bed and breakfast plan (B&B) | Room tariff with continental breakfast |
| AP | All inclusive/Full board/En pension | Room tariff with all three major meals |
| MAP | Half board/Demi-pension | Room tariff with breakfast & either lunch or dinner |
| BP | Quarter board/Petite pension | Room tariff with American breakfast |
| ASP | Business plan | Room tariff with breakfast & evening cocktails |

## Room Configuration Based Rates

✓ Room configuration is concerned with the physical characteristics (or makeup) of the guest room. This is the room configuration only which assists in determining the final room rate and also differentiate one hotel room from another room. This room configuration can be classified as standard, suite and promoted configured rooms.

✓ *Standard configured rooms* are the paradigm product of every hotel and restrains maximum number of rooms whereas *promoted configured rooms* are standard rooms with additional amenities & facilities (such as additional soft and hard goods) and enclose limited number of rooms. *Suite configured rooms* are the profligacy collection of rooms in the hotel and restrain restricted number of rooms.

## Standard & Level of Services

- ✓ *Standard* is related to level of quality or attainment which in hotels is supported by the amenities & facilities provided by the hotels to their guests. Amenities & facilities are offered to guests or placed in guest rooms for the comfort and convenience of the guests. It is also a key room rate determinant because the cost of these amenities and facilities affect the final price of guest room. Amenities & facilities are designed to increase a hotel's appeal, luxury, enhance a guest's stay and encourage guests to return. So its quality should be excellent.

- ✓ As all hotels offer homogeneous products thus hotels find themselves in many types of competitive situations which affect the pricing policy of the room rates and other hotel products. This is the level of service which distinguishes one hotel from another and create the base for charging different set of room rates.

| Quick Scroll on Standard & Level of Services | | |
|---|---|---|
| Full service | Expensive | As all luxurious amenities, facilities & services proffer with hotel rooms |
| Limited service | Moderate | As only basic amenities, services & facilities proffer with hotel rooms |
| No frill | Economy | As all luxuries are removed from the basic product or hotel rooms |

### Activity

Apart from above given basis for tariff fixation, there are also sundry bases for determining room rates. Some of these are given below and you need to explain three points in each.

| Location | Room size | No. of pax | Ancillary revenue | City events | Season | Booking horizon | Check-in/ check-out |
|---|---|---|---|---|---|---|---|
| ? | ? | ? | ? | ? | ? | ? | ? |
| ? | ? | ? | ? | ? | ? | ? | ? |
| ? | ? | ? | ? | ? | ? | ? | ? |

## 6.3   METHODS OF DETERMINING ROOM RATES

**Hubbart formula** (room rate above break-even/an approach to determine ARR or ADR)

It is a target pricing method which involves constructing an income statement from the *bottom-up approach* to determine the room rates and required return on investment (ROI). This formula was developed in 1940 and was published in 1952. Ray Hubbart was the founder of Hubbart formula. He was the chairman of the committee appointed by the AH&MA (American Hotel & Motel Association) for developing or computing a break-even room rate. *Break-even* is an economic term which means hotel will neither be in loss nor in profit. Often, this room rate approach is also known as average room rate/revenue.

The Hubbart formula requires that management forecast estimated operating expenses (EOE), desired return on investment (DROI), revenue from other sources {(RFOS) or

**Original Hubbart Formula**

$$\frac{(EOE+DROI)-RFOS}{RNRS}$$

**Standard Formula For ARR/ADR**

$$\frac{\text{Total Room Revenue}}{\text{Number of Room Sold}}$$

except room revenue} and required number of room sale (RNRS) for the next accounting period. Originally, the Hubbart formula was very extensive to calculate but nowadays Modified Hubbart formula is used to calculate average room rate.

Average room rate/revenue (or average daily rate or average house rate) is a middling possible room rate for any given period, i.e., day/season/year. The main purpose of this formula is to achieve a desired return on investment (ROI). To simplify it, the formula can be divided into five main stages/steps to calculate the final room rates, which includes:

Step 1    Compile all operating expenses like room expenses, food & beverage expense, repair & maintenance, advertising & promotion, fuel & lighting, salaries & wages and so forth running/operating expenses.

Step 2    Determine the desired return on investment (ROI/DROI). This can be calculated by multiplying the capital investment with desired rate of return for a particular period of time.

Step 3    Calculate the expected revenue generation from other revenue points/ sources (except room revenue) like revenue from food & beverage outlets, revenue from barber shop, laundry, gift shop, gym and so forth point of sale outlets.

Step 4    Now calculate the required number of room sales in order to achieve the desired revenue target. Thereafter, *add* estimated operating expenses with desired return on investment then deduct revenue from non-room services (or add step 1 with step 2 then subtract step 3). Resultant figure will show the required room revenue.

Step 5    This is the last stage/step where you require calculating the ARR (average room revenue/rate) and this can be calculated by dividing the required room revenue by required number of room sales.

| Activity |
| --- |
| ✓ Give three foremost reasons why a hotel requires to calculate average room rate on daily basis, even when room demand and occupancy level almost look alike during a particular season. |
| ✓ Among the hotels that you know or you have recently visited (like during hotel visit, vocational training or industrial training), how do they calculate their average room rate? Make a list of average room rates (of different weeks) as per your capacity. |

**Rule of Thumb Approach** (or 1:1000 room rate approach or cost rate approach)

It is a very simple and old formula to calculate or fix room rate and is usually concerned with the need of the hotel management. It represents the direct relationship between the cost of investment in the hotel establishment and the required average room rate. Under rule of thumb approach, Re. 1 is an average room rate for every Rs. 1000 cost of investment. Suppose, a hotel XYZ has constructed 40 rooms at the total of cost of Rs. 160000000. Here,

| **Rule of Thumb Formula** |
|:---:|
| (Re. 01: Rs. 1000) |
| $\dfrac{\text{Re. 01}}{\text{Rs. 1000}} \times \dfrac{\text{Cost}}{\text{Total rooms}}$ |

hotel XYZ requires to obtain Rs. 4,000 as an average room rate. Therefore, this formula is also known as 1:1000 formula.

It is a traditional method to calculate average room rate and widely used among hotels due to its simplicity. But in current scenario, as the room construction cost has risen at about the same pace as room rates, thereby rendering it invalid. This formula is also known as *cost rate formula/cost based rate*. The major drawbacks of this formula are: firstly room rates may not be completely dependent on initial construction expenses and secondly the formula produces an estimate of an ARR that may be difficult to strive for and/or attain. The formula does not have anything to do with the market price and the guest's willingness to pay the rates. It is largely concerned with the needs of hotel; not with guest's desire and expectation. Thus, nowadays many hotels do not apply it.

| Activity |
|---|
| Consider the factors which influence or constrain the room rate of any given room and how the resulting room rate will affect ancillary activities of the establishment. Imagine that you are a revenue manager or front office manager and held with the responsibility of determining room rate for the coming shoulder season. Now enlist what are the factors which you would take into account when deciding what room rates to be charged for the coming shoulder season. |

## 6.4  ROOM RATE TYPES

The term *room rate* refers to the room rent or room tariff. There are different grades of rooms sold at different rates, these different rates are collectively called as "room tariff" (i.e. rate charges per room). *Room tariff* is computed for a "revenue day" which usually begins from 12 noon and will remain for next 24 hours. It is not obligatory that every hotel should start its revenue day from 12 noon only; many hotels also have a policy of 24 hours. It means their revenue day starts from the time of guest check-in and will remain for next 24 hours. Remember, room charges are levied for a *revenue day* and nowadays room may be sold for half a day as well. For these special rates are applicable, such rates are referred to as "half day" rates. Due to such room rate policies, a hotel can sell more rooms than available room inventory. A brief description of major room rate types is given below.

**Premium rates** (*optional price policy*, i.e., rack rates + price of additional in-room offers)

✓ These room rates are more than offered rack rates because it associates additional amenities and facilities. Here, hotels give choice to select additional services (besides meal plans) at the time of registration or during accepting guaranteed reservation.

✓ These additional amenities and facilities can range from baby sitters, in-room entertainment, valet services, room with full mini bar choices, whiteware items to an array of personally preferred bathroom supplies and software items.

✓ In accordance with coalition of additional amenities/services/facilities, the premium room rate will be calculated or determined.

### Activity

✓ There are several reasons why a hotel may choose to offer premium room rates. Can you give two good reasons?

✓ Among the hotels that you know or have gone through their websites, what rack rate and premium rate are they charging on different meal plans? Make a list and see how many you can name.

| S. No. | Name of Hotel | Rack Rate | | | | Premium Rate | | | |
|---|---|---|---|---|---|---|---|---|---|
| | | CP | EP | AP | MAP | CP | EP | AP | MAP |
| | | | | | | | | | |
| | | | | | | | | | |
| | | | | | | | | | |
| | | | | | | | | | |

**Rack rate** (*ceiling price* or maximum possible room rate, i.e., break-even rate + maximum DPM)

✓ Maximum possible rate for any given room that is offered to general market segment/s. There are different rack rates for the same room; it is due to inclusive, semi-inclusive or exclusive of meals.

✓ Remember, every special rate (or discounted rate) is calculated on rack rate. Consequently, it can be termed base rate of every special room rate types including premium room rate.

✓ Therefore, hotel should carefully determine its rack rate as it is a base for all other rate types, simultaneously property cannot charge more than this room rate. It is also important because in many Indian states like Delhi, hotel requires to pay tax on rack rate instead of charged/discounted rate.

## Best Available Rate (BAR)

✓ Lowest possible room rate which is not restricted for any specific market segment/s, thus every segment can avail it. BAR can change several times in a week to several times a day.

✓ Thus, BAR is also termed as best flexible rate (BFR). Apart from it, many hotels also offer *best guaranteed rate* (BGR) which they publish on their websites and update regularly.

✓ BGR claims as a minimum room rate offered by hotel for a homogeneous product in comparison to any other hotel in same location.

## Average Room Rate

✓ Actually, it is not a room rate type but it is a *middling rate* of all sold out rooms. Average room rate is also known as *average daily rate* and often known as *average house rate*. This rate is the average of all the rooms sold on a given night. But remember, average room rate is not the highest, nor the lowest booked rate.

✓ This rate changes on daily basis as hotel rooms are sold to varying guests at varying prices on daily basis. Consequently, total room revenue collection varies on each night.

✓ Average room rate is a foundation stone that is primarily used to determine a starting point in developing/modifying hotel's room rate structure.

## Special rates (room rates for varying selected market segments, i.e., rack rate – adjustable DPM)

✓ These are discounted rates that are offered to specific selected market segment/s only. Remember the term special rate is always preferred in place of discounted rate because discount is an inferior word.

✓ These special rates are usually planned to attract multiple market segments, especially during shoulder and low demand periods. This rate category is offered in order to avoid any further room rate bargaining.

✓ Discount is a kind of adjustment with differential profit margin (DPM); it is neither increased nor decreased total cost/room cost (or product cost) but affects profitability, thus it should be offered only as and when necessary.

| Group rate | • A special room rate agreed between group leader and hotel for group room reservations. Generally, discounted rate is offered to group leader, in which a flat rate is charged for each allocated room irrespective of its location within the hotel. The tariff is usually exclusive of suites. Wash factor should also count while dealing with group. |
|---|---|
| Consortia rate | • A special room rate negotiated between a hotel and a travel agency. A consortium is a conglomerate of travel agencies (or travel agent chains), for example, American Express, Navigant, Carlson Wagon Lit, BCD travel or HRG. |
| Crew rate | • A special room rate agreed between the airline and hotel for their airline crew room reservations only. Generally, not all rooms but several rooms have been contracted out to various airlines for the purpose of generating some fixed revenue and maintaining the full occupancy level. In addition, policy regarding PSO and MAO is also pre-determined. |
| Membership rate | • A special room rate offered to the members of several honoured organizations like FHRAI, AHLA and so forth. The rate is lower (and some rebate is also given on food and beverage products) because hotels get benefit in terms of international publicity and loyalty from the members of these organizations. |
| Half-day rate | • A special room rate that is usually offered by transient and commercial hotels for the day use of room (as room is not used for night, hence the offered rate is usually half of rack rate). This rate is for specific/limited time only – this time is usually from morning 6 O' clock to evening 6 O' clock. |
| Hurdle rate (break even room rate + minimum DPM) | • It is a minimum possible room rate (or floor price) below which a hotel cannot sell its room. This rate is calculated on a daily basis thus it may fluctuate regularly. The hurdle rate is determined by considering factors like expected demand, operating expenses, competitors, guest flow, etc. But it is above break-even room rate. |
| Family rate | • A special room rate offered to the family clientele. Generally, a discounted price is offered by the establishments when two or more members of the same immediate family make use of hotel services. |
| Incentive rate | • A special room rate that is offered to companies which promise to give regular business through their employees. Here, representing company gives rewards in the form of free accommodation (additional services may also be included) to their star/performer employees (or employees who have achieved company stated sales target). |
| Package rate | • Package rates generally comprise room rates for the number of nights (depending upon the rate of package) along with some other ancillary services & facilities. This package is often called "bundling" that is mainly offered during off-season. Nowadays, dynamic package rate policy is dominating the market. |
| Per-diem rate | • This special room rate typically applies to government employees who travel/stay in hotel during their official tour. It usually covers hotel's room rate, meals, and other out-of-pocket expenses. This style of room rate policy prevails in western countries. |
| Early bird rate | • It indicates a promotional rate that is available for only advance booking, especially when lead period is very short. It is often a discounted rate with fencing. |
| Other room rates | • It includes inclusive rate, exclusive rate, driving rate, corporate client rate, advance purchase rate, weekend rates, competitive rates, off-season rates, peak season rates, crib rates, quoted rates, and so forth. |

| | Activity | | | | | | | | | | |
|---|---|---|---|---|---|---|---|---|---|---|---|
| | On the basis of the below given guest profile, indicate which room rate you will offer to your prospective guest while dealing face-to-face or selling rooms over the telephone. Tick the appropriate boxes. | | | | | | | | | | |

| S. No. | Guest Profile | RR | IR | MR | GR | HR | FR | PR | CR | WR | ER |
|---|---|---|---|---|---|---|---|---|---|---|---|
| 1 | A delegate comes to attend afternoon meeting, thus wants room for a few hours only. | | | | | | | | | | |
| 2 | A member of FHRAI comes to your property. | | | | | | | | | | |
| 3 | An employee of a company (company listed on your approved list) comes to your hotel. | | | | | | | | | | |
| 4 | Tour operator gives you business & demand appropriate room rate. | | | | | | | | | | |
| 5 | A street guest directly comes to your hotel and wants a room. | | | | | | | | | | |
| 6 | A government employee takes LTC and now he wants to enjoy with his family in your hotel. | | | | | | | | | | |
| 7 | Scandinavian airlines wants to take several rooms for their crew members for a period of one year. | | | | | | | | | | |
| 8 | While dealing on telephone, you have observed that the guest is money conscious. | | | | | | | | | | |
| 9 | An old man wants to take efficiency room or any other room but on European plan, as he is a patient, thus needs dietary food. | | | | | | | | | | |
| 10 | You plan to organize musical night on Saturday in your hotel, as next day is holiday. | | | | | | | | | | |

**Note.** RR (rack rate), IR (incentive rate), MR (membership rate), GR (group rate), HR (half-day rate), FR (family rate), PR (package rate), CR (crew rate), WR (weekend rate), ER (exclusive rate).

## CONCLUSION

Accurate room tariff is a base for generating business, attaining profit targets, winning market competition and achieving maximum guest satisfaction level. Wrong calculation of room tariff, whether over or under, is always destructive for the establishment because overcharge will reduce your guest turnover rate whereas undercharge will lead you towards revenue loss. Most hotels have premium rate plans by which they can charge more than written rack rates. Remember, printed room rates also create base for tax fixation, VAT and other government charges. In many Indian states, tax is levied on charged room rate (like it may on discounted rates) whereas in Delhi tax is levied on rack rate (even when room was sold at discounted price). In fact, discount attracts market segment/s but at the same time discount is a kind of negative revenue adjustment from differential profit margin (DPM), simultaneously it may create negative image among guests (because discount indicates that there is lack of demand in market for your products and services). Therefore, hotel requires to carefully determine its discount policy. Nowadays many hotels, use the term special room rate in place of discounted rates.

| Terms (with Chapter Exercise) | |
|---|---|
| Market segmentation | Process of customer division (or division of marketplace) in order to choose profitable customer group/s. |
| Transient market | It is a collective term for all market segments irrespective of group segment. It indicates all those market segments like incentive, family, walk-in, airline crews, corporate, company clientele and so forth, except group segment. |
| Group market | Bulk business segment which generates huge profit and revenue within limited time period, especially during off season. |
| Revenue day | Duration/period for which hotel sells its room to guests. It usually begins between 12 noon – 2 p.m. and ends the next day at the same time (this shows a revenue day of 24 hours). But in case of day use room, it usually begins at 6 a.m. and ends at 6 p.m. (it shows a half revenue day of 12 hours.) |
| All exclusive | ? |
| Meal plan | Room tariff plan that is based on inclusion (such CP, AP, MAP and BP) or exclusion (i.e., EP) of meals. Remember, meal plans never include beverage rates (or bar beverages). |

| | |
|---|---|
| Premium room rate | Room rate in which hotel offers something extra, consequently charge above standard rates. For instance, a room with a special view or with extra in-room goods like *soft goods*. |
| All inclusive | ? |
| Rate change | The room tariff of already occupied room is altered. |
| Quantifiable | Term used for expressing the quantity (or productivity level or number of units) of product. |
| ? | Category of a special room rate or a negative adjustment with differential profit margin. |
| Software items | Guest room fixtures that are normally considered a part of depreciable fixed assets such as curtains, draperies, pillows, mattresses and other similar items but never include bed and bathroom linens. |
| Qualifiable | It is a term used for expressing the quality (or degree of quality) of any product or service. |
| Incognito stay | Confidential guest or his/her stay and is mainly demanded by renowned personalities like celebrities, film stars, etc. |
| Meal plan | Meal based room rate strategy, i.e., whether room rate/rack rate is inclusive of meal charges or exclusive of meal charges. |
| Crib rate | Rate charged for children below 5 years. |
| Room tariff card | Room rate sheet (or room rent card) that shows the price of different types of rooms (or different configured rooms) along with extra bed charges and applicable taxes and VAT (CGST and SGST). |
| ? | Special rate (or discount rate) for the company representatives because hotel has legal tie-up with these companies. Therefore, the name of these companies will always be on the Hotel's Corporate Account List. |
| Flat rate | ? |
| Quoted rate | It refers to that the room rate offered to any guest will be same as provided at the time of confirmation, regardless of room availability. |
| ? | It is the maximum possible room rate for any type of room. But it may vary with type of meal plan offers with room rate like rate on MAP plan will always be expensive than rate on EP for same configured room in that hotel. |

| Special rate | It is also known as discounted rate that is usually offered to specific target market segment/s. Therefore, room rate of similar room type will vary in selected market segment/s like incentive guest will get less discount than group guests. |
|---|---|
| ? | It refers to a bundle which comprises various services, facilities & products all together. It is offered for sale at low (or discount) price and mainly available during off-season and shoulder season. |
| Bottom-up approach | Strategy/plan in which hotel's sales division/reservation desk or other related desk tries for upselling (or tries to sell more expensive room in place of less expensive room). |
| Price competition | ? |
| Average Daily Rate | It is also known as *average room rate* and *average house rate* that may change on daily basis. It is middling of all offered room rates on sold out rooms. |
| Hubbart formula | It is a formula developed by Mr. Ray Hubbart to calculate average room rate (ARR). It is also known as bottom-up approach of room rate. |
| Market condition approach | It is an approach to charge room rate on the basis of elasticity of market demand as well as market competition. |
| Rule of thumb approach | It is a traditional room rate determination approach in which property charges every Re. 1 for every Rs. 1000 of cost of investment. Thus, it is also known as 1:1000 approach. |

# 7

# Reception Section

*After reading this chapter, students will be able to...*

- define reception section and explain its functions;
- describe the necessary preparations for transient and group guest check-in;
- explain the check-out procedure for different guests;
- express the role of reception in giving business to other departments/sections of hotel; and
- express how reception section works as a public face of hotel.

## 7.1 RECEPTION SECTION

It is one of the major sections of front office department (and of room division). Reception section is also known as *front desk* or *command centre* that is usually situated near the main entrance of the hotel, therefore it is considered as the very first area of guest interaction with hotel after entering the hotel building. It is an area where guests are received (check-in), register and depart (check-out) the hotel establishment (or front desk). Due to its all time guest oriented work profile, the reception section is open for 24 hours.

Being the first point of contact with the guest, the front desk plays an imperative role in the success of hotel operation. The front desk manages the in-flow and out-flows of guests on daily basis. Therefore, most of the guests adjudicate the entire hotel image by the surrounding of reception area and by the attitude & eloquence of front desk staff. Major functions of the front office may seem simple but these are tremendous and convoluted. The layout of reception section is available in enormous attractive styles but mostly straight, L-shaped, C-shaped, curve or S-shaped are found in hotels. The ideal height of reception desk should be 112 cm (44 inches) whereas width of reception counter should be 76 cm (30 inches). The back side of reception desk is mostly highlighted with wall clocks which show time of different countries.

The reception section is manned with reception manager, reception supervisor and receptionist/s. The reception/front desk section is headed by the "reception manager" or "front desk manager". The reception manager needs to work in close coordination with other front office staff. He/she is mainly responsible for the smooth running of entire front desk. Reception supervisor is also known as "front desk supervisor" who reports to reception manager. He usually organizes the duty rotas and intermittently handles guest complaints like when receptionist is not able to handle the situation/guest. In certain cases (like VIP, CIP, etc.), he/she also escorts the guest towards his/her allotted room, meanwhile he/she provides information about the various hotel facilities & services. The person who is responsible for receiving, registering, rooming thereafter assist in departing (or check-in and check-out activities) the hotel guest is known as *front desk agent(s)* or *receptionist*. Females are mostly preferred for this position.

**Note.** Front desk agent (or reception clerk/receptionist) and front desk cashier must work in extreme close coordination because work profile of both is amalgamated. Resultant, often they work side-by-side on the same counter or their counters will be very much close to each other. Thus, often the term front desk is used to address any of them.

**Major Equipment of Reception**

- Computer system
- Stamps (time/date)
- Photocopy machine
- Fax machine
- Telephone system
- Holding rack/s
- Registration
- Stationery

## 7.2  FUNCTIONS OF RECEPTION SECTION

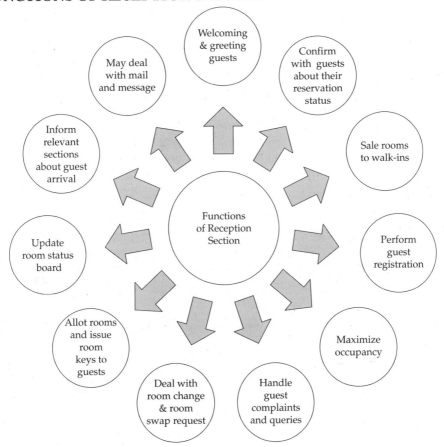

## 7.3 ROLE OF RECEPTION SECTION IN DIFFERENT STAGES OF GUEST CYCLE

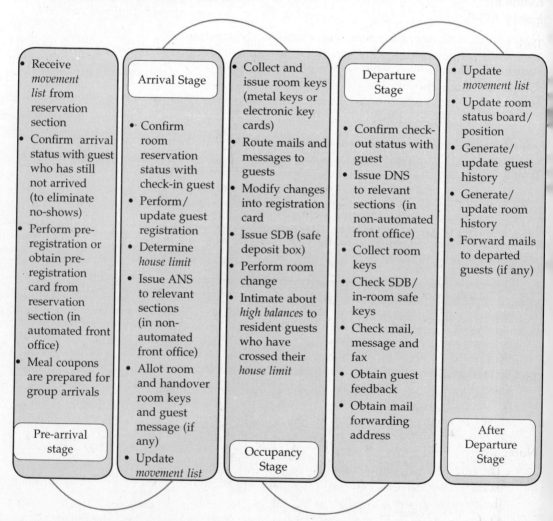

**Pre-arrival stage**
- Receive *movement list* from reservation section
- Confirm arrival status with guest who has still not arrived (to eliminate no-shows)
- Perform pre-registration or obtain pre-registration card from reservation section (in automated front office)
- Meal coupons are prepared for group arrivals

**Arrival Stage**
- Confirm room reservation status with check-in guest
- Perform/update guest registration
- Determine *house limit*
- Issue ANS to relevant sections (in non-automated front office)
- Allot room and handover room keys and guest message (if any)
- Update *movement list*

**Occupancy Stage**
- Collect and issue room keys (metal keys or electronic key cards)
- Route mails and messages to guests
- Modify changes into registration card
- Issue SDB (safe deposit box)
- Perform room change
- Intimate about *high balances* to resident guests who have crossed their *house limit*

**Departure Stage**
- Confirm check-out status with guest
- Issue DNS to relevant sections (in non-automated front office)
- Collect room keys
- Check SDB/in-room safe keys
- Check mail, message and fax
- Obtain guest feedback
- Obtain mail forwarding address

**After Departure Stage**
- Update *movement list*
- Update room status board/position
- Generate/update guest history
- Generate/update room history
- Forward mails to departed guests (if any)

A step-by-step description of some major tasks of front desk section that predominantly occurs during different stages of guest cycle is given below.

## 7.4   PREPARATION FOR GUEST CHECK-IN

### Preparation for Transient Guest Check-in (Preparation for FITs Check-in)

Front desk's main function is to receive the guest at the time of arrival thereafter registering then rooming a guest. It is the arrival stage where the skill of a front desk agent is put to test and his ability to pick right room for a guest is checked. In manual front office, generally the housekeeper submits a list of all available rooms for the day. From this list the front desk agent updates the room status chart/board and blocks the rooms for those guests who have been pre-registered, especially for VIP/SPATT/Alien guests. Major tasks of front desk agent are:

Task 1   After joining the shift, the first major task of front desk agent is to get *briefing* from the night staff. Secondly, front desk agent will refer to *front office memorandum* or *log book* for any further instructions, any incomplete work, etc.

Task 2   Now, front desk agent checks the expected departure and expected arrival for the day, thereafter calculates the actual *room position status* and current *room count*. It is calculated as previous *room count* + expected due-ins – expected due-outs (or expected departure + available vacant room – expected arrivals).

Task 3   After calculating room position status, front desk agent blocks rooms for VIPs, groups and airline crew members. Next indicate HWC (handle with care) on reservation slips wherever necessary like for VIPs.

Task 4   Now, front desk agent requires preparing of amenity vouchers in accordance with guest status or request of the expected arrivals (like fruit basket or flower bouquet for the VIPs) then sends these vouchers to the room service or housekeeping.

Task 5   During the peak season, many hotels pre-register their regular transient guests (or a guest who holds room reservation). In order to pre-register a guest, hotel normally collects the required guest registration data from the reservation record created at the time of reservation.

Task 6   Nowadays, in electronic front office (or in PMS) pre-registration/registration window (or form) automatically retrieves information from reservation window (or form). Generally, both a room type and rate category is assigned as a pre-registration activity for a guest with reservation.

**Note.**
  ✓ The information about *vacant rooms* could be obtained from the *room rack* whereas expected departures information could be obtained from the *movement list* or from the folded slips of room rack.
  ✓ *Movement list* is also called expected arrivals cum expected departures list. Expected arrival information can also be obtained from the *current reservation rack*.
  ✓ In case of pre-registration of forthcoming guest, most of the hotels also *pre-assign* the room. This pre-room assignment should be done in accordance with guest's room preference.

| Factors Affecting Room Position | |
|---|---|
| OOO | It would decrease the room count/lettable rooms position/room inventory |
| OOS | It would decrease the room count/lettable rooms position/room inventory |
| Early departure | It would increase the room count/lettable rooms position/room inventory |
| Overstays | It would decrease the room count/lettable rooms position/room inventory |

- *Plus* room position shows that rooms are available for sale (or available rooms for sales are more than the number of expected arrivals).

- *Minus* room position shows that rooms are not available for sale (or available rooms for sales are less than the number of expected arrivals.)

## Preparation for group guest check-in (preparation for GITs check-in)

Group check-in process is different from transient check-in. Group check-in preparation often entails information regarding welcome drink, cold towels, airport pick-up, pre-registration of members, single C form when all group members are from same nation, rooming list, remote check-in counter (it depends on the size of group), billing instructions, meal coupons, *pre-key* formality (like make ready *key cards* in *key jackets*) and so forth. Groups are usually welcomed by guest relation desk instead of front desk. Major tasks of front desk agent/GRE are:

Task 1     First, on the morning of group arrival the needed rooms should be *blocked off*, update room rack, intimate to bell desk/concierge, determine where to register a group, sort out room keys, check group's complementary items like welcome drink, recognize/tick name of group (or group leader), country of origin of group and so forth.

Task 2     Second, prepare individual/group registration card/s (or perform pre-registration of GITs) and keep all cards ready for signature. By coordinating with reservation section, front desk agent will confirm that whether prospective group needs to be registered at front desk counter or at *satellite check-in counter*.

Task 3     Small groups register at front desk counter whereas in case of large group check-in, front desk should create a *remote check-in counter* (or *satellite check-in counter*) in lobby (or business centre may also use for it) to handle group arrivals efficiently and effectively. This remote check-in counter keeps the front desk counter uncrowded.

Task 4     In small groups, front desk agent will directly handover room keys to each group member whereas in large groups *key-pack process* is followed. In *key-pack process*, room keys along with welcome letters are kept in an envelope. These key-envelopes (also known as pre-key/key-pack process) are usually directly handed over to group leader.

Task 5    Group leader will further distribute *key-envelopes* to all group members. This key-envelope distribution by group leader can be done, in bus, while group is on the way to hotel or group leader may also distribute in hotel lobby.

| Activity |
|---|
| Suppose you are a front desk agent and your next day's movement list shows the arrival of a foreign group (Australian cricket team) and one VIP guest (your corporate manager). Now prepare a 10 points opening task-list to deal with both clienteles. |

| Opening Task-list for Australian Cricket Team | Opening Task-list for Corporate Manager |
|---|---|
| ✓ Welcome drink will be offered during team arrival | ✓ GRE will receive him at airport terminal |
| ✓ | ✓ |
| ✓ | ✓ |
| ✓ | ✓ |
| ✓ | ✓ |
| ✓ | ✓ |
| ✓ | ✓ |
| ✓ | ✓ |
| ✓ | ✓ |
| ✓ | ✓ |

## 7.5  GUEST CHECK-IN PROCEDURE

## Transient Guest Check-In Procedure

It is concerned with greeting and welcoming fleeting guest, confirming that he/she holds reservation with hotel or he/she is a *chance guest*. If the guest holds reservation then move towards registration formalities, otherwise provide room details, tariff structure and whatever information guest desires to know. Finally, allot room, handover room keys and call bellboy to escort guest towards allotted room. In manual front office (or where metal keys are used) before handing over room key to guest, front desk agent needs to raise *key card* and fill it with required details. In metal key system, it is compulsory for guest to show *key card* every time the guest requests his room key from front desk agent. If property uses *card key* then there is no need to raise *key card*. A step-by-step process for transient guest registration is given here.

**Key Objects**

- Tariff card
- Registration card
- Key card
- Vouchers
- Stationery
- Luggage tag

**Stage 1**

- On guest arrival, a valet parking attendant should greet, welcome, open car door, park vehicle, collect car docket and finally handover car keys and car docket to guest. Meanwhile doorman calls the bell desk to collect guest luggage. Afterwards doorman will open the door, warmly greet the guest and welcome him/her. Bellboy will bring all guest luggages at bell desk area from luggage entrance door.

**Stage 2**

- Next guest will be directed towards reception desk where receptionist (or front desk agent) will again greet and welcome the guest. Thereafter, she confirms that whether guest holds reservation with hotel or he is a chance guest, then move forward accordingly.

**Stage 3**

- If guest holds reservation then receptionist proceeds towards registration formalities. On the other side, chance guest must be informed about whether room is available or not. If room is available then receptionist will give all room related information to guest. And when guest agrees to take room, then she proceeds towards registration. If room is not available then receptionist or front desk agent can *farm out* the guest.

**Stage 4**

- Now receptionist (or front desk agent) will complete *guest registration card* (fill C form in case of foreigner) and obtain guest signature on it. If guest has pre-registered then take only guest signature. The allotted room number should be ticked off on the *movement list* (or arrival & departure lists) and enters the guest name into the *tabular ledger*. If guest seems *skipper* or with *scanty baggage*, bellboy should immediately intimate to receptionist.

**Stage 5**

- Check the guest details in registration form for any doubt thereafter handover the room keys (the guest room key is usually given to bellboy) with any message or letters waiting for him. In case of guest with scanty baggage, receptionist needs to put remark on guest registration card.

**Stage 6**

- During registration process, receptionist should also give information regarding the location of restaurant, bar, coffee shop, emergency exits and so on. Now, front desk will intimate all relevant departments about guests checked-in. Bellboy will escort the guest towards the allotted room.

**Note**

✓ The process of identifying whether guest holds reservation with hotel or he/she is a *chance guest* is not a simple task because if front desk agent will not be careful and tactful in his approach then agent may offend the guest and make him/her feel uncomfortable in case he/she does not hold reservation with hotel.

✓ Phrases like "Are we holding a reservation for you" will make the guest uncomfortable as this would indirectly give an impression to the chance guest that the hotel allows only those guests who have booked the room in advance.

## Group Guest Check-in Procedure

In majority of group bookings, group leader demands complementary pick-up facility, predominantly during large group bookings. Hotel mostly assigns the responsibility of group pick-up to guest relation agent who goes along with a team of bellboys to pick up the group. Often, airport representative also assists in this task, especially when group belongs to a foreign country. Generally, a mini-bus is sent to pick up the group. Remember, group members are always pre-registered. A step-by-step process for group check-in is given below:

| Key Objects |
| --- |
| - Rooming list |
| - Key envelopes |
| - Registration card |
| - Errand card |
| - Vouchers |
| - Stationery |
| - Luggage tag |

Step 1    When group arrives at the hotel, all group members are welcomed by some responsible front office staff like lobby manager/front office manager/guest relation executive or any other senior member.

Step 2    Generally, group members are welcomed by offering flower bouquets or garlands. Many hotels in India also welcome their group arrivals in traditional Indian style (especially when sports team arrives), i.e., by offering aarti and tilak. The guest relation desk is particularly responsible for all hospitable services. Luggage of all group members is brought by bellboys from luggage door.

Step 3    Afterwards, group leader is directly moved towards the front-desk-counter or remote-check-in counter for registration, whereas group members are escorted towards the waiting area in hotel lobby where welcome drink, cold towel, etc. are offered to each group member. Meanwhile, bellboys will tag all luggage in accordance with *rooming list*.

Step 4    At registration counter, front desk agent will reconfirm all details about the group members with group leader. Thereafter, pre-registered card (or partially filled registration card) is given to group leader for verification and signature. *Rooming list* is always attached with this card. If any variation is found in rooming list then it must be rectified and immediately notified to bell desk.

Step 5      After the completion of registration, front desk agent will handover all *key envelopes* (or *key jacket*) which contain room keys and welcome letters to group leader who will further distribute *key envelopes* to each group member. Now group members are escorted towards their allotted rooms, simultaneously all relevant sections/departments must be notified. Next, *bag delivery* will be done in all GIT rooms.

Step 6      Lastly, front desk agent will count number of group members that arrive, thereafter compare it with group reservation, the number of absentees will be counted as group *slippage*. This *slippage* will be confirmed with group leader, and after confirmation it will be added back into current vacant rooms. So that front desk can take timely action to fill these rooms or to attain *sold out*.

**Note**

✓ At registration desk, either one registration card can be prepared for whole group (and signed by the group leader) or individual guest registration cards can be prepared then signed by all group members. It depends on legal formalities. When single registration card is made, on behalf of entire group, then generally a *rooming list* copy is attached.

✓ If group members are of the same nationality then only one guest registration card (for group leader) is prepared, but if the group comprises members from different nationalities then the number of guest registration cards will vary with total nationality break-up of the group. The same guideline will also apply to C form.

✓ GRE or whoever is going to pick up the group can carry key envelopes in order to handover to group leader who will distribute them among the group members, in pick-up van, while group is coming towards the hotel.

## 7.6   GUEST CHECK-OUT PROCEDURE (AT RECEPTION DESK)

Different hotels provide different methods for guest check-out. Some hotels use manual check-out process whereas others use express check-out system. Modern luxury hotels use self-check-out and video-check-out system. The basic purpose of every check-out system is to prevent skippers, smooth check-out process and to provide memorable farewell to departing guest. Remember, check-in process may differ for transient and group guests but check-out process for both are virtually alike. A standard check-out process for transient and group guest is given below.

### Transient Guest Departure Procedure

Generally, front desk agent confirms about the departure status with guest a day before an expected date of departure. If a guest wants to *overstay* then agent checks room status position and allows accordingly. Next, front desk agent will intimate about overstay to cashier, bell desk and other relevant departments and sections. Otherwise, cashier will prepare a *grand bill* (or *guest folio*) of departing guest along with supportive documents like expense vouchers.

On the day of departure, the bellboy on his way to the guest room informs the front desk agent about the departure taking place. The front desk agent carries out the following steps (in manual/non-automated front office):

Step 1    Immediately checks the list of expected departures for the day to assess room availability status/room position status for the day.

Step 2    Now reception desk agent makes an entry in the departure notification register/slip (DNS) and despatches to housekeeping, bell desk, room service, telephone operator and other relevant sections.

Step 3    Simultaneously, agent will inform housekeeping department about guest departure, so that the concerned housekeeper or room boy can attend to the departing room.

Step 4    Next, reception will telephonically intimate the bell desk. Thereafter, bellboy will attend to departing guest, run departure errand card, collect room keys and submits to the front desk. So that front desk agent can make an entry in the *movement register* (or arrival/departure register).

Step 5    Front desk agent should also remember to collect keys of SDB, obtain *mail forwarding address* (when guest instructs to divert mail/message/fax) and feedback on comment card from departing guest.

Step 6    Lastly, the room rack and information rack is updated. If an alphabetical register is being maintained then departure details are noted against the guest's name. Remember, bill settlement will be done at the cashier desk.

## Group Guest Departure Procedure

Guest relation desk confirms/intimates to group leader about their departure, a day before the expected date of departure. There is minor/no chance of group overstay. Consequently, front desk agent intimates to all relevant departments and sections about group departure, so that they can make their bills ready and send them timely to front desk cashier. Generally, there are two types of bills prepared for group clientele, i.e., *master folio* and *incidental folio*. Group leader is responsible for the settlement of master folio whereas group members are responsible for the settlement of their incidental folios. Group guest check-out process comprises the following major tasks at front desk:

Task 1    The front desk agent issues a *group departure notification slip* about half an hour in advance of actual departure time to the housekeeping, room service, telephone operator, coffee shop and other relevant departments and sections.

Task 2    Next team of bellboys will move towards the guest rooms to attend to their luggage. Bellboys keep guest luggage in safe custody and bell captain will enter details in *luggage control sheet*. Meanwhile, group leader and members are directed to settle their bills at front desk cashier.

Task 3   Thereafter, bellboys will collect room keys from departing guests and submit to the front desk, so that front desk agent can make an entry in the *movement register* (or arrival/departure register).

Task 4   Front desk agent should also remember to collect keys of SDB, obtain *mail forwarding address* (when group leader/group member instructs to divert mail/message/fax) and guest feedback on comment card from group leader. Apart from front desk agent, bellboy should also ensure the collection of keys.

Task 5   Lastly, the room rack and information rack is updated. If an alphabetical register is being maintained then departure details are noted against the guest name. Remember, bill settlement is done at the cashier desk.

**Note.** In automated front office (or electronic front office), room and guest statistics are automatically updated at every moment as and when any guest arrives, departs, changes room/meal plan, is going to overstay and so forth. Therefore, front desk agent does not require preparing and sending any ANS and DNS to any department or section.

## 7.7   OTHER TASKS

| Steps / Tasks | Registration (It is a legal procedure which starts before actual guest arrival but is completed only after guest signature on registration card) | Room Change (It occurs when room is not assigned in accordance with guest's room preference or guest faces any problem in occupied room or room is wrongly allotted to guest and so forth) | Upgradation/Modification (It occurs when guest changes room, meal plan, number of pax either increases or decreases in any room, departure date changes and so forth) |
|---|---|---|---|
| 1 | Confirm reservation status with incoming guest. | Enquire/ask guest why he/she wants to change the room. | Open guest registration form or open modify account into PMS. |
| 2 | Guests who hold reservation will directly delegate GRC. | If reason is genuine then find out the alternate room for guest. | Enter upgraded information into system and save it. |
| 3 | GRC may be pre-filled (by reservation agent) or front desk agent may fill on the spot. | Inform to bell desk to send a bellboy who will assist the guest in room change. | Simultaneously modify all other related information like room upgradation always demands rate modification. |

| | | | |
|---|---|---|---|
| 4 | In automated front office, front desk agent will verify/modify pre-filled registration card then it will automatically update into each department/section. | In automated front office, room change status will automatically get updated into each department/section. | In automated front office, room statistics will automatically be modified in each department/section. |
| 5 | Any modification of information on GRC during stay period will also automatically update into each department/section. | If problem is major and room is not available then front desk agent may allot higher grade room (but avoid lower grade room) or may also farm out new booking to deal with room change request. | Modification in meal plan and number of pax affects RevPAC whereas departure date modification affects RevPAR, ARR, RevPPAR and RevPOT (Revenue Potential). |
| 6 | Guest will verify details then put his/her signature on filled GRC. | Update room status board, guest registration card and so forth. | Print modified guest registration card (GRC) and file into record. |

**Note**

- When confirming reservation status with guest, if guest is walk-in then front desk agent will first refer room status position in order to find out whether room is available for sale or not. If room is available then agent will move towards registration formalities, otherwise he/she may suggest alternate accommodation to guest (or bounce/farm out the booking).
- In automated front office, front desk agent is merely required to insert reservation confirmation number (or related credential) then the window will automatically retrieve all information for guest with advance reservation.
- In non-automated front office department, front desk agent is required to generate a *room change notification slip* which will be prepared in multiple copies and circulated to all relevant sections and departments.
- RevPOT indicates to the maximum revenue generation capacity of a hotel from all available room sales.

| RevPAR (it stands for Revenue per Available Room) | RevPAC (it stands for Revenue per Available Customer) | RevPPAR (it stands for Revenue Potential per Available Room) | ARR (it stands for Average Room Rate or Average Daily/House Rate) |
|---|---|---|---|
| It means revenue share of each available room from total generated room revenue. | It means revenue generation share of each resident guest from generated total revenue. | It means maximum revenue generation ability of each hotel room. | It means revenue share of each room sold from total generated room revenue. |

| RevPAR will be equal to ARR at 0% and 100% room occupancy. | RevPAC is affected as and when revenue is generated from any POS/revenue centre. | RevPPAR is affected with day use room, as rooms can be sold twice a day. | ARR varies as room occupancy percentage varies. |
|---|---|---|---|
| Its formula $$= \frac{\text{Total room revenue}}{\text{Room Inventory}}$$ | Its formula $$= \frac{\text{Total room revenue}}{\text{House Count}}$$ | Its formula $$= \frac{\text{Maximum room revenue}}{\text{Room Turnover}}$$ | Its formula $$= \frac{\text{Total room revenue}}{\text{Room Count}}$$ |

## Verification of Room Discrepancy

Night auditor is responsible for preparing room discrepancy report. He/she prepares it when *room count* and *room statistics* shown by housekeeping (via housekeeper's report) differs from front office department report. Thereafter, night auditor sends this report to heads of both departments. In general, lobby manager or front office manager is responsible for sorting out room discrepancy. Remember, if room discrepancy is not solved at appropriate time then it is possible that front desk can execute *double rooming*. A step-by-step procedure for sorting out room discrepancy is given below.

Step 1    Time stamp the housekeeper's report. Compare it with the room rack for room status, if discrepancies found then check *guest folio/room folio*.

Step 2    If this check does not reveal the discrepancy then ask the housekeeper to double check the room status. Meanwhile, night auditor will prepare a discrepancy report indicating differences between the room rack and housekeeper's report.

Step 3    If the housekeeper's second check does not solve, send the prepared room discrepancy report to front office manager and executive housekeeper. Night auditor may also refer the room service record to verify the room discrepancy. Otherwise, a personal inspection of the room will be carried out.

Step 4    If the rack shows that the room is occupied but it is vacant and no guest folio is found in the folio rack then night auditor should check previous day's checkouts. A room status may also suffer from *sleeper* dilemma. It means guest has left the room after proper settlement of room account but front desk agent/cashier forgets to update room statistics.

| Activity |
|---|
| As a front desk agent, you have been informed by Mr Pumpkin, a resident guest on own account, who receives daily newspaper, assorted chocolate tray (on alternate days) and fruit basket that he has decided to leave early (one day earlier than filled date of departure). Fill in the reason/s column below showing why the following departments and sections need to be informed. The first one has been done for your reference. |

| S. No. | Departments/ Sections | Reasons |
|--------|----------------------|---------|
| 1 | Room service | To stop delivering assorted chocolate tray and fruit basket |
| 2 | Telephone | To............................................................................................ |
| 3 | Housekeeping | To............................................................................................ |
| 4 | Restaurant | To............................................................................................ |
| 5 | Concierge | To............................................................................................ |
| 6 | Front desk cashier | To............................................................................................ |
| 7 | Bell desk | To............................................................................................ |
| 8 | Night auditor | To............................................................................................ |

## CONCLUSION

First of all, we should be aware that reception desk is not a front office department; instead it is a subdivision of front office department. After entering any hotel, reception is the first desk with which every guest interacts. Therefore, it is not wrong to say that reception desk is responsible for generating first positive *moment of truth*. Apart from guest handling hub, reception also works as a direct room selling junction for the hotel. With the help of movement list, reception desk prepares its welcome plan for *due-in* rooms (or expected arrivals) and makes all arrangements ready for *due-out* rooms (or expected departure guests). Remember, welcome as well as farewell process of transient guests is different from group guests. Room *discrepancy report* generated by night auditor is sent for clarification to front office manager, executive housekeeper or other responsible employees like lobby manager via reception desk.

| Chapter Terms (with Chapter Exercise) | |
|---|---|
| Inspection | Process of room check done by a supervisor looking for problems in the cleaning. |
| Current reservation rack | ? |
| Block room/room block | Condition whereby specific number of rooms are kept aside in advance. It is mainly concerned with group reservation. |

| | |
|---|---|
| Book room/Room book | ? |
| Chance guest | Walk-in guest. |
| ? | Room pre-allotted to a new arrival. |
| Farm out | It means transferring guest to alternate property in lieu of non-availability of accommodation in hotel's own property. It is also known as *bounced reservation* or *walking of guest*. |
| Slippage | ? |
| Room discrepancy | Room status variance whereby status shown by housekeeping department is found different from front office. |
| Rooming list | It is a list of names and room sharing details for a group reservation which is submitted by an inbound tour operator or group leader to an accommodation establishment prior to the group's arrival. |
| ? | *Guest bill.* |
| Movement list | Expected arrival-cum-expected departure list which is prepared by reservation section on daily basis for next day. |
| Double rooming | Situation in which front desk agent assigns the same room to two different guests. It mainly happens due to unresolved room discrepancy of previous night. |
| Tabular ledger | ? |
| Sold out | Situation when all hotel rooms are sold. Simply, it indicates that hotel rooms are totally sold. |
| ? | Guest who properly (or after account settlement) check-outs from the hotel but front desk still shows that room is occupied by the same guest. |
| Blocked off | ? |
| Moment of truth | Series of activities/events which take place between a prospective guest and hotel from pre-arrival/arrival stage and will remain till departure stage. This moment of truth works as a foundation stone for future repeat business and develops buyers' confidence. |

| Room rack/card index/room index/visible index/numerical index | Manual front office equipment which contains an array of metal file slots (like pockets and often known as *pigeon holes*) that are designed to hold room rack slips bearing the name of the guest, expected date and time of arrival, number of rooms, room type and rate, expected date of departure and other supportive details like reserved by, billing instructions, special instructions. It is used to indicate room status. |
|---|---|
| Key envelope | Group registration process whereby front desk agent pre-assigns room to individual group member. Consequently, puts assigned rooms' keys and welcome letters into envelope thereafter handing it over to group leader who will further distribute among group members. Often, it is also known as *hotel passport* which is directly related with electronic key card. |
| Satellite check-in | It is also known as *remote check-in* and is offered to group arrival whereby separate check-in counter is provided to group for their registration and key handover. |
| Key card | Document that is issued to guests along with metal keys at the time of assigning room. Guest is required to show this card each time when he/she collects room key from the front desk agent. |
| Room folio | Individual guest bill at front desk and is often also termed as grand folio. |
| Card key | Electronic key card. Remember, it is different from *key card*. |
| Room position status/room statistics | Actual room position, i.e., how many rooms are occupied, vacant (vacant/dirty or, vacant/clean), OOO/OOS, pre-allotted/assigned and so forth. It may be a kind of board or electronic display which gives a snapshot of current actual room status at a glance. |

# 8
CHAPTER

# Cash Section
(Special reference to account settlement and check-out procedure)

<hr/>

**OBJECTIVES**

*After reading this chapter, students will be able to...*

- describe the functions of cash section;
- explain different types of accounts and for whom these accounts are opened;
- elucidate different modes of settling guest accounts/bills;
- express guest check-out procedure and process; and
- clarify different modes of guest check-out.

## 8.1 CASH/CASHIER SECTION

It is the financial section of front office department which is primarily responsible for opening accounts, tracking payments, tagging advance deposit into room accounts, handling high balance accounts, accepting interim payments, zeroing out to the accounts of departing guest and so forth monetary transactions. Specifically, it deals with daily cash inflow and outflow. Simultaneously, it handles entire guest check-out process in coordination with the front desk.

This section posts all guest charges and credit transactions into respective guest folios, accurately and in time, so that guest accounts can be settled efficiently and quickly. This section also handles foreign exchange. In certain hotels, it also provides SDB (or lockers) to resident guests to keep their valuables and their safety is the responsibility of this section.

Cash section comprises head cashier/cash supervisor who works under the direct supervision of accounts manager (in large hotels) or front office manager (in medium and small hotels). Cashier head/supervisor is assisted by *credit manager* (in large hotels) and front desk cashiers who are directly involved in operational work of cash collection, account opening & maintenance and bill settlement.

Front desk cashier is also responsible for accepting initial *cash bank* from accounts department at the beginning of the shift, thereafter for *cash remittance* with head cashier.

This cash remittance will be deposited in closed and sealed envelope in an apposite manner (known as *close cashier*). In case of any monetary variance between the money placed in the front office cash envelope in comparison to cashier's stated *net cash receipts* (by formulating *audit formula* = Previous balance + Debit transactions – Credit transactions) inside the envelope then it should be noted as *overage, shortage, turn-in* or *due back*. Remember, neither overage nor shortage is considered good, from the perspective of financial integrity, for the organization because both show that the implemented control system is not effective or has some ambiguity.

## Activity

Match the below given descriptions with their correct terms.

| S. No. | Description | Term used |
|---|---|---|
| 1 | It is an imbalance that occurs when the total cash & cheques in the cash register drawer is less than the initial bank (or initial cash bank) plus net cash receipts. | Overage |
| 2 | It is an imbalance that occurs when the total cash, cheques, vouchers & other negotiable items (like paid out vouchers) in cash the register drawer is greater than the initial bank plus net cash receipts. | Due back |
| 3 | It is a situation that occurs when a cashier pays out more than he/she receives; the difference is ............... to the cashier's cash bank. It usually occurs when a cashier accepts so many cheques and large bills during a shift due to which he/she cannot restore the initial bank at the end of the shift without using the cheques or large bills. | Shortage |
| 4 | ............................ is the amount of cash, cheques, vouchers & other negotiable items in the cashier's drawer less the amount of the initial cash bank and the paid outs added. Its formula is: Received cash, cheques, voucher & other negotiable items – Initial cash bank + Paid outs. | Net cash receipt |
| 5 | It indicates to the cash deposited by individual department cashier with general cashier at the end of the day. | Turn-in |

## 8.2    FUNCTIONS OF CASH SECTION

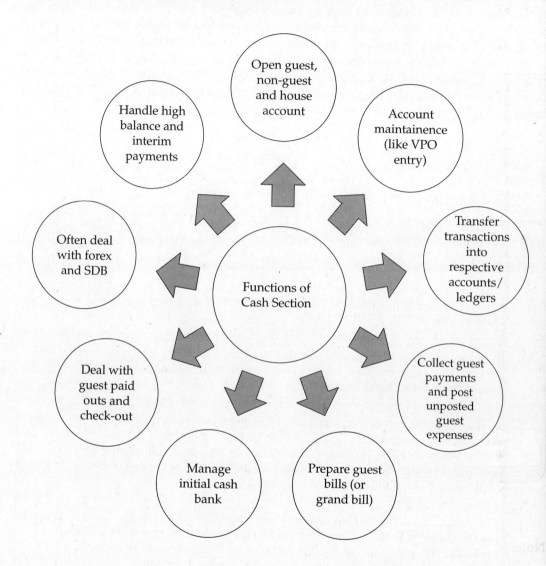

## 8.3 ROLE OF CASHIER SECTION IN DIFFERENT STAGES OF GUEST CYCLE

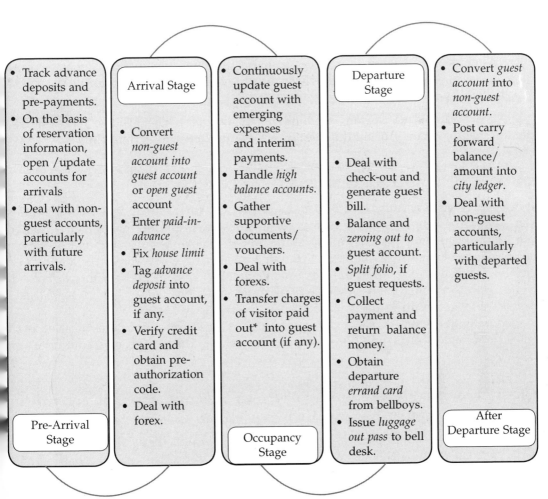

**Pre-Arrival Stage**

- Track advance deposits and pre-payments.
- On the basis of reservation information, open /update accounts for arrivals
- Deal with non-guest accounts, particularly with future arrivals.

**Arrival Stage**

- Convert *non-guest account into guest account* or *open guest account*
- Enter *paid-in-advance*
- Fix *house limit*
- Tag *advance deposit* into guest account, if any.
- Verify credit card and obtain pre-authorization code.
- Deal with forex.

**Occupancy Stage**

- Continuously update guest account with emerging expenses and interim payments.
- Handle *high balance accounts.*
- Gather supportive documents/ vouchers.
- Deal with forexs.
- Transfer charges of visitor paid out* into guest account (if any).

**Departure Stage**

- Deal with check-out and generate guest bill.
- Balance and *zeroing out to* guest account.
- *Split folio,* if guest requests.
- Collect payment and return balance money.
- Obtain departure *errand card* from bellboys.
- Issue *luggage out pass* to bell desk.

**After Departure Stage**

- Convert *guest account* into *non-guest account.*
- Post carry forward balance/ amount into *city ledger.*
- Deal with non-guest accounts, particularly with departed guests.

## Note

- *Visitor paid out* is a supportive voucher prepared when a hotel's in-house guest (or resident guest) requests a service/product that hotel does not provide. So in this situation, hotel (its concierge or travel desk) arranges the guest's demanded services/products from outside sources and makes the payment on behalf of guest. For example, car rental, arranging tickets for cinemas or theatre, etc. Here, a VPO is prepared on which incurred expenses along with service charges are written then guest signature will be obtained, and finally deposited to front desk cashier. Next cashier will transfer charges into respective guest account.

- While dealing with guest-with-reservation, front desk cashier converts his *non-guest account* into *guest account* as guest has already paid *advance deposit* to hotel in lieu of guaranteed reservation. In this situation, account for guest has already been created and during check-in process it only requires to be converted.

- Similarly, when front desk cashier deals with checking-out guest then he requires to convert guest account into non-guest account for those guests who defer their payment. Otherwise guest account is brought to *zero balance*.
- When dealing with walk-in guest, front desk cashier directly opens guest account as paid-in-advance (PIA) and will receive the charges from guest during the check-in process.

A step-by-step description of some major tasks of cashier desk/section that predominantly occurs during different stages of guest cycle is given below.

## 8.4   OPENING AND SETTLEMENT OF ACCOUNT/BILL

It is the process of account creation, maintenance and final settlement. *Account creation* starts with guest registration process whereas guest accounts are maintained/updated during the occupancy period. Finally, at the time of check-out guest accounts are *zeroed out* either via instant payment or through transferring carry forward balance into city ledger; it depends upon the settlement mode. In case of guest-with-guaranteed-reservation, account opens in pre-arrival stage and deposited advance (or *advance deposit*) is tagged with respective account during the arrival stage.

Generally, there are three kinds of account, i.e., guest account, non-guest account (plus combined account) and house account. A brief description of each account is given below.

## Guest Accounts (also known as pay own account, resident account or incidental account)

✓ Guest himself/herself is responsible to settle down his/her account. It is a record of financial transactions, which occur between a resident guest and the hotel. Guest accounts are created when guests register at the front desk, specifically for walk-in guests.

✓ But in case of guest-with-guaranteed-reservation, account (i.e., *non-guest account*) usually opens during pre-arrival stage. And during check-in process, non-guest account is converted into guest account. All guest account transactions are shown by *guest folio*, supported by vouchers and backed up by *guest ledger*.

✓ In group arrival, guest account may be further categorized as incidental and master account. *Incidental account* represents to an individual group member whereas *master account* represents a group leader. All collective charges/transactions of master account are shown in *master folio*, supported by vouchers and backed up by *master/group ledger*.

✓ Group leader settles down master folio whereas for incidental expenses, individual group members are responsible. Payment mode for both accounts depends o hotel's policy.

# Non-Guest Accounts (also known as non-resident account or city account)

✓ It mainly represents future business/guests (who have not yet registered into hotel) plus past business/guests (who have already departed from the hotel). The basic reason behind creating and maintaining non-guest account is to track the transactions of non-resident guests.

✓ Non-guest accounts are also created when a former guest fails to settle his or her account at the time of departure. During departure stage, as guest's status is changed from resident to non-resident guest. Here, unsettled guest accounts are first *zeroed out*, afterwards converted into non-guest account for *deferred payment*.

✓ The responsibility of carry forward balance of unsettled guest accounts (or zero out account) transfer from front office to accounts division which means accounts division will be responsible to collect *deferred payment*.

✓ Incongruent to guest accounts which is compiled on daily basis, non-guest accounts are normally billed on a monthly basis (or billed at specified time period) by the hotel's accounting division. All non-guest account transactions are supported by range of *folios* and *vouchers* and backed up by *non-guest ledger* or *city ledger.*

| | Chapter Activity | |
|---|---|---|
| | Fill in the below given spaces with appropriate differences, some have been given for your reference. | |
| **S. No.** | **Guest ledger** (a ledger to record monetary transactions of resident-guests or guests currently staying in the hotel rooms) | **Non-guest ledger** (a ledger to record payment transactions of non-resident guests like future arrivals, skippers & deferred payments) |
| 1 | ? | ? |
| 2 | Also known as transient ledger, front office ledger, rooms' ledger or visitors' tabular ledger (VTL). | Also known as city ledger, city ledger account, debtors' ledger or customers' ledger. |
| 3 | It is a subsidiary ledger that lists the *account receivables* of guests who are currently staying in the hotel. | ? |
| 4 | ? | It represents skipper, no-shows, bad debts, future and past guests. Sometime *house accounts* too. |
| 5 | It is the total set of all guest folios of resident guests hence any debit entry to the guest folio will not only increase the guests' balance but will also increase the net outstanding balance of the guest ledger. | ? |
| 6 | ? | Maintained by account section; not by front desk cashier. |

# Combined Account (guest account + non guest account)

✓ It is the combination of guest account and non-guest account. These kinds of accounts usually require open and operate while dealing with company clients, incentive guest or any client who came via a third party reservation. During check-in, front desk agent does not forget to tell about diverted charges which will be settled down by the third party.

✓ For instance, a company client who shows a *letter of authorization* and after verification with company approved list, front desk agent allots a room. Here, guest is responsible for the settlement of his/her incidental charges/expenses whereas representing company takes the responsibility to settle down remaining specified charges.

✓ Therefore, often at the time of departure resident guest demands to split his bill/folio; guest bill/folio can split vertically as well as horizontally. Now the vertically or horizontally prepared folio/bill is known as *split folio* or split bill. The term *folio* is used for bill.

✓ *Split bill/folio* may also be divided as individual guest folio and company folio on the same piece of paper. The settlement of individual guest folio will be the sole responsibility of individual guest whereas *company folio* will be settled by the representing company.

# House Account (also known as management account)

✓ It is mainly opened to settle the accounts of those rooms which are allotted as complementary room or free of charge (FOC). All monetary transactions of such rooms are settled/written-off in house accounts. Generally, complementary rooms are offered to hotel corporate employees, group leader, CIPs, owner or guest who is staying on owner's reference.

✓ The reason behind offering complementary rooms is to promote future business/room sale, such as complementary room to group leader/tour leader. House account is also raised to track the daily non-profitable transactions that occur within the hotel organization including transactions based on NC-KOTs.

✓ Most hotels use house accounts to record supply chain commissions and other obligations. This commission is mainly offered to travel agencies, tour operators, group leaders and other third parties who promise to generate and prop-up business for hotel.

**Note**

• At the registration phase, non-guest account (or account of guest-with-guaranteed-reservation) converts into guest account because status of non-resident guest converts into resident guest. Similarly, at departure phase account of those guests (or guest accounts) who are diverting their carry forward balance towards third party also convert from guest to non-guest account. This carry forward balance will transfer into city ledger and account status of departed guest will convert from guest account to non-guest account.

## 8.5 ACCOUNT/BILL SETTLEMENT MODES

It is the major responsibility of a front desk cashier to collect the payment from departing guest, guest flourishing with high balances, etc. Generally, guest pays during reservation phase then registration, during occupancy and finally at the time of check-out (in certain cases, payment also received after guest departure). Some widely used methods of guest payment in hotel industry include cash, credit cards, cheques (like travellers and personal cheque), travel agent vouchers, third party billing and so forth. A brief description of some major modes of payments are given below.

### Cash Based Settlement

Cash based settlement involves settlement of accounts either via paper money or metal coins. Apart from direct cash, use of debit card, travellers' cheque, bankers' cheque, bank draft and other similar negotiable instruments are also considered a part of cash based settlement options. This cash can be in national currency as well as in foreign currency. A brief description of both is given below.

### In Local Currency

✓ The cashier accepts and counts cash in front of the guest. Any change and a receipt are given to the guest. Local currency means settlement in national currency which includes payment made via paper currency or metal coins. Every organization is bounded by legal guidelines to accept payment via both currency notes and coins.

### In Foreign Currency

✓ Here, first cashier converts the foreign currency into local currency as per the day's currency exchange rate then deducts the billed amount or vice versa. The remaining balance should be refunded in local currency only. Front desk cashier should record the entry into foreign exchange voucher and must issue an encashment certificate to *alien guest*.

**Note.** Foreign currency notes/coins is ready-money but in the economy of alien guests' own country; not for visiting country. Because, no foreign currency can promulgate into the open market of visiting country, it must be first exchanged from authorized dealer then used.

### Credit Based Settlement

It indicates to all those modes of account settlements in which payment is deferred and usually received after a specified period of time from the third party. For example, settlement via cheques (such as traveller's cheque and personal cheque), vouchers (such as travel agent voucher and airline voucher, MAO and PSO), BTC, charge card (or *plastic money*) like credit card and other similar negotiable instruments.

Collectively, settlement parties for deferred payment are known as third party and billing style is known as third party billing whereas diverted payment is known as deferred payment. The account used to track such transactions is termed account receivable. The *account receivable* period may vary with type of settlement, therefore hotel's accounts division opens *account age receivable* statement which comprises current, delinquent and overdue account. A brief description of major credit based settlement modes is given below.

## Foreign Currency

✓ It is concerned with acceptance of payment in overseas currency. It is important that a cashier should be familiar with those currencies that are accepted in the hotel because not all nations' currencies are acceptable, simultaneously cashier should also ensure that accepted currency should not be counterfeit.

✓ Cashier should update himself with daily *buying rate* and *selling rate* of foreign currency, on routine basis. A step-by-step process for accepting payment via foreign currency is given below.

Step 1    Enquire currency from the guest and determine whether it is redeemable or not. If it is redeemable then check its legitimacy.

Step 2    If legitimate then fill in all currency and guest related credentials (like name, nationality, passport number, etc.) in *Encashment Certificate* in lieu of currency exchange.

Step 3    Now convert the foreign currency into local currency (cashier must be aware with foreign exchange rate & applicable buying rate) and deduct incurred charges.

Step 4    Handover the original copy of Encashment Certificate to the *alien*/foreign national and return outstanding balance, after deducting billed amount, in local currency.

Step 5    Attach the second copy of the Encashment Certificate with exchanged foreign notes and leave the third copy in book itself for future reference or for evidence.

Step 6    At last, cashier must fill in details about the transaction in RLM 1 register, cashier report (updated on daily basis) as well as in *foreign exchange control sheet*.

## Personal Cheque

✓ It is concerned with acceptance of payment through guest's personal cheque. Some hotels allow such method of payment while others have totally restricted policy against this method of payment. Personal cheques are verified from the issuing bank. For the approval of personal cheque, hotel needs the following points of information: guest's driving licence/passport, guaranteed card (accompanied with cheque), address, telephone number, and so forth.

✓ These need to be written on the back side of the personal cheque. In contrast to personal cheques, traveller's cheque and certified cheque do not need any approval because they are solely issued by exchange of payment when issued. A step-by-step process for handling personal cheque is given below.

Step 1    A cashier should be aware that only one cheque is used for one transaction and the given cheque should be under-signed by the guest in front of cashier.

Step 2    Cashier requires to check or match the bank code like IFSC, guest signature, account payee details (or name of hotel), etc., and all should agree with accompanied guaranteed card.

Step 3    Cashier must check the expiry date of the guaranteed card, verify & ensure that guaranteed card limit covers the billed amount and check the date and year mentioned on the cheque.

Step 4    Cashier must check the similarity between amount shown in words & figures and if there is any kind of overwriting or corrections then it must be properly corrected and signed.

Step 5    Cashier should check the correct name of the guests (there should not be any spelling mistake in the hotel name/cheque prepared in the favour of) before acceptance.

Step 6    Cashier should accept only MICR (Magnetic Ink Character Reader) cheques and must be account payee only. Do not accept undated, future dated (or post-dated) or third party cheques.

**Note**
- Front desk agent must ensure well in advance (or during check-in) whether guest is allowed to use personal cheque for account settlement or not, it must be as per hotel's policy.
- Generally, front desk agent should ensure that personal cheques of overseas bank as well as cheques drawn on a new account should not be commonly accepted. In some hotels, guests are required to fill-in an application form in lieu of using personal cheque.

## Traveller's Cheque/Check

✓ It is concerned with a pre-printed, fixed-amount cheque designed to allow the person signing it to make an unconditional payment to someone else as a result of having paid the issuer for that privilege. Remember, traveller's cheques are issued in fixed denominations by financial institutions like Amex.

✓ The process of validation in the hotel requires the cashier to watch a second sign in front of him, then cashier should make sure that both signatures are identical. A step-by-step procedure for accepting traveller's cheque is given below:

Step 1    After accepting the traveller's cheque from the arriving or departing guests, cashier must check the denomination of the currency.

Step 2    On the basis of denomination, a cashier requires to calculate the correct exchange value of the traveller's cheque.

Step 3    After calculating cheque's value, cashier requires to check the date and payee's detail on the traveller's cheque. He may also demand passport for the purpose of double check.

Step 4    Now, take the guest signature and compare it with previous signature. If signature matches then fine, otherwise request guest to sign again, next appeal to produce valid document for verification or inform a senior officer.

Step 5    At last, a cashier should check the cheque number against the *current stop list*. If cheque is authentic then cashier will write down the room number behind the cheque for further reference.

Step 6    Finally, accept the traveller's cheque, deduct billed amount thereafter return outstanding money in local currency. If traveller's cheque is from overseas bank then make an entry in RLM 2 register, foreign exchange voucher and generate *encashment certificate*. Original copy of encashment certificate must be given to guest.

## Activity

Differentiate between traveller's cheque and personal cheque on the basis of themes given below.

| S. No. | Base | Traveller's Cheque | Personal Cheque |
|---|---|---|---|
| 1 | Validity | More than personal cheque mode | ? |
| 2 | Bank account | ? | Must to have bank account to clear the cheque |
| 3 | Account payee | ? | ? |
| 4 | Denomination | ? | May be fixed/decided on the spot at POS terminal |
| 5 | Discredit | Never dishonoured due to advance payment | ? |
| 6 | Acceptability | ? | Limited acceptability |

# Credit Card

✓ Payment through credit card is the most common method of payment. Some of the well known brands are American Express (or Amex which introduced the concept of traveller's cheque into hotel industry), Diner's Club (it introduced the credit card into hotel industry in year 1950s), Visa, Master Card, Maestro, Barclays, JCB International (Japan Credit Bureau), etc. The front desk agent obtains the card at the time of registration and takes an imprint or directly swipes credit card via *hot line machine* (or EDC machine or PDQ machine—Process Data Quickly machine).

✓ While swiping, hot line machine will generate a *pre-authorization code* or may also generate *denial code*. It depends on credit card's authenticity and available balance. This process is called card approval. A step-by-step procedure for accepting a credit card is given below:

Step 1    After collecting the guest's credit card, a cashier must verify the name of credit card company, validity date (or expiry date), cardholder's name and signature on the reverse side of the credit card. Cashier must also check the company's current stop list.

Step 2    Now cashier will swipe the credit card for on-line authorization (OLA) through EFTPOS or hot line machine (or EDC machine or PDQ machine) for authorization. Here, the system will generate pre-authorization code (card acceptance code) or denial code (card rejection code).

Step 3    Cashier must check the *floor limit* of the credit card, so the excess balance or payment should be charged in cash from the guest. Floor limit may also assist in determining *house limit* to the registering guest.

Step 4    Now, cashier will prepare sales voucher, usually in triplicate (one original + two carbon copies) and enter the authorization number on it, if any.

Step 5    Cashier must take the guest signature on the generated/imprinted sales voucher. Cashier requires verifying the guest signature on voucher with credit card signature.

Step 6    All required activity has been finished over here; now cashier requires destroying the used carbon (placed between triplicate copies), so no one else can misuse it.

**Note**

• In semi-automated front office, cashier uses the credit card imprinter for obtaining the impression of credit card number and validity date on the sale voucher/charge slip; next obtain the guest signature on it. A copy of imprinted sales voucher along with grand bill and other charges is to be given to the departing guest.

- If pre-authorization has been given by the credit card company then cashier can directly process the credit card on the off-line mode with approval code and enter the billed amount into it.
- During check-out, guest should present same credit card for which pre-authorization has been taken, but if guest wants to charge the expense to another credit card then it must be verified on-line before settling the account.

## Travel Agents Voucher

✓ It indicates to a voucher that is issued by a travel agency which means the responsibility of payment lies with travel agency. Travel agent/agency prepares its voucher in triplicate system. The first copy is given to guest, second copy sent to hotel's reservation section and last copy is retained within the voucher book.

✓ Before the day of arrival, reservation department sends the copy of received travel agent voucher to front desk. So, the front desk agent could enable the guest's identity during check-in process. A step-by-step process for handling payment via travel agent's voucher is given below.

Step 1    After check-in, front desk cashier/agent should collect the voucher from the guest and check with a list of approved travel agents, thereafter cross-refer received copy with guest copy.

Step 2    Cashier requires checking the date validity, meal plan, credit status, number of rooms, number of pax and so forth relevant details.

Step 3    At the time of check-out, the cashier must ensure that guest pays for all items which are not covered by the voucher. Now, enter the amount in *master account* and ensure that guest cannot notice it because the room rate paid by guest to travel agent is different from the amount that is allowed to travel agent.

Step 4    After check-out, the *master account* should be attached with collected travel agent voucher and sent to the account & administrative department for further processing.

Step 5    Generally, at the end of the month these accounts along with received vouchers and a statement of the total account is sent to the travel agent/agency for payment.

**Note.** Often it has been found that many travel agencies/agents are not a part of hotel's intermediary supply chain system, thus they are not listed on the hotel's approved list of travel agents. In case of such bookings, travel agencies/agents have to deposit an advance before the actual arrival of transient guest or group. It is also possible that travel agent can request to pay via cheque then cashier is required to send it to credit manager in accounts section for approval.

# Bill to Company (BTC)

- ✓ Often, it is also known as *corporate billing, deferred billing* or *third party billing*. It usually happens with those guests whose payment will be given by their representing company. This is only possible when there is a legal agreement (or tie-up) between the hotel and the guest's representing company.

- ✓ In BTC settlement mode, a cashier will transfer the amount to the debtor's ledger (or *city ledger*) in accounts section then accounts section sends invoice to the company at regular intervals (usually on monthly basis). This style of settlement, at regular intervals, helps the hotel to settle various small debts with one cheque or bank transfer. A step-by-step process for handling payment via BTC is given below.

Step 1   At the time of check-in, front desk should enquire for mode of payment, if guest states that he/she wants to pay via corporate account then front desk must obtain the *letter of authorization* from guest. Next check the representing company & other relevant details with its own list of approved companies.

Step 2   Now open the correct account; cashier may be required to open *company folio* (for those expenses which will be settled by representing company), guest-account (for incidental expenses) or to both. Account creation will be done in accordance with *letter of authorization* (LoA) that is received from the guest.

Step 3   Cashier should clarify to the guest about the settlement system, for instance what charges will be settled by his/her representing company and for what charges guest is solely responsible. Now, cashier will establish *house limit* (or credit limit) to cover incidental charges.

Step 4   During the occupancy stage, cashier transfers the accumulated charges into respective accounts. Generally, room charges plus taxes are mostly charged into *company folio* whereas meal charges are transferred into guest's incidental account (or transferred as per the instruction mentioned in received LoA).

Step 5   During the occupancy period, continue monitoring the *house limit* and when guest reaches a limit, intimate the guest and collect interim payment for rendering services continuously.

Step 6   During the departure stage, obtain guest signature on the grand bill. Collect payment for which guest himself is responsible and bring the incidental account to zero balance.

Step 7   Transfer the carry-forward balance of *company/master folio* (in respect of room charges and taxes) into the city ledger in accounts section. At the end of the month (or after stipulated period of time), accounts section sends the guest's signed bills with letter of authorization (or its reference number with cover letter) to representing companies for final settlement. Next update *account age statement*.

**Note**

- In many circumstances, cashier settles the BTC/corporate bills via *split folio* provision. In it, cashier splits the guest folio (or grand bill) into two parts, either horizontally or vertically, thereafter post charges accordingly.
- Generally, incidental charges are transferred on one side of *split folio* whereas diverted charges post on another side of the split folio. It depends on settlement condition shown in LoA.

| Activity | | | |
|---|---|---|---|
| Under FERA 1973 thereafter in FEMA 1999, RBI has provision of FLM (Full Money Changer License) and RLM license for the purpose of forex. Hotels are authorized under RLM (Restricted Money Changer License) to deal in forex then record each transaction into respective RLM register. You need to fill in below given blank spaces with an appropriate description. | | | |

| RLM Series | Description | RLM Series | Description |
|---|---|---|---|
| RLM 1 | ? | RLM 5 | ? |
| RLM 2 | ? | RLM 6 | Monthly statement of purchase and sale of foreign currency. |
| RLM 3 | ? | RLM 7 | ? |
| RLM 4 | To record sale of foreign currency | | |

## 8.6   GUEST CHECK-OUT PROCEDURE (AT CASHIER DESK)

It is always desirable that cashier should be aware and updated, well in advance, with expected check-out guests. With the help of *movement list* (by referring to expected departure column), every shift cashier must search guest folios of that guest who probably are to check-out the next day. Consequently, arrange all supportive documents to be presented so that upon guest's request each document can be checked. A step-by-step process to handle the check-out guest is given below.

**Stage 1**
- After getting guest intimation for check-out, a cashier should ensure that all charges & payments made have been posted to the guest's account (if possible check with other departments too, mainly with room service & restaurant). Next keep grand bill (along with supportive bills of POS outlets) ready to be presented for final settlement.

**Stage 2**
- Cashier must also ensure that all deposited valuables (kept in safe deposit box) are collected by the guest before departing. Simultaneously, do not forget to collect key of SDB and obtain guest signature in SDB handover column. Meanwhile cashier will also intimate the bell desk, housekeeping and maintenance department.

**Stage 3**
- Bell desk will send a bellboy to guest room for *down bell activity*. In many hotels, cashier is also required to provide *luggage clearance pass* to the bell captain. So that bell desk can update/check off its expected departure list when luggage will be cleared.

**Stage 4**
- Similarly, housekeeping department will also send a room boy into the departing room for room checking and *room make up*. Next, housekeeping's control desk will tick off (in departure list) against the departed guest's room number, thereafter make it available for sale as soon as possible.

**Stage 5**
- In some hotels, maintenance department may also be required to send an employee to check in-room electronic gadgets. Next, cashier will present the generated bill to the guest for payment then depart the guest with a warm farewell and invitation to come again. Remember, do not forget to collect room keys.

**Stage 6**
- Finally, cashier will notify the reception desk about the successful check-out of the guest in order to avoid *sleeper* situation. Reception desk will update the room status position after receiving the room keys from the bell desk or guest. Next, it will prepare a *room history card* and updates *guest history card*.

**Note**

- In automated front office (or PMS driven front office), room history and guest history automatically generates/updates as soon as status of guest converts from resident to non-resident. Simultaneously, PMS also eliminates the probability of *sleeper* like room status.

## Transient Guest Departure Procedure

Step 1  The bellboy on his way to guest room also informs the cashier about the departure of the guest. So, cashier can check for voucher/s required to be posted into guest account then prints out the guest's grand bill.

Step 2  Cashier calls up room service, coffee shop and relevant sections for any last minute billing. Afterwards, cashier updates the bill and checks guests billing instruction.

Step 3  He takes out all supporting vouchers and checks against bill entries and arranges them date-wise. When the guest comes to cash desk for bill settlement, cashier presents the bill to the guest (to verify) then ratify the charges. Finally, reconfirms the billing instructions with guest.

Step 4  If guest wishes to see supporting vouchers, they are also presented to him and the guest is helped in locating the charges on the main bill (or *grand bill*) plus any query raised is answered.

Step 5  When the guest has approved of the charges, the cashier finalizes the bill after passing any allowance or discount as per the instructions or authorization of the front office manager or duty manager.

Step 6  On receiving the cash, cashier makes out a receipt and notes in the appropriate column of the grand bill the receipt number and date. He then issues the *baggage out pass* to the bellboy.

Step 7  Cashier then staples the receipt on the original copy of the grand bill and folds both gently and places them in an envelope on which he writes legibly "Your Receipt".

Step 8  The supporting vouchers are properly kept in another envelope and on this cashier writes "Supporting Vouchers". Finally, both envelopes are handed over to the departing guest.

| Activity |
|----------|

*Voucher* is a supportive document that reinforces the applied charges as well as the guest's payment. It is available in different forms. Match the below given vouchers' description with their correct names/types.

| S. No. | Voucher Description | Voucher |
|--------|---------------------|---------|
| 1 | It is usually raised when wrong entry is made into guest's account as well as when hotel gives any compensation to their guest in lieu of poor services offered. | Charge voucher |
| 2 | It is usually raised when hotel staff is paid on behalf of guest in lieu of securing payment from guest like movie tickets for their resident guest. | Cash advance |
| 3 | It refers to Visitor Paid Out voucher. | Cash voucher |
| 4 | It is a supportive document or voucher that is used to shore-up the various cash payments between the guest and hotel. | Guest disbursement voucher |
| 5 | It is usually raised when any wrong entry is made into any guest or non-guest account like wrong amount, wrong side posting, transfer into wrong account and so forth. | Allowance voucher |
| 6 | It is usually raised when any section/outlet wants to transfer something into another section/outlet. | Correction voucher |
| 7 | A supportive document that shows the details of the amount to be charged to a guest's account, like drinks in the bar. | Transfer voucher |

## Group Guest Departure Procedure

Task 1    The cashier takes out the print of *group folio* (or master folio) to present to the group leader and *incidental folios* for GITs (or for group members). In non-automated front office, front desk cashier should check with point of sale outlets like room service and coffee shop for any last minute billing.

Task 2    If there is any outstanding charge(s) from point of sale outlets for departing guest then cashier must post it before printing folio. Next, cashier should take out all supporting vouchers (in respect of incidental charges) then separate them room wise and make a small room wise summary to facilitate the cash collection.

Task 3    When group leader approaches the cashier desk for account settlement, cashier should present the group bill (or *master folio*) for approval. Simultaneously, the list of the rooms that have to pay their incidental charges is also given to the group leader.

Task 4    Group leader assists the cashier in collecting all incidental charges from the GITs (or individual group members). Next, after ratifying the master folio the group leader is requested to sign the bill. The carry forward balance of master folio will transfer to *city ledger*.

Task 5    Now cashier will issue the *luggage out pass* to the group leader/bell captain. At the end of the shift, cashier will send all original unpaid bills to the accounts section which will then be tracked out with the help of *account age statement*. Thereafter, accounts section will claim these bills from respective group leaders, tour leader or representing company.

**Note**

- In case group business is given by travel agent/agency then front desk cashier needs to attach travel agent voucher with signed *master folio* thus noting down the voucher number, name of travel agent/agency, travel agent's/ agency's reference file number and so forth details.
- Simultaneously, if entertained group business is supported by meal coupons then front desk cashier also requires collecting them from respective point of sale outlet/s then attach them all with master folio for claim and further references.

| Activity |
|---|

Suppose you are a front desk cashier and your next day's movement list shows the departure of a foreign group (Australian Cricket Team) and one VIP guest (your corporate manager). Now prepare a 10 points closing task-list to deal with both clienteles.

| Closing Task-list for Australian Cricket Team | Closing Task-list for Corporate Manager |
|---|---|
| ✓  Balance incidental and master folio | ✓   Balance to house account |
| ✓ | ✓ |
| ✓ | ✓ |
| ✓ | ✓ |
| ✓ | ✓ |
| ✓ | ✓ |
| ✓ | ✓ |
| ✓ | ✓ |
| ✓ | ✓ |
| ✓ | ✓ |

## 8.7 MODES OF CHECK-OUT

The term *mode of check-out* refers to the method or approach of guest checkout; it can be non-automated, semi-automated or fully-automated. Each method has its own pros and cons and their applicability depend upon guest flow, size of organization, standard of hotel, financial viability, etc. But in all of these check-out methods front desk must ensure that guests should never opt for *overstay* without prior concern else problem may arise with new arrivals. In addition, front desk must update room status position (except SCO) as soon as guest departs else *sleeper* room status like problem may arise. A brief description of different modes of check-out is given below.

### Non-automated (supported by *guest's weekly bill*)

✓ In manual check-out system, all bills and supportive vouchers and documents are manually prepared by the cashier. A day before or early morning on the day of expected departure, front desk/cashier desk prepares the multiple copies of *departure notification slip* and sends to different departments/sections, in lieu of any outstanding charges that need to be transferred into respective guest's account.

✓ At the time of departure, cashier hands over the handwritten bill (or *guest's weekly bill*) to guest for verification, if guest has any doubt then cashier will rectify with the help of raised supportive vouchers. Afterwards, cashier will obtain payment from guest and properly settle his account. Lastly, collect room keys, keys of SDB, instruct bell desk (through *luggage out pass*) to release guest luggage, update room status board plus prepare/update guest history.

### Semi-automated (supported by *express check-out* form)

✓ A condition whereby guest leaves the hotel without checking-out/passing through cashier desk. Generally, express check-out form is used in semi-automated process whereby settlement can be done either in cash or through *charge card*. However, guest needs to check then furnish his credit card details on the *express check-out form*, prior to his date of departure or during the check-in also.

✓ Express check-out is an activity which involves compilation and distribution of guest accounts (or *guest folio*/bill) to all those guests who are expected to depart next morning.

✓ After leaving the room, guest needs to deposit the signed express check-out form along with room key at cashier/reception/guest relation desk (or may drop in drop box) then proceed through remaining check-out formalities.

### Fully-automated (powered by *self check-out* and *video check-out terminals*)

✓ It is also called "computerized check-out" that is usually a mechanized system of checking out which may either be front desk supported or guest can directly check-out through self-operating terminal. Generally, *self-check-out* terminals are used by only those guests who pay their bills through credit card; cash is usually not eligible for this facility. In *video check-out*, guest can use in-room television for the purpose of account settlement and check-out.

✓ In fully automated system, front desk staff may also use computer system for guest check-in and check-out. Front desk agent enters all the guest given information into the computer system, afterwards guest account is automatically generated and at the time of departure, an updated guest folio will be generated.

| Where can guest check-out in fully-automated system? |
|---|
| At front desk (or cashier desk directed check-out) |
| ✓ It is specifically known as "computerized check-out" because front desk cashier uses computer system to settle guest account then check-out. These computer systems may be powered by software packages like IDS, Opera, Fidelio ShawMan, Hogatex, etc. During check-out hours, front desk cashier makes the total of guest folio/bill (or bring to guest account at *zero balance*) and gives print command for final bill generation (or guest folio/grand bill). |
| ✓ Thereafter, guest needs to go through bill details to find out any discrepancy. In this entire process, guest requires doing nothing, except signature on the generated *guest folio* (signature is required when guest wants to settle his/her account other than cash mode). If guest pays in cash then there is no need to take his signature on guest folio. |
| At self check-out terminal (or guest directed check-out) |
| ✓ It is related with a mechanized check-out process in which guest can either use *self check-out machine* installed in hotel foyer or use *video check-out* system that is available in guest rooms. This system directly allows guests to review their folio, well in advance, and settle their account through credit card. Identical to front desk directed computerized check-out system; both of these systems must also be powered by software packages. |
| ✓ By adopting this way of check-out, the front desk/cashier desk staff reduces a lot of burden from their shoulders and can utilize this time in effective dealing with other guests. At the same time, the hotel guests also not needed to stand in long queues. |

## 8.8 PROCESSES OF GUEST CHECK-OUT

Hotels may offer either front-desk-directed or guest-directed-check-out provision. In front-desk directed check-out, cashier is responsible for performing check-out course of activities whereas guest-directed check-out system is a kind of self-guided check-out system. A step-by-step process for check-out under guest-directed-system (ECO as well as SCO based) is given below.

| | Express check-out (Guest directed semi-automatic system) | Self check-out (Front desk or guest directed automatic system) |
|---|---|---|
| Step 1 | ECO form is usually handed over to guest during registration process plus signature must also be obtained on imprinted credit card voucher. | SCO provision can be supported by either self-check-out machine (at lobby) or in-room television system. |
| Step 2 | A day before, hotel sends a copy of guest folio/grand folio in order to verify the charges. | Guest checks his/her incidental account balance either in self-check-out machine or in-room television. |
| Step 3 | In return, guest fills express check-out form and submits to the front desk/cashier. | Guest verifies his/her account details. In-room television check-out is also known as *video check-out*. |
| Step 4 | Guest puts his signature on the guest folio and deposits with front desk cashier. | Guest uses his/her credit card to settle his account or to pay incidental expenses. |
| Step 5 | On the day of departure, guest drop his/her room key into drop box and leaves the property. | On the day of departure, guest drop his/her room key into drop box and leaves the property. |
| Step 6 | After departure, cashier fills in carry forward balance into an imprinted credit card voucher. He sends a copy of voucher to guest also. | When check-out gets completed, machine automatically prints an itemized statement of account and dispenses to guest for reference. |
| Step 7 | Front desk must update room status position as soon as possible. | Machine automatically updates room status position and creates/updates guest history file. |

## 8.9 STANDARD CONVERSATION BETWEEN CHECKING-OUT GUEST AND CASHIER

**Situation** – Mr. Blue is going to check-out today. During check-in phase, he told the front desk to settle his account via credit card. Now cashier Mr. Red is processing the check-out of guest plus ensures that guest has not consumed anything this morning. Guest is also requesting to arrange a vehicle to drop him at airport. Pretended standard conversation phrases between Mr. Red and Mr. Blue is given below:

| Settlement Instructions | |
|---|---|
| Guest | Mr. Blue |
| Cashier /agent | Mr. Red |
| Room No. | 101 |
| Payment mode | Credit card |
| Extra services | No extras |
| Bill dispute | Nothing |

Mr. Red     Good morning sir, how may I assist you?

Mr. Blue    Good morning sir, I would like to check-out today.

Mr. Red     Ok sir. May I know your name and room number please?

| Mr. Blue | Oh yes, my name is Mr. Blue and I am from Room No. 101. |
|---|---|
| Mr. Red | Wait a minute Mr. Blue, let me check your account. |
| Mr. Blue | Ok |
| Mr. Red | So, Mr. Blue you are from Room No. 101 and you would like to check-out today? |
| Mr. Blue | Yes |
| Mr. Red | Sir, did you have breakfast this morning? |
| Mr. Blue | No |
| Mr. Red | Sir, have you used any of our services today? |
| Mr. Blue | No, not at all. |
| Mr. Red | Ok sir, I'll compute your bill for kind reference. Here is your bill. Would you like to check it, sir? |
| Mr. Blue | Oh please, let me check. |
| Mr. Red | Sure sir, no problem. |
| Mr. Blue | Ok, I think everything is fine. Now, I would like to pay my bill. |
| Mr. Red | May I have your credit card, please? |
| Mr. Blue | Here it is! |
| Mr. Red | Please sign your name here, sir. |
| Mr. Blue | Oh, of course. |
| Mr. Red | Would you be going back to your room, Mr. Blue? |
| Mr. Blue | No, I already checked-out from my room. |
| Mr. Red | May I please have your room key? |
| Mr. Blue | Oh sure! |
| Mr. Red | Sir, would you like us to arrange for a drop to airport? |
| Mr. Blue | No, not at all. |
| Mr. Red | Thank you for staying with us Mr. Blue. Have a nice journey. And hope to welcome you back again soon. |

## Additional Standard Phrases During Check-out

| *What to check with guest during check-out* | *Standard phrases used with guest during check-out* |
|---|---|
| - Confirm check-out with expected check-out guest | - Good morning Mr. X, will you be checking-out today? |

| | |
|---|---|
| - Confirm that guest has enjoyed his stay | - Hope you enjoyed your stay with us Mr. X. |
| - Confirm mini bar consumption with guest | - Is there any mini bar consumption? |
| - Confirm guest has emptied to his/her SDB | - Have you emptied your safe deposit box? |
| - Confirm guest has last minute phone calls | - Did you have any phone calls from your room? |
| - When phone calls made and guest wants to verify | - That's for the phone calls you made. |
| - Try to sell additional service to guest | - Would you like us to arrange for a drop to airport? |
| - Offer guest folio to check applied charges | - Here is a copy for you to review. |
| - If guest has any doubt in guest folio | - Which item did you have a question about, I shall clear it for you. |
| - If guest folio has mistake | - I apologize for the inconvenience, Mr. X. |
| - When demand payment via credit card | - Will you be using your credit card to settle the account? |
| - When there is not enough amount in credit card | - I do apologize, Mr. X the amount of your stay was not approved on your credit card. |
| - When requesting guest to give credit card | - May I take an imprint of your credit card? |
| - To offer copy of sales voucher to guest | - Here is a copy of your receipt. |
| - When guest changes his/her payment mode | - I'll enter the new method of payment for your records. |
| - When refusing guest payment via personal cheque | - I am sorry. We don't accept personal cheques. |

| | |
|---|---|
| - If guest asks why not you accepting personal cheque | - It is the policy of the hotel ...... |
| - When guest requests you to send bill to his/her address | - Yes but you'll have to give me your address. |
| - Confirm that guest again wants to go in his room | - Would you be going back to your room? |
| - Taking back room key | - May I please have your room key? |
| - Ask guest for luggage assistance | - Will you need any assistance with your luggage? |
| - Send bellboy for luggage assistance | - Please allow me to get a bell person to assist you. |
| - Seeing off the guest | - Thank you for staying with us. Have a nice journey. |
| - For getting business again from the guest | - Hope to welcome you back again soon! |

## CONCLUSION

Cashier desk is responsible for all cash inflow and outflow. Initially, the desk is responsible for collecting *cash bank* thereafter to open guest, non-guest and house account for different guests in accordance with the situation. Apart from cash based settlement policy, cashier should also be perfect in settling guest bill via other methods, for instance, foreign currency, charge card, traveller's cheque, travel agent voucher and so forth. Like reception desk is responsible for welcoming the guest, similarly cashier desk is responsible for settling guest account thereafter warm guest farewell. At the end of the shift, cashier should deposit all collected cash and other negotiable instruments with head cashier. Simultaneously, cashier should track and monitor *due back, overage, shortage, net cash receipt*, VPO, *allowance*, discount, *cash bank* and so forth transactions. Cashier should have complete knowledge of foreign exchange regulatory act, governing licensing laws and must update with daily buying and selling rates of foreign currency. He/she should be comfortable in working with manual, semi-automatic and fully automatic systems/mechanisms.

| | **Chapter Terms**<br>**(with Chapter Exercise)** |
|---|---|
| ? | It is an itemized record of payments, charges and credits. |
| Voucher | ? |
| Management account | Hotel's house account that is used to record the transactions of NC (or no-charge) guests. |
| Delinquent account | Account receivable (or account aging) statement that contains record of amount that is yet to be received. Under this system, the payment is usually received within a period of 60-90 days. |
| Incidental account | Incidental charges thus often also known as *incidental folio*. These are the charges of an individual guest for which he/she is solely responsible. |
| Overdue account | It is also one of the components (or columns) of account receivable statement. Under this system, the payment is usually received within a period of 40-60 days. |
| Current account | It is too a part of account aging statement (or column). Under this system, the payment is usually received within a period of 30-40 days. |
| Account aging | ? |
| ? | Also known as account age/account aging. |
| EFTPOS | ? |
| Traveller's cheque | It is a kind of payment option like *banker's cheque* which was pioneered by Amex (American Express). It is a cheque, worth the stated amount and may be brought for cash by the guest. The guest must sign the cheque twice, in the presence of the issuing cashier and again in the presence of the paying cashier. |
| Travel agent voucher | It is a voucher issued by travel agency to a guest (travel agent may also send another copy directly to concerned hotel). In it, hotel accepts travel agent's voucher from guest and provides accommodation and other services but will collect payment from travel agency. |
| Master account | It is a group leader account whereby leader is responsible for paying all posted expenses. It is also known as *group folio* or *master folio*. |
| SDB | *Safe deposit box* and mainly this facility is offered (at front office) to all resident guests for the safe keeping of their valuables and confidential items. |

| | |
|---|---|
| Company account/ folio | In this chapter, the term company account is used for transferring third party charges which are settled by guest's representing company. Such transactions must be supported by LoA. |
| Video check-out | It is a guest-directed-check-out facility whereby guest can verify his/her account on in-room television then directly check-out without passing through cashier desk (guest settles his/her account prior a day of departure). It is also known as *in-room-check-out*. |
| ? | A guest-directed-check-out option whereby guest settles his/her account via machine and proceeds to check-out without passing through cashier desk. |
| Express check-out | Semi-automatic check-out option whereby resident guest fills the express check-out form and submits it to front desk cashier with payment then check-out without passing through cashier desk. |
| Cash bank | It is also known as *house bank*, *cash float*, initial cash bank or simply *bank*. It is a fixed amount that is allotted to cashiers for dealing in foreign exchange, VPOs, cash payments or refund of cash deposit. |
| ? | It is a kind of machine that gives the impression of credit card on the sales voucher. |
| Cash remittance | It means depositing collected cash with head cashier. The specific day's collected cash (national plus foreign currency) and other negotiable items must be submitted to head cashier/general cashier on daily basis. |
| Overage | ? |
| Shortage | ? |
| Close cashier | It refers to the system whereby respective outlet cashiers close their departmental account, thereafter print reports for management review. |

# 9 Telephone Section

OBJECTIVES

*After reading this chapter, students will be able to...*

- explain the work profile of telephone section;
- differentiate between PBX and PABX and explain their bases of differences;
- elucidate wake-up call and how it is placed in guest room; and
- explain the importance and effect of correct telephone protocol.

## 9.1 TELEPHONE SECTION

Telephone section is also known as *switchboard*. This section is usually situated in the back area of front office department and principally responsible for connecting incoming and outgoing telephonic calls (or landline phones calls) of guest as well as of hotel employees. A vital role of telephone section is to act as the communication centre, especially in the event of emergency, for instance during the bomb threats, fire alarm, etc.

Traditionally, this section equipped with PBX (private branch exchange) which or that is used to communicate within the hotel as well as outside the hotel. The PBX or "switchboard" manages all incoming and outgoing communications. Along with phone calls, faxes and emails are also routed through PBX. Nowadays, PBX is replaced by EPBX or EPABX (Electronic Private Branch Exchange or Electronic Private Automatic Branch Exchange). The EPBX reduces the cost of manpower (up to some extent) plus also makes easy the transferring of incoming and outgoing calls. It can easily track call details like call duration, call amount, call timing, caller number and so forth. EPABX is generally used in large hotels because it offers additional facilities like worldwide dialling from the guest room, automatic recording of the charge in the bill office, powered by an automatic wake-up call system, voice mail, convey pre-recorded message and a message taking system without the assistance of switchboard operator.

| | **Activity** | |
|---|---|---|
| Fill in the below given spaces with suitable differences between PBX and PABX systems. | | |
| **Criteria /Base** | **PBX**<br>(a conventional system<br>used in small hotels) | **PABX**<br>(an updated version of PBX used in<br>medium and large hotels) |
| Dialling | Do not offer direct dialling which means every call goes through telephone operator. | Offer direct dialling facility into guest rooms without moving through telephone operator. |
| No. of extensions | ? | A number of extensions can be connected, as & when required. |
| Operator | Manually operated | ? |
| Target | ? | It is fit for large organizations |
| Price | It is less expensive than PABX (installation charges) | ? |
| Call information | ? | Bill is automatically generated/ transferred by the machine (it has separate meter for all extensions) into respective guest rooms. |
| Service connections | ? | Service connections cannot be minimized. |
| Traffic & workload | Traffic & workload of operator cannot be streamlined. | ? |

## 9.2   FUNCTIONS OF TELEPHONE SECTION

Due to incessant work, the telephone section should be manned at all the time or 24 hours a day/night because it is also a location for fire checking equipment. The telephone section's hierarchy includes telephone supervisor under whom telephone operators work for both, day & night operation. The telephone section is headed by telephone supervisor who is responsible for supervising the activities of telephone operators. The telephone section staff processes all incoming and outgoing calls through the switchboard/system, places wake-up calls, records messages, prepares *traffic sheet* and plays crucial role during emergency. These telephone operators must have good communication skills, courtesy and good command over their language.

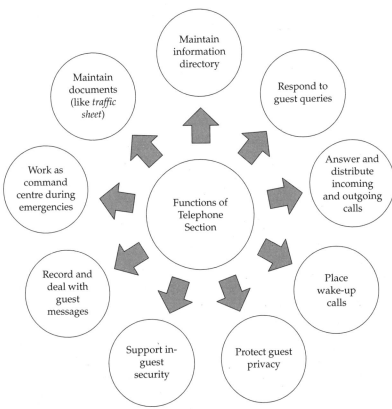

## 9.3 ROLE OF TELEPHONE SECTION IN DIFFERENT STAGES OF GUEST CYCLE

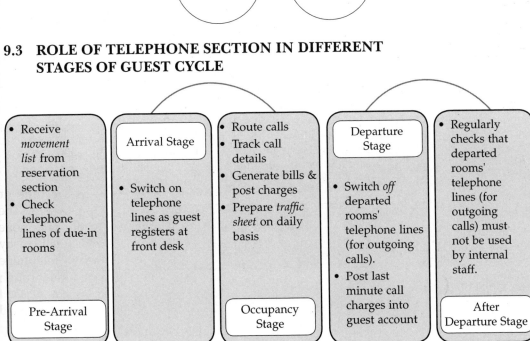

A step-by-step description of some major tasks of telephone desk/section that predominantly occurs during different stages of guest cycle is given below.

## 9.4 WAKE-UP CALL PROCEDURE

It is concerned with all those telephone calls usually placed by telephone section (it may also be placed by front desk or guest relation desk—depending upon individual hotel's SOPs) in the guest room in order to awaken the guest. In simple terms, wake-up call is an in-house telephone call to a sleeping guest at a specified time (given by the guest) to wake him up. It is generally made in the morning but a guest may also request for it at any time of the day, for instance as in case of airline crew members. It is the duty of telephone department to place wake-up call but if the call is made for airline crews then lobby personnel will first call local airline operators and obtain the timing of the flight and then they should forward the message to telephone operator to place wake-up call. Thus, telephone operator (in coordination with front desk agent) can also refer to *airline crew check-in sheet* for this purpose.

In case of airline crew members and group, generally the wake-up call is given 45 minutes or 1 hour prior than pick-up time. In small hotels, it is the duty of front desk agent to place wake-up calls. Due to the development in most modern technologies, today instead of requesting to hotel operator for placing wake-up call, guests can simply dial a special hotel extension number on their in-room telephone then follow the given instructions provided by the system to place/request a wake-up call time. Many hotels also provide/combine a wake-up call with room service, especially when wake-up call is for early morning, allowing the guest to order breakfast upon receiving a wake-up call.

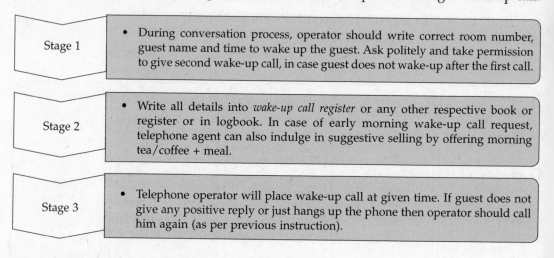

**Stage 1**
- During conversation process, operator should write correct room number, guest name and time to wake up the guest. Ask politely and take permission to give second wake-up call, in case guest does not wake-up after the first call.

**Stage 2**
- Write all details into *wake-up call register* or any other respective book or register or in logbook. In case of early morning wake-up call request, telephone agent can also indulge in suggestive selling by offering morning tea/coffee + meal.

**Stage 3**
- Telephone operator will place wake-up call at given time. If guest does not give any positive reply or just hangs up the phone then operator should call him again (as per previous instruction).

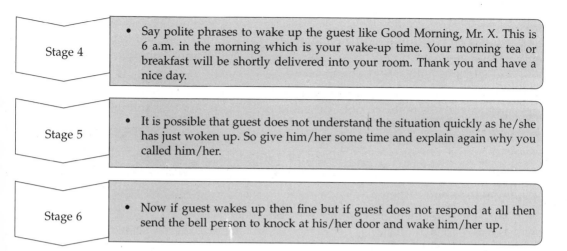

| Stage 4 | • Say polite phrases to wake up the guest like Good Morning, Mr. X. This is 6 a.m. in the morning which is your wake-up time. Your morning tea or breakfast will be shortly delivered into your room. Thank you and have a nice day. |
| Stage 5 | • It is possible that guest does not understand the situation quickly as he/she has just woken up. So give him/her some time and explain again why you called him/her. |
| Stage 6 | • Now if guest wakes up then fine but if guest does not respond at all then send the bell person to knock at his/her door and wake him/her up. |

## 9.5 TELEPHONE PROTOCOL

*Telephone protocol* refers to telephone etiquettes and code of behaviour of an employee who handles telephone calls in any department/section of the hotel organization. Generally, in hospitality industry it is most important for everyone because the term hospitality itself is related with protocol as it demands generosity. Several major components of telephone protocol and their brief description is given below.

### Pick-up Telephone and Address Guest

✓ Telephone agent should pick up the call within three rings, next greet the guest then introduce himself/herself and organization promptly, thereafter identify the caller (it means whether guest is male or female).

✓ People easily become friendly and feel happy if you call them by their names. So, first ensure how you should address a guest and then call him/her.

### Continuous Response

✓ While talking on telephone, telephone operator should regularly respond because it shows your caller that you are eagerly listening to him. So, sometimes making noises like "hmm", "yes", "ok", "I understand" can make the conversation spontaneous.

✓ On the other side, if telephone operator keeps silent during conversation your caller may be confused whether you are listening or not. So, don't confuse your caller because communication is always a two-way process.

## Pay Attention

✓ Perfect call handling demands intense attention. If you are talking over telephone, simultaneously doing other work/s then you cannot completely concentrate/ understand both things equally.

✓ So, if another work is urgent then telephone operator should finish the other job first. Here, it is better to hold the caller or tell him/her that you will call back after a certain period of time.

| Widely Used Telephone Phrases |
| --- |

**While Addressing**

✓ Sometimes just by hearing the tone you may not be able to identify whether your caller is male or female. In such circumstances, it is better to ask "How should I address you?" or "May I have your name please?" If caller requests you to connect his/her call with any of the resident guest or hotel employee then you should never directly connect the phone call, instead first take permission then proceed further.

**While Transferring Call**

✓ If you need to transfer the call then inform the guest in advance that you are transferring the call to the concerned person. The ideal phrases for it "Please allow me to transfer your call to Mr. X. Could you please hold down for a minute?" After that if the caller allows you and says 'Yes' or 'Ok', then transfer the call.

**If Concerned Person is Not Present**

✓ If you find that extension is not reachable or dead then say "Thanks for holding. But I am afraid Mr. Y is not available. Would you like to leave a message for him or call back later?" When you leave the telephone, there may be forwarding messages. So, keep your reliever informed about this. People rely on you to pass their messages. Therefore, always ensure that it is to be done on priority basis.

**If Extension is Busy**

✓ If you find that extension is busy then first of all apologize to guest for holding then say "Thanks for holding. Mr. Y is busy and still on line. Would you like to leave a message for him or call back later?"

**Ending any Conversation**

✓ While finishing a conversation say "Thank you Mr. X for calling. Have a nice day." Remember during the entire conversation period, telephone attendant must listen to guest carefully.

**Common Phrases**

✓ Some common phrases are: "May I have your name please?", "May I have your contact number please?", "Mr. X please let me repeat the message.....Is that all right?", "Mr. X could you please hold down for a minute?"

# Be Friendly and Keep Yourself Handy

✓ Make your tone as friendly as possible plus try to smile while talking over the telephone. The way the telephone operator talks, uses pitch variation, pace of talking and tone of voice all reflects the personality traits of an individual plus organization too. Therefore, telephone operator should try to visualize the caller and speak clearly in order to make conversation more efficient and effective.

✓ The term *keep handy* means, a person who is dealing/working on telephone desk must always keep *writing kit* with him/her. This handy aid/kit includes writing pad (or message book), pen/pencil, list of telephone numbers & directory, room tariff information (or room rent card) and so forth reference materials.

# Never Neglect any Call

✓ Don't neglect any call. Take all calls as a business opportunity for your organization. In certain cases, telephone operator may put the guest on hold. So, if you have had to put the person on hold, or leave the line for whatever reason, do not say "Hello" when you come back on the line. It is better to say either – "Sorry to have kept you waiting sir/madam…" or "Sir/madam, you asked about the ……."

✓ When you returned to the hold line (or attend to previously held telephone call) and found that the caller is talking to someone else (background voice may help you in identifying the situation) then do not interrupt him/her with "Hello….. Hello". Simply say "Excuse me sir/madam" (as you would do in a face-to-face conversation.).

---

### Important

✓ When you put the caller on hold, "you should never talk (or ask any questions) to your colleague without pressing the hold button – as the caller may hear you".

✓ If the caller enquires for something for which you do not have the right answer then you can say, "I will check on that sir/madam. Would you please hold the line one moment?"

---

# Give Call Back

✓ If you or the caller cannot listen to each other then offer him/her to call back. On the other hand, if you have to call the caller back to give some information (like about reservation confirmation number), make sure you should introduce yourself fully, thereafter provide required information. For example, "Good morning Mr. Singh, this is Manoj Yadav from (hotel name). You had asked for information regarding ……"

| Few Favourable & Unfavourable Telephonic Conversation Phrases for Practice | | |
|---|---|---|
| **Criteria** | **Unfavourable Practice** | **Favourable Practice** |
| Starting | Start a conversation with only "Hello" or "Department Name" or "Hotel's Name". | Start with greeting and then follow the procedure we discussed earlier in this chapter. |
| Response | Avoid taking responsibility in the way like "I am not working in HR department". | If you are not the person whom caller is looking for then transfer to the appropriate person but after taking permission. |
| Keep handy | Requesting to hold down because you don't find your pen or pad. | You should always keep handy a pen, pad (or message book) or pencil. |
| Continuous reply | Being silent while talking for a long time. | Must make some noises or response like "Hmm", "Ok", "Yes", "I understand" etc. |
| Ideal or magic words | Use simple language. | Use some magic words like "Certainly", "You are right", "I do understand", "Thank you" etc. |
| Message taking | Taking wrong or incomplete message. | Pay deep attention while taking message and then repeat and be sure you have taken the proper message. |
| Follow-up | Don't take follow-up. | While transferring any call or message, be careful to take follow-ups. |
| Distribution of information | Give personal information about guest. | You can't share guest information or room number with any unknown person. You should only transfer a call or take message on behalf of him. |

| Activity | |
|---|---|
| Briefly explain the below given calling ways. | |
| **Calling modes** | **Description** |
| Calling cards | |
| Direct dialling | |
| Transfer call | |
| IDD | |
| BTR | |
| Toll free | |

## 9.6   STANDARD PHRASES FOR CONVERSATION BETWEEN TELEPHONE OPERATOR AND GUEST

Do you need anything else sir.
I am sorry to have kept you waiting sir.
Excuse me for interrupting.
Feel free to call me, sir.
My name is Mr. X, if there is anything I can do for you, please let me know sir.

## CONCLUSION

Telephone section is responsible for routing incoming and outgoing calls. Conventionally, telephone section must always have switchboard operator (or telephone operator) but nowadays due to the development of technology *call management system* can easily manage most of the activities of telephone section like voice call, message recording provision, call accounting system and so forth. PBX or also known as *switchboard* is a traditional call management system whereas PABX/EPBX/EPABX is a perfect prototype of today's call management systems. Nowadays, call accounting system (CAS) immaculately manages all tasks of telephone section. But still whoever works at any telephone system (in front or back area) must need to follow *telephone protocols* because it indicates the hospitality of the organization. For example, reservation section and sales & marketing division mostly relies on telephone system for generating business. Apart from regular telephone handling, operator should also be aware of wake-up call procedure and how it should work during emergency situations.

| Terms (with Chapter Exercise) | |
|---|---|
| Switchboard | It is a manually operated telephone system, also known as PBX (Private Branch Exchange). |
| EPBX | ? |
| Telephone protocol | ? |
| CAS | *Call accounting system* that is a computer interfaced system used to monitor all calls, especially outgoing calls and provide complete call details like extension, duration, charges, etc. |
| ? | Widely known as call accounting system. |
| Wake-up call | Telephonic call placed by telephone operator or front desk agent into guest room (at designated time) in order to wake him/her. Often, early morning wake-up request associates with *door knob menu* card/order. |
| Calling cards | These are the cards like a mobile's top up card. These are billed to a code number on a calling card issued by either a telephone company or by a private billing company. |

| | |
|---|---|
| Direct dialling | Facility of direct dialling from inside the room. |
| IDD | International Direct Dialling. Nowadays, due to the technological advancement, a guest can directly dial to any country throughout the world; the IDD is a prefix number to be called. |
| Keep handy | Person who is working at telephone desk must keep/carry required stationery and writing material in order to record messages. |
| Toll free | Telephone/call system whereby a prospecting guest can obtain whatever information he/she desires. These calls can be directly dialled from a guest room as either local calls or long distance calls. |
| BTR | *Bill-to-room.* In it, the call charges are transferred into respective room account/guest account, either manually or electronically. |
| Message book | It is a writing pad used to record messages for residing guests or for internal employees which later on transfers to the respective guest or employee. |
| Magic words | Collection of courtesy words. These magic or courtesy words are used as a part of protocol or etiquettes. |
| Follow-up | Call back provision after something has been taken care of to ensure satisfaction. |

# 10 CHAPTER

# Reservation Section

---

**OBJECTIVES**

*After reading this chapter, students will be able to...*

- explain the functions of reservation section and process of accepting room reservation;
- confer about different forms of reservation and tools for determining room availability;
- determine overbooking/adds-on reservation over-room inventory; and
- identify different systems, sources & modes of reservation.

## 10.1  RESERVATION SECTION

A *reservation section* is a hub for room reservation and normally situated near by the reception desk but usually not in the view of transitory guests. All requests for room reservation from various sources are received and processed from here only. The term *reservation* means taking an advance booking or reserving the room (or accommodation) either *in-person* or through conciliators like travel agent, tour operator, etc. It involves a particular type of guest room being reserved for a particular person or *block* for group of persons for a requested period of time. As room reservation request can be received at any moment of time, so the reservation section must be manned throughout 24 hours. The reservation can be received either through telephone, fax, letter, travel agent or directly from the guest.

The reservation section is headed by reservation manager (or reservation supervisor). The other reservation staff includes reservation clerks or agents who are mainly responsible for dealing with telephonic enquires related to room reservation and for modifying or cancelling earlier accepted room reservations. The reservation manager would be the controller of this section and would organize the staff duty schedules, establish & maintain high standards of work and make decisions on whether bookings should be accepted or not. The reservation agents or clerks are responsible for handling queries and for generating room booking data in terms of reservation requests, its modification, confirmation and cancellation for the purpose of recordkeeping and future reference. They keep records of the number of bookings taken for each night, and records all the relevant personal and financial details of each room booking. They can also do upselling of an accommodation by suggesting upgraded rooms and by indicating the benefits of purchasing inclusive meal rate plan over exclusive meal plan for any room. Reservation staff also intimates about the conditions for guaranteed reservation, *lead time*, generating reservation confirmation letter, reservation cancellation/modification letter. Some major functions of reservation section are given below.

## 10.2    FUNCTIONS OF RESERVATION SECTION

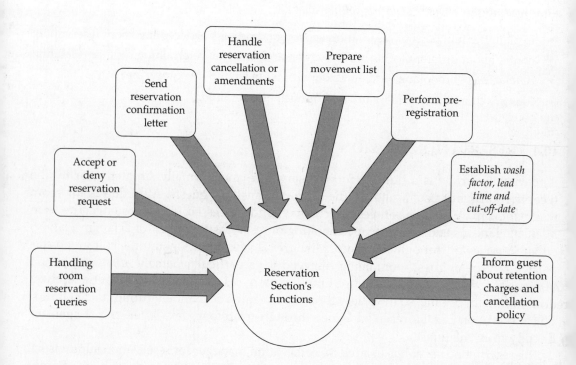

## 10.3 ROLE OF RESERVATION SECTION IN DIFFERENT STAGES OF GUEST CYCLE

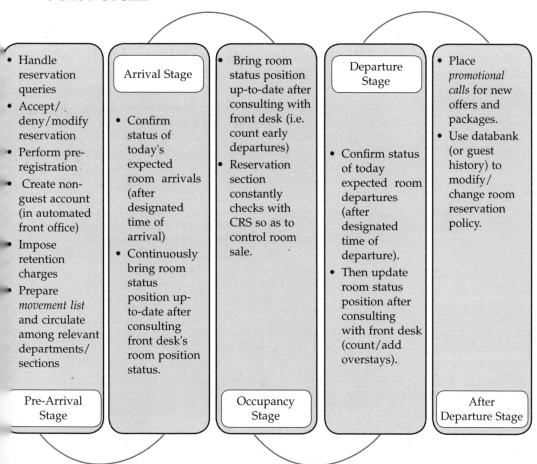

• Handle reservation queries
• Accept/ deny/modify reservation
• Perform pre-registration
• Create non-guest account (in automated front office)
• Impose retention charges
• Prepare *movement list* and circulate among relevant departments/ sections

Pre-Arrival Stage

Arrival Stage

• Confirm status of today's expected room arrivals (after designated time of arrival)
• Continuously bring room status position up-to-date after consulting front desk's room position status.

• Bring room status position up-to-date after consulting with front desk (i.e. count early departures)
• Reservation section constantly checks with CRS so as to control room sale.

Occupancy Stage

Departure Stage

• Confirm status of today expected room departures (after designated time of departure).
• Then update room status position after consulting with front desk (count/add overstays).

• Place *promotional calls* for new offers and packages.
• Use databank (or guest history) to modify/ change room reservation policy.

After Departure Stage

A step-by-step description of some major tasks of reservation desk/section that predominantly occurs during different stages of guest cycle is given below.

## 10.4 PROCESS OF ROOM RESERVATION

It is the procedure (or step-by-step process) by which room reservation requests are handled. In hotels, reservation section is responsible for dealing with all kinds of reservation requests. This section is headed by reservation manager/supervisor and operational work is performed by reservation clerks/agents. Guest reservation requests can be received either through telephone, fax, letter, third party or directly from the guest (or in-person). At the time of handling reservation request, reservation agent should obtain all required personal as well as financial information from the guest. Reservation agent should try to do upselling of an accommodation unit and ask guests to confirm or guarantee their reservations. The different stages involved in handling room reservation is briefly given here.

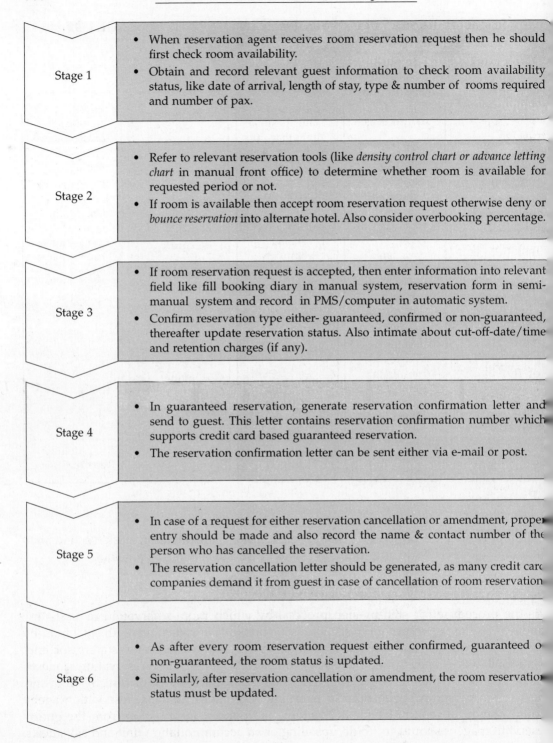

**Stage 1**
- When reservation agent receives room reservation request then he should first check room availability.
- Obtain and record relevant guest information to check room availability status, like date of arrival, length of stay, type & number of rooms required and number of pax.

**Stage 2**
- Refer to relevant reservation tools (like *density control chart or advance letting chart* in manual front office) to determine whether room is available for requested period or not.
- If room is available then accept room reservation request otherwise deny or *bounce reservation* into alternate hotel. Also consider overbooking percentage.

**Stage 3**
- If room reservation request is accepted, then enter information into relevant field like fill booking diary in manual system, reservation form in semi-manual system and record in PMS/computer in automatic system.
- Confirm reservation type either- guaranteed, confirmed or non-guaranteed, thereafter update reservation status. Also intimate about cut-off-date/time and retention charges (if any).

**Stage 4**
- In guaranteed reservation, generate reservation confirmation letter and send to guest. This letter contains reservation confirmation number which supports credit card based guaranteed reservation.
- The reservation confirmation letter can be sent either via e-mail or post.

**Stage 5**
- In case of a request for either reservation cancellation or amendment, proper entry should be made and also record the name & contact number of the person who has cancelled the reservation.
- The reservation cancellation letter should be generated, as many credit card companies demand it from guest in case of cancellation of room reservation.

**Stage 6**
- As after every room reservation request either confirmed, guaranteed or non-guaranteed, the room status is updated.
- Similarly, after reservation cancellation or amendment, the room reservation status must be updated.

**Task Force for Reservation**

Task 1   Handle group and transient reservation & generate confirmation

Task 2   Modify reservations on guest request (cancel or amend reservation)

Task 3   Prepare for group (by referring *rooming list*) and transient arrival

Task 4   Prepare *movement list* (expected arrivals and expected departures)

Task 5   Create and update reservation records or electronic files

Task 6   Coordinate with sales and marketing division and supply chain/third party such as travel agencies to generate business

> **Reservation Section Equipment**
>
> - Whitney rack
> - Shannon slips
> - Reservation chart
> - Office equipment
> - Office stationery

## 10.5   PREPARATION FOR GUEST RESERVATION

Apart from reservation handling, reservation section also requires preparing *movement list* on every evening for the next day's *due-ins* and *due-outs*. Thereafter, sends the copies of movement list to various relevant departments and sections. The term *movement list* is concerned with a common list that shows both next day's expected arrivals and expected departures. In this list, reservation agent also indicates about any VIP guest, complementary room and other relevant remarks like HWC. It is a routine task of reservation section. A brief description about transient and group guest preparation is given below.

### Preparation for Transient Guest Reservation

Task 1   Handle guest queries; provide with all room relevant information to prospective guest, confirm whether room reservation is guaranteed or non-guaranteed, specify *cut-off-date/time*, bestow different room rate structure that varies with room configuration, *lead time* and retention charges/refund policy.

Task 2   Accept, deny or bounce guest reservation request on the basis of availability of rooms kept aside for transient market segment. Before accepting room reservation, agent may also refer to past bill settlement history of the guest. This history will help in determining whether to grant *house limit* or not.

Task 3   Reservation agent should confirm with guest about the bill settlement policy (or about *incidental charges*) and inform about cancellation policy (*retention charges*). Finally, he sends reservation confirmation letter to guest. Reservation agent too is responsible for any modification (or cancellation or amendment) in earlier taken room reservation, if guest requests.

Task 4    Now before guest arrival, reservation section will prepare a *movement list* and forward to various relevant front office sections (such as front desk, telephone section and bell desk) and other departments (particularly to housekeeping and room service).

Task 5    In many hotels, reservation section allots room (in accordance with guest's room preference) to their guest-with-guaranteed reservation as well as also performs their pre-registration. Remember, this pre-registration activity is not possible for guests who hold non-guaranteed reservation as well as for chance guests.

Task 6    Reservation agent should obtain complete information to reserve a room like booker name, guest name, number of pax, payment mode, expected date & time of arrival and departure.

**Note**

✓ VIP, CIP, MIP, HWC or SPATT are the acronyms often written on reservation forms or Whitney slips or on the expected arrival list to indicate the front office staff that special care has to be taken while dealing with such guests.

✓ This is particularly important for housekeeping department as they have to arrange flowers for such guests, sometimes extra amenities too. Thus, housekeeping department has to pay special attention, in terms of room inspection to the block room/s of such clientele before their actual arrival.

✓ Generally, room service arranges fruit baskets, chocolate tray or Champagne/wine bottle for these guests in their rooms before their actual arrival. These are often given as complementary items. Whether to place one complementary item or all it depends upon guest's designation/grade like VIP 1, 2 or 3.

## Preparation for Group Guests Reservation

Task 1    Handle queries of group leader; provide with all room relevant information to prospective group, calculate *wash factor*, decide *cut-off-date/time*, give different room rate structure that varies with number of room bookings and *lead time*, confirms/expects how much ancillary revenue group will generate. Generally, reservation supervisor handles group business.

Task 2    Accept or deny group reservation after determining number of rooms requested, *group booking pace*, group settlement history, anticipated group business, *displacement factor*. Rooms will be given from group room *blocks* or rooms that are kept aside for group market segment.

Task 3    Reservation supervisor should also confirm with group leader about the group guests' bill settlement policy (or about *master folio charges* and *incidental charges*). Finally, send reservation confirmation letter to group leader. Reservation supervisor is also responsible for any modification (or cancellation or amendment) in earlier accepted group room reservation, if group leader requests.

Task 4    Now reservation section will prepare *group arrival list*. Activities concerning the group arrival start a few days before the actual arrival of the group. In the back office area reservation section performs pre-registration and allots rooms in accordance with received *rooming list*.

Task 5    Some hotels even allocate the rooms plus keys (through *pre-key* and *key-pack* process) to all members of the group who are mentioned on the *rooming list*. But this is possible only when all necessary information like name, passport details, nationality, types of rooms, room preference & sharing and so forth details are available with reservation section.

## Note

✓ At the end of the day (and at specified time) when reservation section closes for the day, the expected arrival list and the reservation rack for the next day's arrivals is prepared and sent to the front desk section to facilitate the arrivals.

✓ Reservation agent also refers to its prepared *correspondence set* (or *corros*) to check next day's expected arrivals. It mostly refers to use in those hotels which use *Whitney rack system* and is hardly found nowadays.

✓ *Movement list* contains the detail of next day's due-ins and due-outs with special reference to VIPs and crew members' (of different airlines) expected time of arrival and departure on that day. The expected arrival list is prepared in alphabetical order.

| Activity | |
|---|---|
| Differentiate between the following: | |

| S. No. | Advance Deposit (also known as *reservation advance*) | Advance Payment (also known as *paid-in-advance*) |
|---|---|---|
| 1 | ? | Usually taken from chance guests |
| 2 | ? | ? |
| 3 | Usually demanded at the time of reservation | ? |
| 4 | ? | Usually demanded at the time of registration |
| 5 | ? | ? |
| 6 | Usually taken from guest-with-guaranteed-booking | Also termed as paid-in-advance (or PIA) |

## 10.6 TYPES OF ROOM RESERVATION

It is concerned with nature of room reservation which may vary with the number of guests (such as individual or group guest reservation), guest type (VIP, CIP or SPATT), payment confirmation (guaranteed or non-guaranteed), stage of authorization (tentative, waitlisted or confirmed) and so forth. Type of room reservation also indicates the specific market segment which supports in constructing and modifying room rate and discount strategy for them. A brief description of major forms of room reservation is given below.

| | |
|---|---|
| Tentative reservation | • It is a non-guaranteed or simply provisional reservation, thus hotel gives cut-off-date/time to guest to confirm the reservation else it will be treated as cancelled and room released for normal sale. |
| Waitlisted reservation | • It is linger-on or hold-on reservation. In simple terms, it is a situation of overbooking where hotel accepts more room reservation than available room inventory. Here, reservation acceptance or rejection is affected with rooms' cancellation/modification, confirmation, wash factor, etc. |
| Confirmed reservation | • It is also known as 06:00 P.M. release reservation where hotel holds guest's room reservation request (often this facility is offered to regular guests) until 06:00 P.M. on the day of arrival. If guest does not arrive till 06:00 P.M., the room will be released for normal sale. |
| Guaranteed reservation | • It gives guarantee to the hotel that whether guest arrives or not, payment will be made. The guarantee may be given/supported by either credit card company, travel agent voucher, traveller's cheque and so forth negotiable instruments. |
| Transient reservation | • It is concerned with individual guest's room reservation. Transient reservation request can be received from both business guest segment and pleasure guest segment. Transient market segment is also termed as non-group business segment. |
| En-block reservation | • It is concerned with *group reservation*. Here, hotel must consider *wash factor*, market demand, group history, ancillary revenue, number of rooms required then offer suitable room rate (or special room rate). |
| Blanket reservation | • In some cases, rooms are blocked for the whole group (blanket reservation) and the allocation of rooms is done only at the last minute when group arrives. *Blanket reservation* is done particularly when the members of the group arrive individually. |

| Bounced reservation | • A reservation to which hotel cannot honour thus transfer the guest request into another property. Such reservation is also called "Farm Out or Walking Reservation". In *referral hotel group*, such room bookings are diverted towards member hotels. |
|---|---|

| Star reservation | • Room reservation of any VIP or VVIP guest or their representative. Often these guests also want their stay to be kept confidential. Thus, guests from this market segment are also known as *incognito guests*. |
|---|---|

### Activity

Differentiate between the following:

| S. No. | Confirmed Reservation | Guaranteed Reservation |
|---|---|---|
| 1 | ? | Usually powered by advance deposit policy |
| 2 | ? | ? |
| 3 | It is also known as 06:00 p.m. release reservation | ? |
| 4 | ? | Almost any guest can go for guaranteed reservation |

## Bases on which Reservation can be Guaranteed

It is concerned with how and by which mode of payment a guest can legalize his/her room reservation. Earlier, guest could use limited payment options like mostly by *advance deposit* (or direct cash deposit) but nowadays there are numerous ways to guarantee the booking such as by means of credit card, cheque, third party guarantee plus cash. Remember all guaranteed reservations are always preferred to be bound with certain rules and policies such as reservation cancellation terms and conditions. A brief description of some major bases for claiming guaranteed reservation are listed below.

| Advance deposit | • A room reservation guarantee is supported by *advance deposit* and if cancellation request is received after cut-off-date/time, retention charges will be applicable as per policy. This advance deposit is also known as *anticipated amount*. |
|---|---|

| Credit card | • Reservation guarantee is given by credit card company on behalf of prospective guests. Therefore, credit card company often demands reservation confirmation letter from guest to support the guaranteed room reservation. |
|---|---|

| Contractual deal | • Contractual agreement between a hotel and representing company regarding the use of certain number of rooms. Thus, hotel bills to the company. Often *room blocks* for such booking is termed *contract rooms* and whether room will be used or not, the representing company will pay for contracted number of rooms/bookings. |
|---|---|

| Voucher | • Supportive documents like travel agent voucher supported reservation guarantee. In this, travel agent/agency takes the responsibility of payment on behalf of the guest. Thus, hotel bills to travel agent/agency. |
|---|---|

| Cheques | • Supportive document especially traveller's cheque and bankers cheque. Here, a company/bank that has issued a cheque takes the responsibility of payment on behalf of prospective guest. |
|---|---|

### Reservation Cancellation

Like reservation acceptance, reservation cancellation also plays a crucial role in generating business in near future. In proper cancellation system, a *cancellation code* is generated which supports the transaction and often demands by credit card company to prop-up hotel in systematic cancellation of earlier accepted guaranteed room reservation. Cancellation for different types of room reservation is given below.

| S. No. | Reservation Type | Cancellation Description |
|---|---|---|
| 1 | Tentative | Cancelled upon guest's request/intimation/after cut-off-date. |
| 2 | Waitlisted | Cancelled upon property's reservation status/overbooking policy. |
| 3 | Confirmed | Cancelled after 06:00 p.m. on the day of arrival, if guest does not arrive. |
| 4 | Guaranteed | Cancelled after receiving cancellation request & generates cancellation letter. |
| 5 | Non-guaranteed | Cancelled after cut-off-date/cut-off-time. |
| 6 | Transient | Cancellation depends on individual guest's request. |
| 7 | En-block | Cancellation depends on *wash factor* and group leader's request. |

## 10.7 TOOLS FOR DETERMINING ROOM AVAILABILITY

There are numerous techniques/tools available for forecasting/determining the future room availability. But the basic motive of all these tools (whether traditional or modern) is to ease management to achieve *perfect room sale*, determine overbooking percentage and to generate perfect sale (or balanced business) from each targeted market segment, i.e., transient and group market. Remember, to achieve perfect room sale designation reservation section develops *room-sales-mix-ratio* (or *guest-sales-mix-ratio*) of its available *room inventory*.

This room-sales-mix-ratio facilitates reservation section in determining *vacancy forecasting* (or room available for sale forecasting), *occupancy forecasting* (or room sale forecasting) and even in *room revenue forecasting*. Room-sales-mix-ratio also helps in establishing the *room blocks* for each targeted market segment, thereafter assists in revising/updating room rates, framing discount policy, adjusting room blocks (or shifting unsold rooms among high demand sensitive targeted market segment) and analyzing customer flow. Up to some extent, the below mentioned tools are similar in terms of uses and their importance. A brief description of these tools is given below.

| S. No. | **Vacancy Forecasting** (it means *room available for sale forecasting*) | **Occupancy Forecasting** (it means *room sale forecasting*) | **Revenue Forecasting** (it means *room revenue forecasting*) |
|---|---|---|---|
| Formula | Room Inventory<br>+ Expected no-shows<br>+ Expected under-stays<br>+ Expected cancellations<br>+ Expected departures<br>- Expected arrivals/reservations<br>- Expected OOO/OOS rooms<br>- Expected stay-over + overstay<br>- Expected walk-ins | Room Count<br>+ Expected arrivals/reservations<br>+ Expected early arrivals<br>+ Expected stay-over + overstays<br>- Expected cancellations<br>- Expected OOO/OOS rooms<br>- Expected no-shows<br>- Expected departures<br>- Expected under-stays | Forecasted average daily rate (or ADR)<br>×<br>Forecasted occupancy percentage<br>×<br>Forecasted duration (or time period) |
| Example | Room Inventory 100 (RI)<br>　+ EN　　10<br>　+ EU　　05<br>　+ EC　　12<br>　+ ED　　25　　152<br>　- EA　　25<br>　- EO　　05<br>　- ES　　15<br>　- EW　　15　　**60**<br>F. vacant rooms 92 (or 92%) | Room Count  75 (out of 100 RI)<br>　+ EA　　25<br>　+ EE　　04<br>　+ ES　　15　　119<br>　- EC　　12<br>　- EO　　05<br>　- EN　　10<br>　- ED　　25<br>　- EU　　05　　57<br>F. occupied rooms 62 (or 62%) | Forecasted ADR Rs. 5500<br>×<br>Forecasted occupancy percentage 75%<br>×<br>Forecasted duration 365 days<br>=<br>Forecasted room revenue Rs. 15,05,625/- |

| Advance letting chart | • A reservation section maintains 12 ALCs for a year (one ALC for each month) and keeps them in a serial order.<br>• It is mainly used in small hotels. It shows the guest's name, proposed length of stay and sometimes reservation type also. |
|---|---|
| Density control chart | • It is a modified version of ALC, to sort out its weaknesses. Reservation section maintains 12 DCCs for a year (one DCC for each month) and keeps them in a serial order.<br>• It is mainly used in medium hotels. It does not only show reservation against each room but also helps in framing their overbooking status. |
| Room status board | • It may be prepared manually as well as electronically. It is also known as "room letting board".<br>• It just indicates the mere position of available rooms and the future letting position at a glance. In manual system, usually a pencil is used to fill up and update the room status board. |
| Reservation journal | • It is mainly used by small hotels (like B&B hotels) and resorts, especially in those properties which have high proportion of one night stay.<br>• It is a *bound book* in which guest entries are updated on daily basis. It indicates the room position (whether room is occupied or vacant) as well as provides relevant guest details. |
| Loose leaf record | • This system has pre-printed pages/cards which shows the rooms and their current status (like vacant, occupied or out-of-order). Room reservation requests records against the respective *revenue day*.<br>• It is required to be maintained on daily basis, the closing *house-count* of previous day will be treated as opening house-count for the next day. |
| Computerized reservation system | • Here, reservation section records and updates room reservation request directly on computer system. Standalone software with limited modules can become a part of computerized reservation system.<br>• But when these computer systems powered by dominating software packages are simultaneously supported by external links then it becomes a part of *central reservation system*. |

## Activity

Give five reasons (why to adopt) for each of the six tools of determining room availability, listed above. Next, for the purpose of comparative study determine which is most suitable for the present scenario. Also suggest appropriate property where each tool will fit.

| S. No. | Tools for determining room availability | Suitable property | Reasons (why to adopt this tool) | | | | |
|---|---|---|---|---|---|---|---|
| | | | 1 | 2 | 3 | 4 | 5 |
| 1 | Advance letting chart | | | | | | |
| 2 | Density control chart | | | | | | |
| 3 | Room status board | | | | | | |
| 4 | Reservation journal | | | | | | |
| 5 | Loose leaf record | | | | | | |
| 6 | Computerized reservation system | | | | | | |

## 10.8 SYSTEMS OF RESERVATION

The term *system of reservation* refers to the scheme of recording reservation details. In simple words, it means how to record, maintain and update guests' reservation outline. Generally, reservation outline comprises two kinds of information, personal and financial information of guest. In general, there are three systems of room reservation, i.e. manual, semi-automatic and fully automatic. A brief description of each is given ahead:

| Activity |
|---|

Fill in the below given blank spaces with appropriate description that reflects personal & financial information about guest, thus reservation agent will always ask about it during the room reservation process.

| S. No. | Personal Information | Financial Information |
|---|---|---|
| 1 | ? | How guest wants to settle his/her bill? |
| 2 | ? | ? |
| 3 | Correspondence address of guest | ? |
| 4 | ? | Which credit card guest wants to use? |
| 5 | ? | ? |
| 6 | Correct name and expected date of arrival | ? |

# Manual System of Reservation (Supported by Handwritten Material)

✓ In these reservation systems each and every work is performed manually thus reservation section indents stationery, reservation register/diary (such as *hotel diary, cardex system* and *card system*) and so forth. There is no use of pre-printed formats and any of the electronic devices (even calculator also).

✓ Formats for handling reservation request, booking guest room and rooming guest (or room allotment sheet), all are prepared manually. Even reservation confirmation letter/cancellation letter/modification letter are written manually. A brief description of several famous manual systems of reservation is given below.

| Hotel diary | • It is a bound book that is mainly used by small hotels to record room reservation requests. It consists of 365/366 pages; each page represents *revenue day* (or single day) of the year. |
|---|---|
| Cardex system | • It is a library cardex system that is mainly used by small and medium size hotels. It consists of 12 drawers/pockets which hold 30/31 cardex sheets. Each drawer/pocket represents a *revenue month* whereas cardex sheets represent a reservation status of each *revenue day* (or one day). |
| Card system | • It comprises 4 compartments/metal rack carriers, i.e., 1st rack represents current month, 2nd represents following months, 3rd and 4th racks are used for keeping expired cards. Current month's rack contains 31 metal pockets (one pocket for each revenue day of the month) whereas 2nd rack contains 11 metal dividers for remaining 11 months of the year. |

# Semi-automatic System of Reservation (Powered by Pre-printed Formats)

✓ It refers to those reservation systems in which all reservation oriented activities are performed with the help of certain devices/peripheral devices which are not fully mechanical and simultaneously also not fully manual. Reservation section also uses pre-printed formats to fill relevant details of prospective reservation.

✓ In this kind of reservation system, reservation section uses computer system, printer, Whitney system and pre-printed formats. Remember, standalone software programmes can become a component of semi-automatic reservation system but never PMS. A brief description of two basic semi-automatic reservation systems is given ahead.

| | |
|---|---|
| Whitney rack System (supported by *Whitney front desk system* for registration) | • It is a patented system that incorporates all guest details from the time a guest makes a reservation until departure. The system may be used in part or as a complete system. The Whitney rack system was developed by the New York based duplicating and checks company, i.e., American Whitney Corporation, in the year 1970.<br>• During reservation, guest details are recorded in *Whitney slip* (also known as *shannon slips*), thereafter placed in Whitney slots. Different coloured slips can be used to indicate different market segments. |
| Computerized Reservation System | • It is recording of guest's room reservation request directly on computer screen into respective module. Different standalone software packages which principally cover reservation activities are perfect examples of computerized reservation system. Remember, these are never web based software packages.<br>• In simple terms, use of computer system for the purpose of recording room reservation details is collectively termed computerized reservation system. |

## Automatic System of Reservation (Powered by Different PMS and HMS Software Packages)

✓ It refers to all reservation systems which work electronically or mechanically. In these systems, all room reservation requests, advance deposit, confirmation, cancellation, modification and other reservation oriented tasks work on-line (or on software programmes).

✓ In hotel industry, automatic system of reservation may be linked either with *central reservation system* (in CROs), *global distribution system* and also become an integral part of *instant reservation system*. A brief description of each is given below.

| | |
|---|---|
| CRS (Central Reservation System) | • It is equipped in central reservation offices (CROs) that allows on-line users to gain access into the computer/reservation system of the hotels and provides all relevant information in relation to hotel accommodation and other related supportive services/facilities. It is available in two forms, affiliate and non-affiliate reservation systems.<br>• *Affiliate reservation system* (like Reservahost, MARSHA III, Holidex 2000, Global II, Choice 2001, etc.) is used by chain hotels whereas *non-affiliate reservation system* (or *referral hotel network*) is used by group of independent hotels. |

| GDS (Global Distribution System) | • It is a reservation network which works as an intermediary platform among participating hotels, travel agencies, car rental firms and prospective guests for exchanging information and generating business. It is established to connect the various hotel reservation networks/ systems (or CRS) with the airline reservation system/airline distribution system (ARS/ADS). Remember, CRS of all participating organizations + ARS/ADS = GDS. <br>• GDS provides worldwide distribution of hotel reservation information, airline tickets, automobile rentals and other services required by the travellers. There are a number of GDS systems in operation, the major ones being Amadeus, Galileo, SABRE (a pioneer of GDS concept), Worldspan, Abacus, System One, etc. |
|---|---|
| IRS (Instant Reservation System) | • It is also a kind of CRS which is based on the principle of WAN (Wide Area Network) & usually operated by the IROs (Instant Reservation Office/s). The instant reservation system is usually offered by the large hotel organizations for the hotel of their own chain or franchisee. <br>• In this system, hotel can take reservation for their sister properties but not for itself. Reservation section is responsible for taking booking for its own hotel where IRS is installed. The Instant Reservation System was first introduced in 1960 by the Holiday Inn group of hotels. |

## Activity

In the below given list, indicate the correct answer by putting tick in the appropriate box:

| S. No. | Particulars | CRS (*A) | CRS (**NA) | GDS | IRS |
|---|---|---|---|---|---|
| 1 | Semi-Automatic Business Research Environment (SABRE) | | | | |
| 2 | Reservahost and Holidex 2000 are perfect examples of | | | | |
| 3 | Referral hotels/non-chain hotels reservation system | | | | |
| 4 | A network (or a platform) of CRS of different sectors | | | | |
| 5 | Works on Wide Area Network and started by Holiday Inn | | | | |
| 6 | Global II and Choice 2001 | | | | |

**Note.** Acronym *A indicates *automatic* whereas **NA indicates *non-automatic* system of reservation.

## 10.9   SOURCES OF RESERVATION

It indicates to the channel-of-business (or *business-entry-points*) from where hotel room bookings and other business opportunities knock at the hotel. Remember, source of reservation is different from mode of reservation. Actually, source of reservation involves all associates that create relation between prospective guest and hotel. These business associates

| **Modes of Reservation** |
| :--- |
| - Written or verbal |
| - Manual or electronic |

are often termed *supply channel* (or intermediary channel or business channel) that provides business to hotel by forming positive business image of an organization in the mind of guests, thereafter sell products/services or direct business towards hotel. These sources/associates can be categorized into two broad groups, direct and indirect reservation sources.

| Direct sources | • Direct source of hotel room reservation indicates to materialization of business directly from the targeted market segment/s. These reservation requests may come to hotel in a number of direct ways.<br>• Direct ways/sources include in-person, toll-free numbers, direct telephone lines, letters, faxes, e-mails, telex, telegram, on-line reservation sites and so forth. |
| :--- | :--- |
| Indirect sources | • It is concerned with all those room reservation requests which are generated from intermediary channel/supply chain system. Here, hotel pays business-generation-charges to intermediary players in lieu of generating business.<br>• Indirect sources include travel agencies, tour operators, airlines, corporate offices, hotel booking agencies, GDS, sky scanners and so forth. |

Remember, every hotel has two primary in-house *business-entry-points*, i.e., front desk and reservation desk. Apart from it, grand hotels also have sales and marketing division, corporate sales offices and other contracted out intermediary players for the purpose of business generation.

| Activity | | |
| :--- | :--- | :--- |
| From the below given list, identify which can be described as a direct or indirect source of room reservation. | | |
| **S. No.** | **Room reservation request from** | **Direct** | **Indirect** |
| 1 | Mr. Water Melon comes to your hotel and gives room reservation on behalf of his boss Mr. Honey Dew. | | |

| 2 | Cream Caramel Airline issues either Meal & Accommodation Order (MAO voucher) or Passenger Service Order (PSO voucher) to their layover passengers. | | |
|---|---|---|---|
| 3 | A hotel management college plans a trip to visit several hotels in Goa; resultant college authority contacts a travel agent to arrange vehicle and accommodation. | | |
| 4 | Mr. Potato plans to go to London on a business trip and also needs to return as soon as possible, so he is using GDS to reserve a room as well as to arrange all related services. | | |
| 5 | Mr. Beef Wellington wants to go to Switzerland on honeymoon trip, thus he consults with hotel on toll-free number, thereafter faxes the hotel requesting a room reservation. | | |

## 10.10   OVERBOOKING

Overbooking is also known as "bumping, oversold or elastic booking" and sometimes as "adds (in PMS)". It is concerned with accepting more room reservation than available *room inventory*. Unfortunately on an average basis, 10% guests with reservation do not show up and also do not inform the hotel of their cancellation/no-show. These cancellation/no-show guests create a gigantic revenue loss and customer loss dilemma for hotel because of which it is difficult for hotel to maximize occupancy percentage, subsequently affecting revenue generation capacity of the hotel. Overbooking is a perfect tool that assists in maximizing room occupancy percentage and minimizing vacancy percentage, consequently augments the hotel's ARR, RevPAR and RevPAC.

> **Overbooking Formula**
>
> Room inventory
> - Out of order rooms
> - Confirmed reservation*
> - Guaranteed reservation*
> - Expected stay-overs
> - Expected over-stays
> - Expected walk-ins
> + Expected under-stays
>
> *After deducting expected no-shows

On every night, every unsold room leads the hotel towards a loss of room revenue, simultaneously to the potentiality of supportive revenue points (especially of food and beverage outlets) which can never be recuperated in future (as hotel room is a perishable product). In order to achieve *perfect sell* (or high occupancy percentage) and prevent such revenue loss, in addition guest also does not require to *farm-out* into another property; room division needs to accurately estimate/calculate the overbooking percentage. If a hotel can manage its overselling strategy perfectly then it can achieve its *perfect sell* target. The policy of overbooking should be based on the management of various occupancy categories into which guests are placed. In simple terms, overbooking calculation must be power-driven by degree of effect of last year/s guest turnover. For instance, previous year/s confirmed reservations, guaranteed reservations, stayovers, overstays, understays, walk-ins, no-shows and so forth must be counted while scheming overbooking percentage. Thus, overbooking management is also called as "occupancy management". A step-by-step process for handling overbooking is given below.

| Step 1 | • List the total number of *room inventory* (or lettable rooms) currently available for sale (it is calculated by total number of rooms minus out of order/out of service rooms). |
|--------|--------|
| Step 2 | • List the number of guaranteed no-show reservations, confirmed no-shows, as many guaranteed & confirmed reservation often found as no-shows reservation, thus affect overbooking. Take previous year record or *wash factor* (during group) to calculate it. |
| Step 3 | • List the number of non-guaranteed reservations for that date because it provides a strong base to determine the overbooking level. Again consider no-show factor. |
| Step 4 | • List the estimated number of under-stay guests (or early departures). Once again, past history can be a guide to determine the expected early departures. |
| Step 5 | • List the estimated number of stay-over guests. After that, prepare a list of number of expected walk-ins. To understand the equation exactly, you can refer to the given formula in quick scroll box to calculate overbooking. |

## 10.11 STANDARD CONVERSATION BETWEEN PROSPECTIVE GUEST AND RESERVATION DESK/AGENT

**Situation-** Mr. Jam works as a reservation agent in hotel XYZ. He is dealing with a prospective guest Mr. Jelly who wants to book room for 3 days. Mr. Jelly is not a price sensitive guest and he is coming with his friend. Thus Mr. Jam suggests him to book deluxe twin room and demands advance deposit of Rs. 49,000 to guarantee the reservation. Mr. Jelly wants to use his credit card to pay advance deposit. Mr. Jam is ready to accept credit card and tells Mr. Jelly about cut-off-hours. Pretended standard conversation phrases between Mr. Jam and Mr. Jelly are given here:

| Mr. Jam | Good morning, reservation desk, this is hotel XYZ, I am Mr. Jam on the line, how may I assist you? |
|---|---|
| Mr. Jelly | Good morning, I want to make a reservation for next Saturday the 20<sup>th</sup> of January. |

| Guest | Mr. Jelly Bean |
|---|---|
| Reservation agent | Mr. Jam |
| Booking date | 20-23 Jan, 14 |
| Room type | D. twin room |
| Payment mode | Credit card |
| Extra services | No extras |
| Advance deposit | Yes |
| Assign room No. | No |

**Mr. Jam** Ok sir. How should I address you?

**Mr. Jelly** I am Mr. Jelly.

**Mr. Jam** May I know, did you stay here before, Mr. Jelly?

**Mr. Jelly** No, not at all.

**Mr. Jam** Fine, for how many nights will you require the room Mr. Jelly?

**Mr. Jelly** For three nights.

**Mr. Jam** Right and what type of room do you want?

**Mr. Jelly** I want a twin room.

**Mr. Jam** Ok Mr. Jelly, will you come alone or with family?

**Mr. Jelly** No, I'll come with my friend.

**Mr. Jam** Mr. Jelly, currently we have deluxe twin room for Rs. 24, 500 per night and standard twin room for Rs. 17, 500 per night.

**Mr. Jelly** I'll take the deluxe twin room please.

**Mr. Jam** Ok sir, may I have your last and first name please?

**Mr. Jelly** Oh yes, my name is Bean Jelly.

**Mr. Jam** How should I spell your name sir?

**Mr. Jelly** My last name is Bean, it is B for Benjamin, E for Edward, A for Andrew, N for Nellie then my first name is Jelly, it is J for Jack, E for Edward, double L (L for London), and Y for Yellow.

**Mr. Jam** It is Bean Jelly.

**Mr. Jelly** Yes you are right.

**Mr. Jam** Mr. Jelly Bean, as you know that it is a peak season for Gwalior because Tansen festival is going on. Due to which we have lots of room requests. So I would like to recommend you to guarantee your booking, as we have very high occupancy rate.

**Mr. Jelly** Ok, so how can I give the guaranteed booking?

**Mr. Jam** Sir, you have to pay us a minimum deposit of Rs. 49, 000.

**Mr. Jelly** Can I use my Visa Electron credit card to pay this amount?

| Mr. Jam | Yes sir, we accept Visa Electron. |
|---|---|
| Mr. Jelly | Ok. |
| Mr. Jam | Mr. Jelly Bean I have guaranteed your reservation. Now I also want to make you aware that if you do not come on 20th Jan, then one night room rate will be deducted from your credit card, as it is a part of hotel policy. But if you would like to modify your reservation then you have to inform us 36 hours in advance. I hope you understand our hotel policy. |
| Mr. Jelly | Yes, I understand. |
| Mr. Jam | Ok sir. Now, may I have your contact number and address please, Mr. Jelly Bean? |
| Mr. Jelly | Yes, my contact number is 08989002808 and address is IHM-Campus, Airport Road, Maharajpura, Gwalior. |
| Mr. Jam | Ok sir, may I also have your e-mail address, so that I can send you your reservation confirmation letter? |
| Mr. Jelly | Oh yes, my e-mail address is bean_jelly@rediffmail.com. |
| Mr. Jam | Ok sir, how are you arriving? |
| Mr. Jelly | By flight. |
| Mr. Jam | Please, could you provide your flight number and expected arrival time, Mr. Jelly Bean? |
| Mr. Jelly | Oh yes, I am coming from Kingfisher Airlines and will reach there around 11:00 a.m. |
| Mr. Jam | Thank you sir. Mr. Jelly Bean would you like us to arrange pick-up service for you at the airport. |
| Mr. Jelly | No, thank you very much, I'll manage that. |
| Mr. Jam | Now, Mr. Jelly Bean let me confirm the room reservation details to you. You have booked one deluxe twin room, arrival date is 20th Jan. and expected departure date is 23rd Jan. Your room rent is Rs. 24, 500 per night which includes complimentary breakfast. You do not want any pick-up facility and you are ready to pay advance deposit of Rs. 49, 000 through credit card. Your contact number is 08989002808 and address is IHM-Campus, Airport Road, Maharajpura, Gwalior. |
| Mr. Jelly | Yes, its correct. |
| Mr. Jam | Mr. Jelly Bean kindly note down your reservation confirmation number, that is TH00AG77. We will also send you your reservation confirmation letter at your e-mail address that is bean_jelly@rediffmail.com. |
| Mr. Jelly | Oh, thank you so much. |
| Mr. Jam | Ok sir, would you like us to send you a brochure? |

| Mr. Jelly | No, I don't want. |
|---|---|
| Mr. Jam | Thank you Mr. Jelly Bean for giving us a chance to serve you. So Mr. Jelly Bean see you on 20<sup>th</sup> Jan. Have a nice day. |

Mr. Jam     Thank you Mr. Jelly Bean for giving us a chance to serve you. So Mr. Jelly Bean see you on 20th Jan. Have a nice day.

## Additional Standard Phrases for Conversation

| | |
|---|---|
| - To know with whom you are talking | - May I know to whom I am talking? |
| - To know from which country guest is calling | - Sir, may I know, you are from which country? |
| - To know whether guest wants to make reservation or modify it | - Do you have a reservation/do you want to make one? |
| - When guest wants to modify the reservation but agent doesn't have any record (exceptional case) | - I am afraid we have no record of a reservation for that date in your name. |
| - When cannot find reservation record then ask the name of agent | - Do you remember the name of the reservation agent? |
| - Try to find out reservation through name who booked the room | - In whose name was the reservation made? |
| - Try to find out reservation through date of booking | - When was the reservation made? |
| - When still cannot find any reservation record | - I am sorry, but I have not got any record of that. |
| - Offer for new reservation | - Shall I make a reservation for you? |

## CONCLUSION

Reservation section is responsible for processing room reservation, handling cancellation and amendment requests. The guest's room reservation request can be guaranteed or non-guaranteed. There are different modes of giving guaranteed reservation but nowadays cash and *charge card* are mostly accepted. Reservation request can be received from different sources which can be classified as direct and indirect sources of reservation. *Direct reservation source* means guest directly approaches the hotel for reserving a room such as guest sends a request to reservation section. Whereas *indirect source* includes reservation through third party like travel agency. All reservation requests can be received either via telephone, e-mail, fax or letter. There are different systems for recording room reservation, for instance card system, Whitney, CRS, hotel diary, instant reservation and so forth. It depends on individual hotel which system they want to adopt. Reservation section works in close coordination with sales and marketing division because of identical job profile of both, i.e., to generate maximum sale. Nowadays, hotel also uses *overbooking strategy* to achieve *perfect sell*. In order to accurately calculate overbooking margin, reservation section refers to its previous history, current booking trend and forecast future opportunity. Tools used to determine future room availability also assist in this task.

| Terms (with Chapter Exercise) | |
|---|---|
| Lead time | It is the time gap between room reservation enquiry and its confirmation/guarantee date. |
| Chance guest | ? |
| Adds | Overbooking and the term is often found in PMS. |
| Rooming list | It is the list which contains the names of all group members with necessary details like their names, nationality, passport details, etc. Hotel receives this rooming list, well in advance, from group leader then allots room to each group member. |
| Rooming | Informing guest about the room and its facilities. It mainly occurs while bellboy is escorting guest towards his/her allotted room. |
| Correspondence set | It is a set/record of the reservations that are arranged in an alphabetical order. It contains all information/communication which the hotel and the guest have made from time-to-time like enquiry letter, reservation form, room offer letter, confirmation slip, Whitney/Shannon slip, etc. |
| Perfect sell | ? |
| En block | A group reservation/booking in which number of rooms are kept aside for group arrivals. |
| Farm out (or walking) | It is the situation in which hotel relocates its guest into another hotel because property cannot accommodate him/her due to lack of available rooms (sometimes even when guest holds a confirmed or guaranteed reservation). |
| CRS | A central reservation system which is a kind of reservation network for sharing business. It may further be categorized as affiliate and non-affiliate CRS. |
| ? | Software or system whereby CRS of various standalone sector/industry, specifically travel, tourism and hotel industry, works under the same umbrella. |
| IRS | ? |
| ? | Central reservation system of independent/non-chain hotels and is often also known as non-affiliate reservation system. |
| Affiliate reservation system | ? |

| Cancellation code | Code generated by reservation agent in lieu of cancellation of guest's previously made room reservation. It is often demanded by credit card company in order to properly cancel or refund the amount into respective guest account. |
|---|---|
| Confirmation | ? |
| ? | It is also known as *overselling* or *bumping*. It means selling/receiving room reservation more than the available number of rooms. |
| Anticipated amount | It refers to an advance deposit amount, predominantly advance deposit in lieu of guaranteed room reservation. |
| Advance deposit | It refers to that advance money/prepayment which guest gives in order to guarantee the room reservation request. This advance associates to guest-with-guaranteed-reservation. |
| ? | It refers to the last date of giving reservation confirmation to the hotel, thereafter the temporarily reserved room is released for normal sale. |
| Cut-off-date/time | ? |
| ? | Penalty/charges imposed by hotel against the cancellation of guaranteed reservation; it mostly applies when request for cancellation is received after cut-off-date/time. |
| Whitney rack | ? |
| ? | It is also called "Shannon slip/card" used to record some of the vital information regarding the prospective guest by referring to the room reservation form. |
| ? | An *advance payment* or *cash-in-advance* which is mainly demanded from walk-in guest who prefers to pay his/her account via limited cash amount, thus no *charge privilege* facility is offered to him/her. |
| Confirmed/06 p.m. release reservation | ? |
| House limit | Maximum credit limit (or *charge purchase* or *charge privilege*) given by the hotel to their registering guests. |
| Displacement factor | Simply, it indicates that one guest is accepted in place of another. But in yield management, it occurs when hotel accepts group business at the expense of transient guest. |

# 11

CHAPTER

# Bell Desk

⟨ **OBJECTIVES** ⟩

*After reading this chapter, students will be able to…*

- express the functions of bell desk section and its equipment;
- describe the procedure for handling guest luggage during arrival and departure stage;
- explain step-by-step process for left luggage handling; and
- elucidate the role of bell desk in transient and group check-out process.

## 11.1 BELL DESK

The term *bell desk* is concerned with sub-division of uniformed service department that is mainly responsible for *bell activity*, both *up-bell* and *down-bell* activity. In small hotels, the bell desk is also known as "porter desk". Remember, *concierge* is a French word often termed as porter thus frequently appears on the bell desk in luxury hotels, but not every porter comes under this category because concierge (or *les clefs*) is mainly responsible for giving information in relation to locality and hotel. In large hotels, concierge section is separate from bell desk.

| **Bell Desk Equipment** |
|---|
| - Bell desk |
| - Calling bell |
| - Bell cart |
| - Luggage net |
| - Luggage tag |
| - Various stamps |
| - Stationery & forms |
| - Beach umbrellas |
| - Computer system |
| - Telephone system |

Generally, bell desk section is next to the reception counter in front office or nearby the hotel entrance. In large hotels, bell desk works under *uniformed service department* which comprises all those members of the hotel staff who wear uniforms. This uniform is of a distinctive colour to the hotel but the basic design/pattern is same among all hotels. Uniformed service department comprises bell desk, concierge desk, valet parking attendants and doorman. This section is headed by bell captain who gives directions and instructions to the staff working under him.

In large hotels, there is a separate room for the bell related activities. These large hotels may also provide a separate cabin for bell captain in bell section. In small hotels, there is a small bell counter instead of whole bell room which is usually situated nearby the entrance door (generally in front of front desk) or next to the front desk. This small desk contains all required equipment of bell section from luggage tag to luggage trolley. Behind the bell desk, there is a small room (or may be large lockable cupboard/racks) in which left luggage provision is available for guests who have settled their accounts and checked-out from their rooms but desire to keep their luggage in hotel custody for a limited period of time. Apart from bell activity, bell desk is also responsible for few additional works like paging, delivery of newspaper, rooming guest and so forth.

### Activity

Suppose you are employed as a Senior Bell Captain in a large chain hotel. Make a list of the duties that you think you would be responsible for during the course of one week's work.

## 11.2 FUNCTIONS OF BELL DESK

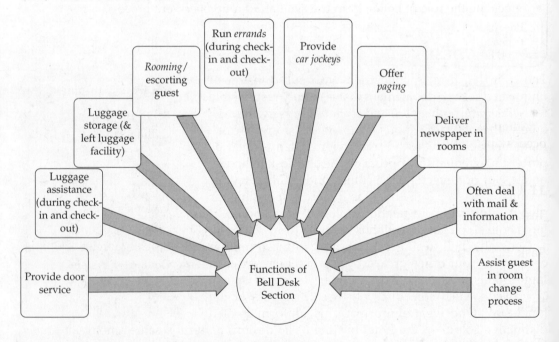

## 11.3 ROLE OF BELL DESK SECTION IN DIFFERENT STAGES OF GUEST CYCLE

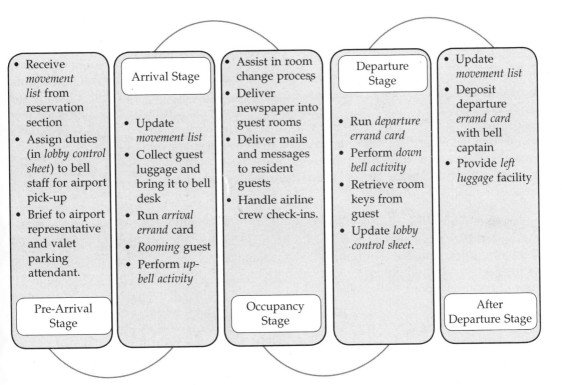

- Receive *movement list* from reservation section
- Assign duties (in *lobby control sheet*) to bell staff for airport pick-up
- Brief to airport representative and valet parking attendant.

**Pre-Arrival Stage**

**Arrival Stage**

- Update *movement list*
- Collect guest luggage and bring it to bell desk
- Run *arrival errand* card
- *Rooming* guest
- Perform *up-bell activity*

- Assist in room change process
- Deliver newspaper into guest rooms
- Deliver mails and messages to resident guests
- Handle airline crew check-ins.

**Occupancy Stage**

**Departure Stage**

- Run *departure errand card*
- Perform *down bell activity*
- Retrieve room keys from guest
- Update *lobby control sheet*.

- Update *movement list*
- Deposit departure *errand card* with bell captain
- Provide *left luggage* facility

**After Departure Stage**

A step-by-step description of some major tasks of bell desk/section that predominantly occurs during different stages of guest cycle is given below-

## 11.4 TASK FORCE OF BELL DESK

Task 1  Unload and load luggage (during check-in and check-out) into vehicle.

Task 2  Provide luggage service (up-bell and down-bell) between the lobby area and the guest room (during arrival and departure)

Task 3  Keep luggage at safe and secure place/maintain the luggage (via proper procedure like put *luggage tag*)

Task 4  Provide left luggage services and *luggage assistance* during room change process

Task 5  Retrieve guest luggage after clarification from cashier

**Bell Desk Staff**

- Bell captain
- Bell attendants
- Commissionaire
- Chauffeur
- Airport agent
- Lift operator
- Page boys

Task 6    Deliver newspaper into guest rooms and sometimes also deliver guest mails and messages.

Task 7    Send filled C forms to the FRRO (nowadays it fills on-line).

## 11.5   PROCEDURE FOR GUEST LUGGAGE HANDLING

It is concerned with step-by-step (or series of activities) process for handling guest luggage. In hotels, it is the responsibility of bell desk section to handle the guest luggage. Remember, it is also important to fill then run *errand card* while dealing with guest luggage. There are separate errand cards for guest arrival and guest departure stage. Luggage handling activity is divided into two groups: during arrival/departure stage and at post-departure stage.

**Equipment**

- Errand card
- Bell cart
- Luggage tag

**Luggage handling** (during arrival and departure stage)

The term *luggage handling* is related with picking then transporting of luggage from the lobby to the guest's room and vice versa as well as counting, tagging and storing of luggage, as and when required. It also includes shifting of luggage from one guest room to another room.

*Luggage handling during check-in (or up-bell activity)*

| Step 1 | • When guest arrives, *commissionaire* buzzes the bell captain desk to call the bellboy for carrying the guest luggage from car. Bellboy greets and welcomes the guest, collects his luggage and brings into lobby (via luggage entrance) at bell desk. |
|---|---|
| Step 2 | • Afterwards, bellboy waits for the guest to register at front desk, meanwhile luggage tagging can be done. In case of scanty luggage, bellboy should immediately inform the lobby manager or front desk. |
| Step 3 | • When front desk allots a room to the guest, it will indicate to bell desk through *arrival errand card*. If there is any damage in guest luggage, bell desk should immediately bring it into the notice of guest. |
| Step 4 | • Now bellboy will lead the guest towards the allotted room along with his luggage. Meanwhile, bellboy can inform the guest about the available facilities and services. |

| Step 5 | • Bellboy will open the guest room door and let the guest enter first. Afterwards bellboy will enter, position guest luggage in the luggage rack or cupboard. Finally, introduce the guest to room light switches, channel music, door lock, AC, etc. |
|---|---|
| Step 6 | • If required, offer other help otherwise wish the guest a pleasant stay. A bellboy should never solicit for tips. At last, report back to the bell desk and deposit *errand card*, on the basis of which bell captain will update *lobby control sheet*. |

## Luggage handling during check-out (or down-bell activity)

| Step 1 | • After receiving guest request (via telephone) to check-out, a bell captain will write down the details carefully on the *departure errand card*, a stack of which is kept at the bell desk itself. He instructs bellboy to proceed to the room. |
|---|---|
| Step 2 | • A bellboy should knock on the room's door and announce himself. Give a look around the room for any guest articles left, any missing room articles, any damage to hotel property. Finally switch *off* the air conditioner/heating, lights, etc. |
| Step 3 | • Collect the room key and depart from the room, allowing the guest to lead the way. Ensure that the guest room is locked. If the guest wants to carry the room key himself, allow him to do so. |
| Step 4 | • Place the guest luggage at the bell desk. Stick on any hotel stickers or publicity tags. Hand over the room key to the reception counter and *departure errand card* to the front office cashier. Wait for the guest to pay the bill. |
| Step 5 | • Bell desk will receive an authorisation (via *luggage out pass* and *departure errand card*) to take the guest luggage out of the hotel which front desk cashier will issue only after the settlement of bill. Also check with receptionist that the room key has been received. |
| Step 6 | • Take the baggage to the car porch and load it to the car or vehicle. Finally, report back to the bell desk and hand over the *departure errand card* with the authorisation signatures to bell captain, so that captain can update lobby control sheet. |

**Note**

- During *up-bell* as well as *down-bell* activity, bellboy should use staff elevator to transfer luggage to/from the room and vice versa. Arrival errand card will be generated during up-bell activity whereas departure errand card will rise during down-bell activity.
- In many hotels, bellboy deposits *departure errand card* to front desk cashier who, in turn, puts his signature on it and gives it to the bellboy in an attempt to release guest luggage whereas many properties use separate *luggage out/clearance pass* (or both) for this purpose.

---

**Activity**

Imagine you are working as a bellboy in a five-star hotel. Describe step-by-step activities that you are supposed to perform at the time of check-in of family guest {family comprises of husband, wife, 2 children (1 child below 5 years and another child above 12 years)}. You may also include suggestions which you think are necessary for up-selling of additional facilities and services.

---

## Left Luggage Handling

The term "left luggage" can be defined as "a luggage left by a hotel guest after checking out from sleeping room and settlement of his account but wishes to collect his luggage later". Generally, these are those guests who want to go for sightseeing or visit other cities and find it inconvenient to carry their luggage with them. For these guests, hotel provides safe and secure left luggage facility which may be chargeable but most of the hotels do not charge for it. A simple procedure for receiving & delivering left luggage from/to guest is given below.

**Key Objects**

- Luggage stickers/tag
- Left luggage register
- Luggage check/ticket
- Stationery

*Procedure for receiving left luggage from guest*

| Step 1 | • Before keeping guest luggage, ensure that a guest who wishes to leave his luggage must settle his/her account. A left luggage attendant can confirm this with front desk cashier. |
|--------|--------|

| Step 2 | • Now, string the luggage ticket on each piece of luggage separately. This luggage ticket contains a number (serial number) which is also printed on the counterfoil (duplicate copy) of the ticket. |
|--------|--------|

| Step 3 | • Enter the important details pertaining to left luggage in the left luggage register like guest name, luggage ticket number, number of pieces of luggage & so forth important details. |
|---|---|
| Step 4 | • Finally, tear off the duplicate copy (or second copy) of each ticket and hand it over to the guest and at last keep the luggage in the left luggage room. |

*Procedure for delivering left luggage to guest*

| Step 1 | • Before handing over the luggage to guest, bell desk requires to retract the duplicate copy (or second copy) of the *luggage ticket* from the guest. |
|---|---|
| Step 2 | • Now, counterpart this duplicate copy with the ticket attached to the luggage in the left luggage room. Counterpart to all required particulars especially serial number & guest's signature. |
| Step 3 | • Enter the date and time of delivery of guest luggage in the "Left Luggage Register" and finally obtain the guest's signature in correspondence to luggage handover. |
| Step 4 | • Safely keep the counterfoil of *luggage ticket* collected from the guest in order to protect hotel from any fake claim. The redeemed tickets and luggage books must be stored at the hotel for a minimum of 2 years. |

**Note.** When number of luggage pieces are more but all are small and belong to same guest then a luggage attendant can tie them all together in *luggage net* and stick a single luggage tag as a control instrument.

## 11.6   GUEST CHECK-OUT PROCEDURE AT BELL DESK

The heading is concerned with series of *bell-activities* of bell desk staff during check-out stage. Like during arrival stage, the handling of transient guests' luggage is different from group guests' luggage, similarly handling of guest luggage during check-out process is also different for transient and group guests. The handling of guest luggage by bell desk during departure stage is known as *down-bell activity*. A brief description of transient and group guest luggage handling by bell desk during departure stage is given below.

## Transient Guest Departure

A guest wishing to check-out normally calls the bell desk or reception. Irrespective of where the call is received the following below mentioned procedure must be followed:

Step 1    The bell captain deploys a bellboy who, in turn, fills up the *departure errand card*. Meanwhile to the room, bellboy should inform the front desk and cash desk about the departure taking place (when guest directly calls to bell desk for check-out).

Step 2    Upon arrival at the guest room door, bellboy should either gently knock at the door or press the door bell and then announce his department.

Step 3    When door opens, bellboy should greet the guest and check with him/her the number of pieces of luggage that are to be taken to the lobby.

Step 4    He picks up the luggage and places it outside the room then collects the room key. In the interim, he should take a quick look around the room to check for any perceptible damage to the room or losses of hotel articles.

Step 5    Bellboy should also check for any left behind items in drawer and closet, afterwards proceed to draw the curtains and switch off the light. Finally, lock the guest room.

Step 6    At the elevator, bellboy should allow the guest to step in first and then position himself near the control panel with the baggage in a corner next to him.

Step 7    During the way to lobby, bellboy may enquire with guest about room experience, meal experience and whether he would like a taxi called for him.

Step 8    Bellboy escorts the guest to the cashier desk and himself proceeds to the bell desk with the luggage. Now he pastes the hotel name tag on the guest luggage and mark a "D" on it.

Step 9    He then comes to the cashier desk and picks up the *luggage out pass* (or baggage out pass) from the cashier. He hands over the room key to the front desk and takes an acknowledgement in return from the front desk agent.

Step 10   Now bellboy asks the doorman to call a taxi (if guest has requested) and after showing *luggage out pass* to the doorman, bellboy places the luggage in the boot of the taxi. He then notes the taxi number on the departure errand card.

Step 11   Finally, bellboy returns to the bell desk to complete the errand card and shows it to the bell captain who signs it and makes an entry in the *lobby control sheet* (or lobby attendant control sheet).

**Note.** After placing luggage in the taxi boot, bellboy should inform the guest and indicate toward the open boot for the guest to check and then closes the boot.

## Group Guest Departure

A group departure is handled by all concerned sections of the front office department in a similar fashion to the FIT departure, except that a team of bellboys is appointed for luggage handling. This team of bellboys are assigned with the responsibility to bring down group members luggage from their rooms as well as keep it in safe custody till baggage clearance pass is not issued by the front desk cashier. In group departure, the stress is on coordination and efficiency; therefore a group's departure demands extra careful supervision. Elicited below are the slightly different and some special procedural tasks that are to be kept in mind while handling any group departure. These tasks may be used in addition to the regular procedure given for the departure of FITs/transient guests.

Task 1    The bell captain must ensure, well in advance, that sufficient staff will be available for group check-out in the shift, thus reshuffle the duty chart if situation arises. He checks the *luggage down time* and the wake-up call time of the group (if group departing early in the morning).

Task 2    Five minutes before luggage down timing, the bell captain allocates the floors and rooms to the bellboys for bringing the luggage down to the hotel lobby.

Task 3    The bellboys collect the luggage from rooms mentioned on their lists (usually group members are advised by the group leader to keep their luggage packed and ready for pick-up outside their doors at the luggage down time).

Task 4    The luggage from all rooms and floors is quietly collected and brought down to the lobby where it is neatly arranged in rows and stickers are pasted on them. The bell captain counts the bag pieces.

Task 5    The bell captain obtains a *luggage out pass* from the cashier after the group leader settles the bill, and ensures that keys from all group members are retrieved.

Task 6    The collected keys are handed over to the front desk agent where the agent scores out (on room status board) the room numbers of those rooms whose keys have been received. After all keys are received, the agent gives keys clearance on the *departure errand card*.

Task 7    The bell captain requests the group leader to count the luggage and also requests to contact those members of group who have not yet given their room keys. The luggage is then loaded.

Task 8    Only one *departure errand card* is made for whole group, however; numbers of all those bellboys who are engaged in group departure are mentioned on it. Finally, an entry is to be made in the *lobby attendant control sheet* on departure.

Though bellboys have been entrusted with the additional responsibility of collecting room keys from departing guests, generally in a group departure this becomes slightly difficult, therefore active assistance is rendered by the cashier, front desk agent, room boy in collecting guest room keys and returning them to the front desk agent before group leaves the hotel.

## 11.7   STANDARD PHRASES FOR CONVERSATION BETWEEN THE BELL HOP AND GUEST

- Taking permission from guest for luggage assistance
  - May I help you with your bags, sir?

- Confirming with guest for his luggage (like in group)
  - Could you tell me which one is your baggage, please?

- Asking guest about how much luggage he is carrying
  - How many bags do you have in all?

- Putting guest on hold meanwhile bringing bell cart
  - Sir just a moment, please, I'll bring a baggage cart.

- Informing guest from where he can collect his luggage
  - Your bags will be at the bell captain's desk.

- Informing group member that their luggage is on the way
  - Your luggage is on the way from the airport.

- Informing the group that their luggage will come soon
  - Your luggage will come after fifteen minutes. And we will bring them over to your rooms.

- Warning guest to take caution while crossing door
  - Please mind your hands in the revolving door.

- After entering, requesting guest for registration
  - Please, could you check-in at the reception counter over there?

- For confirming any valuable/breakable item in luggage
  - Do you have any valuable or breakable items in your bag?

- To confirm that all placed luggage is of guest only
  - Is this all your luggage?

- To keep guest on hold for elevator service
  - Sir just a moment please, the elevator will be here soon.

- Caution intimation to guest while using elevator
  - Please step inside.

- Making guest comfortable by asking courtesy questions
  - How was your trip, sir?

- For giving direction towards the hotel based restaurant

- If you would like to go to the restaurant on the top floor, please change elevator at the fifth floor.

- Permission to keep luggage after entering the room

- Here is your luggage. Where would you like me to put it?

- Informing the guest about taxi fare

- Taxi fares are charged according to the cab model and mileage drive.

## CONCLUSION

Bell desk works under the direct supervision of bell captain and is considered sub-division of uniformed service department. In reality, concierge is the correct French word for bell desk. But nowadays, this concierge word is used for information agent. In all hotels, bell desk is responsible for handling guest luggage during arrival as well as departure. Often bell desk is supportive in guest service during the occupancy period, for instance it assists in room change process, delivering newspaper to guest rooms and so forth. Bell desk is also responsible for word-of-mouth advertisement of hotel, especially while *rooming* the guest during check-in. Apart from luggage assistance, bell desk also offers left luggage facility which is only given to resident guests who have properly settled their accounts. While handling guest luggage, bell desk should remember to tag luggage card.

| Terms (with Chapter Exercise) | |
|---|---|
| Down-bell activity | A collective term for all those bell desk activities which take place between a departing guest and bellboy. It starts from picking luggage from guest room and remains there till farewell. |
| Up-bell activity | A collective term for all those bell desk activities which take place between an arriving guest and bellboy. It starts from picking luggage from car and remains till rooming of the guest. |
| Left luggage | Facility in which departing guest is allowed to keep his/her luggage at the bell desk (in left luggage counter). Remember, this facility is only for departing guests and who have properly settled their accounts. Whether it is chargeable or non-chargeable; it depends on the individual hotel policy. |
| Bell hop | Bellboy. |

| Lobby attendant control sheet | It is a sheet that is used to control the activities of bell desk staff (and other lobby staff), at the same time it also provides information in relation to allotment of duties to various bell staff. Often it is also known as *bell captain's control sheet*, *lobby control sheet* or *front log sheet*. |
|---|---|
| Errand card | Card that contains guest's information in relation to their arrival/departure (time & date), number of pax, number of luggage, etc. It is mainly used by bellboys during arrival (i.e., arrival errand card) and departure stage (departure errand card). |
| Bell cart | A luggage trolley. |
| Paging | It is a system of identifying an unknown guest. It can be manual like paging board with handle or electronic pager or mobile phones. |
| Car jockey | Hotel's car driver or chauffeur |
| Rooming | It means escorting then introducing a guest to a room and its facilities and services. |
| Luggage net | Front office manual equipment, i.e., a *net* that is used to tie all baggage in a single net then paste baggage tag (it contains guest's name, room number, etc.) so that guest's luggage will not misplace which may happen if kept separately. |
| Luggage tag | Also known as baggage tag, is a small card which is tied around the guest's luggage. Tag contains information regarding the guest's name and room number. Often, the number of bags handled should be recorded on the luggage tag. |

# 12
CHAPTER

# Supportive Centres

⟨ **OBJECTIVES** ⟩

*After reading this chapter, students will be able to...*

- identify different supportive centres and their probable location;
- define work profile of each supportive centre;
- describe how lobby supports functioning of all supportive areas;
- explain how concierge aids in guest satisfaction;
- identify different recreational activities and indoor and outdoor games; and
- explain how supportive centres sustain generating ancillary revenue for hotel operation.

## 12.1 SUPPORTIVE CENTRE

A *supportive centre* refers to all those departments/sections/outlets/subdivisions and other similar divisions which may be either *revenue centre* or *cost centre*. Simply, supportive centre may or may not generate revenue for the establishment but fulfils guest expectations, consequently enhances guest satisfaction level. For instance, public telephone booth, newspaper kiosk, salon bar, concierge, spa, gift shop, business centres, game zones and so forth. If property offers recreational activities to their guests then it is also considered a part of supportive services.

Most of these supportive services are mainly found in hotel lobby such as public telephone booth, newspaper kiosk, salon bar and business centre while others may be in the direct access from hotel lobby. Therefore, lobby is also known as *foyer* or hotel entrance.

Nowadays, hotel lobby is considered an important area for supporting services due to which newly opened hotels build their lobby area large enough to offer such supportive services like conducting small meetings and managing get-togethers. For arranging small meetings, hotels offer business centre around lobby area. Here, business people can hold short meetings; meet other business people before moving on to another location, or where friends can meet other friends or family. Every hotel has areas in their lobby where people can meet, for instance a lobby bar, coffee shop or small seating area.

As a hotel employee, it is your job to make sure that people know where they can have these informal meetings. It is also your job profile to make sure that they are comfortable and you should assist them in any way you can.

Most of the hotels also consider *uniformed service department* as a part of supportive centres whereas other hotels regard it is as a division of front office department or room division. Uniformed service department is usually responsible for providing most personalized services and take care of guests and their needs. Therefore, it is also known as *guest service department* and comprises commissionaire, chauffeur, bellboys, elevator/lift attendants, valet parking attendants, cloakroom attendants and other hotel employees who wear hotel's uniform.

---

### Activity

Imagine you are employed as a lobby manager in hotel Blue Mango that is a 100 room boutique property. Your task is to draw a lobby plan which indicates the location of following sections/divisions:

- Guest seating/waiting area   - Front desk/reception section   - Bell desk

- Guest relation desk          - Centre of attraction (like fountain)   - Business centre

- Gift shop                    - Restaurant and coffee shop        - Salon bar

---

A brief description of major supportive centres of front office/room division department predominantly found in hotel lobby is given below.

## 12.2   UNIFORMED SERVICE

The term *uniformed service* is used for subdivisions of room division department including front office employees who wear uniform. It is prominently responsible for providing more personalized services to guests, therefore it is also known as *guest service department*. Staff working in uniformed service department are often classified as *tipped employees* as portion of their income is derived from the guests gratuities. Nowadays, many hotels incorporate the concierge into the GSA.

**Elementary Job Positions in Uniformed Service Dept.**

- Bell attendants
- Door attendants
- Valet parking attendants
- Travel desk attendants
- Concierges

Major functions of uniformed service department are given below.

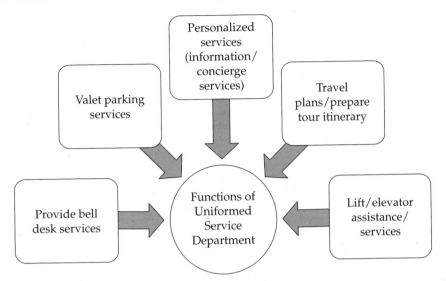

## 2.3  GUEST RELATION DESK

Due to its guest oriented work profile, guest relation desk can also be called a "hospitality desk, welcome desk, help desk or simply guest relation desk". The desk should be located nearby the reception desk or in the front of reception desk. In some circumstances, it may also be built before the reception desk too (or in place of front desk like in automated hotels where front desk is not available due to self-check-in terminals). This front face location is obligatory because of its guest oriented work

**Specific Functions of Desk**

- Welcome & farewell of guest
- Offering garlands & bouquet
- Welcome by Tilak & Aarti
- Offer baby sitting
- Ensure to serve welcome drinks
- Ensure comfortable stay

profile which includes welcoming the guests, offering *tilak, aarti,* flower bouquet or garlands and finally guest farewell. Irrespective of hotel size, i.e., whether hotel is small or large, this section acquires a limited space. Therefore, it may also be made in the form of desk only, behind which guest relation staff can perform their assigned tasks.

The guest relation desk is supervised by guest relation manager and managed by guest relation executives (GREs) along with a team of guest service agents who have a desk in the main lobby. Their main responsibility is to make the guest feel comfortable, welcoming and provide a more personalized service, very often by simply talking to guests who are travelling on their own and perhaps feel lonely as staying in a strange/new town or city. Guest relation executives are also called *guest relation officers*. In certain situations, guest relation staff is also responsible for making *courtesy calls* to the guests in their rooms and in case of VIP guest they are also held responsible for completin formalities of registration either on their own desk or directly in guests room. GREs a also involved in handling many typical situations which are difficult to deal with b usual front office staff. Guest relation agents are primarily employed to create a mo caring and memorable atmosphere for guests.

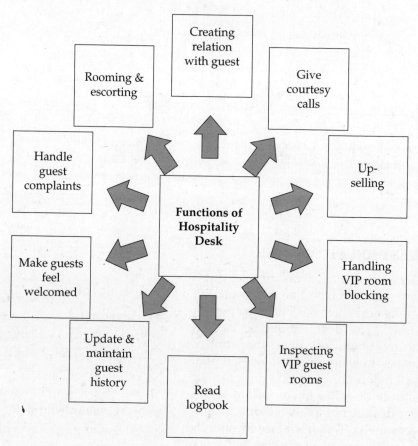

## Standard Phrases for Conversation between GRE and Resident Guest

- For assuring guest that you will definitely look into the matter

- I'll look into the matter.

- For giving confidence to guest

- Don't be afraid sir.

- For assuring guest that the matter will not take too much time

- It would not be too long.

- Assuring women that you will definitely sort out the matter

- Don't worry madam, we'll see to it.

| | |
|---|---|
| Ease guest by presenting yourself in his/ her service | - I am always at your service, sir. |
| When guest is satisfied with your solution | - I am happy everything was to your satisfaction. |
| Giving your concern that it should not happen | - It's a great pity. |
| When your guest is dissatisfied with any of your staff | - I am really sorry sir. |
| Confirming whether your guest is feeling better or not | - Are you ok, sir? |
| When guest is not feeling well | - Would you like to see the doctor? |
| After first aid and when guest feels better | - I hope you are better now. |
| Expressing gratitude and confirming guest has enjoyed his/her stay | - Hope you enjoyed staying with us. |

### Activity

Describe the guest seating arrangement in hotel lobby of any luxury five-star hotel. If possible, also draw the layout of guest seating area.

Or

Describe the guest seating arrangement in hotel lobby of any budget/theme hotel. If possible, draw the layout of guest seating area also.

## .4  CONCIERGE DESK

**Concierge Keep Handy With**

- Hotel Information Brochure
- City Information Guide
- Local Area Maps
- Computer system
- Telephone Directory
- Prepare Tour Itinerary (optional)

is a French word which means *porter desk* at is primarily responsible for handling est luggage, simultaneously providing formation. But nowadays, there are separate sks for performing tasks separately. Porter sk (or bell desk) handles luggage whereas ncierge desk deals with exchange of formation. The concept of concierge desk or formation counter is a common service in ropean countries. The concierge/information section of front office department works a source of information about the hotel itself as well as in relation to locality. Thus, it ould be equipped with all necessary information about the hotel's facilities, services, events taking place in hotel (if any), etc. In addition, it also provides information i relation to important places like temples, drug stores, airport, shopping malls, museum monuments and so forth city/locality attractions.

Concierges may also provide custom services to hotel guests. Desk staff's dutie include making reservations for dining, securing tickets for theatre and sporting event arranging transportation, and providing information on cultural events and loc attractions. Concierges are known for their resourcefulness.

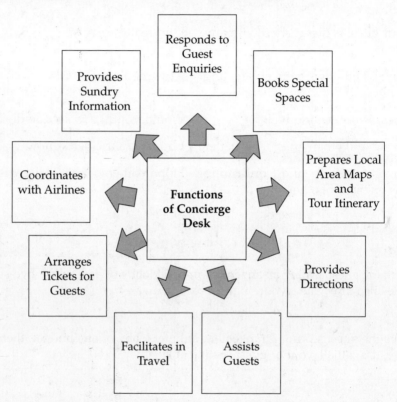

## Standard Phrases for Conversation between Concierge and Resident Gue

- For giving direction towards any lobby
  based outlet
- For giving direction towards floor based
  outlet
- Giving direction with the help of turn
  corner
- Explaining to the guest that it is out of
  our control

- Go down to the lobby.

- Please take the lift to the 5$^{th}$ floor an
  turn right.

- Turn left/right at the first corner.

- This is out of our scope sir.

- Assuring guest that you will try your level best
  - I will try my level best, but can't guarantee you, sir.
- For putting guest on hold, meanwhile search for things
  - If you wait a moment, sir, I'll try to find out.
- For wishing best of journey
  - Goodbye and have a nice trip sir/madam.

---

### Activity

✓ Enlist the differences between hotel ambience and hotel décor. Does each have an impact on customer expectations in terms of service and quality. Explain each one briefly.

---

## 12.5 BUSINESS CENTRE

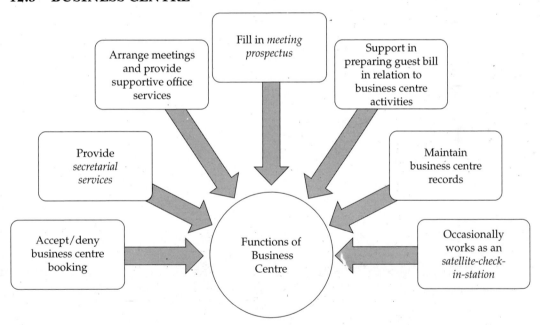

*Business centre* is an area or meeting venue in a hotel (usually in/nearby lobby) where an assortment of office services like word processing, power point presentation and facilities like fax machine, printer, photocopy, etc., are made available to guests or registrants on a commercial basis. It is also called "meeting room or conference room". Many hotels provide a business centre to both their business and pleasure guests. The range of services and facilities provided by business centre may include business equipment, presentation supplies, computers, meeting and function space with wired or wireless internet access, secretarial support, translation services and so forth. Small properties typically offer limited business services with self-serve options whereas luxury hotels offer range of facilities and provisions. Often business centre also works as *satellite-check-in station* particularly for group check-in process.

## 12.6 SPA

It is a Belgian concept concerned with *balneotherapy* (Latin word *balneum* means bath) which is related with health treatment through water. The acronym SPA stands for "Sanum Per Aqua". Spas are the health resorts which provide the therapeutic bath and massage facilities along with other features of luxury hotels. The term *spa* is derived from the name of the town **Spa**, in **Belgium**, whose name is documented back to Roman times, when the location was called *Aquae Spadanae* (a Latin word which means to scatter, sprinkle or moisten). Examples of international SPA resort hotels include Chi The Spa at Shangri-La, Hyatt Resorts and Spa. In India *Ananda Spa* in Himalaya and *Country Spa* hotel chain are two perfect examples. There are different types of spas, a brief description of some of them are as follows:

| | |
|---|---|
| Ayurvedic spa | • This spa is related with all treatments and natural products that are often used as an alternative medicine. Its ranges may be found in Spa resort hotel in the form of bathroom amenities. |
| Club spa | • This spa is related with physical fitness given/directed by a trained professional of a hotel. |
| Dental spa | • This spa is related with traditional style dental treatment (by licensed dentist) with other spa services. |
| Destination spa | • This spa is associated with comprehensive programme in relation to healthy habits like wellness education, healthful cuisine, physical fitness activities, etc. |
| Medical spa | • This spa is associated with comprehensive programme in relation to medical, wellness, care and/or alternative therapies and treatments. |
| Mineral spring spa | • This spa is related with hydrotherapy treatment. It is a source of getting natural minerals. It is done by using thermal or seawater in hydrotherapy treatment process. |

## 2.7 RECREATIONAL ACTIVITIES

*Recreational activities* are those activities which relax the guests or provide diversions to guest from their normal routine life. Nowadays, there is a separate market segment which mostly demands recreational activities only because these guests have more leisure time. Hotel industry focuses on the need, want and demands of this market segment then frame their recreational activities, its related policies and pricing strategy accordingly. In hotel industry, guests from this market segment are known as *recreational travellers*. The customary needs and wants of these travellers are: friendly nature, budget hotel at convenient location, ready to pay for recreational games/entertainment and their duration of stay is usually more in comparison to *business travellers*.

### Activity

Collect various recreational activities which are mostly offered by resort hotel organization then categorize them in accordance with/to resort type. Finally, fill in the given below spaces with appropriate example in each column.

| S. No. | Seaside Resorts | Hill/mountain Resorts |
|--------|-----------------|------------------------|
| 1 | ? | ? |
| 2 | ? | Ice skiing |
| 3 | Scuba diving | ? |
| 4 | ? | ? |
| 5 | ? | Bungee jumping |
| 6 | Rafting | ? |
| 7 | ? | ? |
| 8 | ? | ? |

## 2.8 GAME ZONE

The term *game zone* refers to various amusement and entertainment facilities offered by hotel management to their guests. These games entertain and engage guests, especially their children, during free hours. Simultaneously, it generates small amount of revenue for the hotel organization. Apart from revenue generation, it adds more value (or works as a value addition facility) to hotel benevolence, consequently shore-up the hotel marketing. These games may be provided inside the building premises as well as outside the building. All games available inside the building premises are collectively known as *indoor games* whereas all out-house games are called *outdoor games*.

### Activity

First, enlist different games then categorize them as indoor and outdoor games. Thereafter, fill in the below given spaces with appropriate examples.

| S. No. | Indoor Games | Outdoor Games |
|--------|--------------|---------------|
| 1 | ? | ? |
| 2 | ? | ? |
| 3 | Carom | ? |
| 4 | ? | Lawn tennis |
| 5 | ? | ? |
| 6 | ? | Badminton |
| 7 | Table tennis | ? |
| 8 | ? | ? |

## CONCLUSION

The luxury of any hotel is not only determined by *room configuration* and its supplies and amenities, it is also established on the basis of provided additional supportive services for example, gaming zone, spa, business centre, recreational activities, gymnasium and so forth. These supportive services do not only pull the targeted market segment but also work as a value addition gizmo plus generate additional revenue which will directly affect the hotel's RevPAC and RevPOT. Supportive services work as a bundle of value addition components under one roof. Remember, these supportive services are mostly given by resort properties as they are targeted towards family and group clientele as well as duration of their accommodation is also more in comparison to business hotels. The place or centre which offers these supportive services is collectively known as supportive centre. Almost all subdivisions of uniformed service department like concierge and guest relation desk are counted as best supportive centres.

### Terms
(with Chapter Exercise)

| | |
|--|--|
| Uniformed service | It is a collective term for all those front line employees (like bellboys, doorman, car valet and lift attendant) who wear uniforms. It may be an independent department (or subdivision of front office department/ room division) which directly reports to front office manager/RDM. It is mostly found in European hotels and also known as *guest service department*. |

| | |
|---|---|
| Chauffeur | Car driver (or limousine driver). |
| Leisure time | It means free time, time during which somebody has no obligations or work responsibilities, and therefore is free to engage in enjoyable activities. |
| ? | Doorman, also known as door attendant or link man. |
| Foyer | Property's entrance point like hotel lobby. |
| Supportive centre | It is a collective term for those facilities or zone of facilities which offers additional services and activities, especially recreation oriented. |
| ? | In-house games, for example, carom, table tennis, video games and so forth. |
| Outdoor games | Out-house games, for example, lawn tennis, football, cricket, volleyball, basket ball, etc. |
| Game zone | A place/area which offers different games to resident guests. These offered games may be chargeable or non-chargeable, it depends on individual hotel's policy but are mostly charged for. |
| Recreation | ? |
| ? | A small meeting point/area with almost all business services like office equipment and secretariat services. It is usually found in hotel lobby. |
| Meeting prospectus | A store requisition sheet type document which shows all information or requirement in relation to the panorama of meeting, for example, number of pax, type of seating arrangement, office equipment, business services, etc. |
| Bungee jumping | ? |
| Scuba diving | It is a kind of underwater sport (or mode of underwater diving) in which scuba diver (or guest) uses a self-contained underwater breathing apparatus (scuba) to breathe under water. |
| Concierge | It is a French term for porter desk, but now for information agent who is mainly responsible for providing in-house (or hotel related information) and out-house (or locality related) information to guest. |

| | |
|---|---|
| Rafting | Rafting or river rafting is concerned with water sports in which guest uses an inflatable raft to navigate a river or other water bodies. Remember, rafting is usually done on white water or different degree of rough water. |
| Courtesy call | A *civility call* that usually is placed by guest relation agent/executive into the room of resident guests in order to confirm that guest is comfortable or not. |
| Business traveller | ? |
| Spa | It stands for Sanum Per Aqua, meaning *balneotherapy* which is related with health treatment through water. It is a Belgian concept. |
| Room configuration | Physical characteristics (or make-up) of the guest rooms. At last, it also forms an additional base to determine rack rate. |
| Brochure | It is a kind of information-cum-promotion material for any organization. It provides information about the property and its sister hotels along with address and contact number. But it hardly (or never) discloses the room tariff. |

# 13 CHAPTER

## Mail and Message Section

OBJECTIVES

*After reading this chapter, students will be able to...*

- describe the functions of mail and message section;
- elucidate how incoming-hotel-mail handling is different for different level of employees;
- express the procedure for handling mails of past, present and future guests; and
- explain the correct process for receiving mails like why time and date stamping is necessary.

## 13.1 MAIL AND MESSAGE SECTION

A *mail and message section* is a subdivision of front office department which is mainly accountable for handling guest's as well as non-guest's mails & messages. The section accepts mails and messages on behalf of guests, later on delivers them into guest rooms. It also receives & delivers hotel employees' mails & messages. In case of receiving guest mail and message, first of all agent requires checking the guest status. The guest may be past guests, present guests or future guests. Nowadays, many hotels affix this section with bell desk. This kind of practice reduces the cost of manpower and this cost ultimately turns into additional hotel revenue/profit.

Separate mail and message section is generally found in large establishments and is known as "mini post office" which sorts out mails & messages and delivers them timely to the concerned person. A *mail logbook* is maintained on daily basis in order to track all received mails & messages, so it can have a record to show it to the supervisor. After receiving the mail, the desk agent is expected to time stamp all guest mails when they arrive. Time and date stamp secures the agent from any later on claim which may arise by the guest. After receiving, its entry should be made in *mail signature book* and instantly delivered to the resident guest.

This section is headed by mail & message supervisor with few mail & message handling agents. Mail & message supervisor is responsible for a team of mail & message clerks who are, in turn, responsible for delivering mails and messages of employees and guests. In many medium size hotels, this mail & message section is clubbed with concierge desk. In such situations, working clerks have an additional responsibility to deal with resident guests as well as visitors in order to provide them information about the hotel as well as local attractions. They would also carry a supply of local postage stamps and stationery.

## 13.2    FUNCTIONS OF MAIL AND MESSAGE SECTION

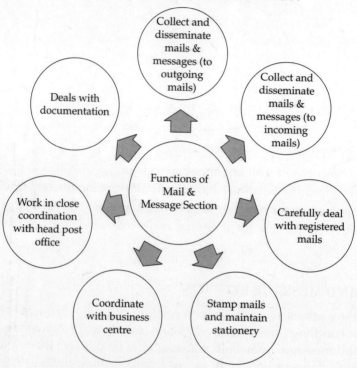

## 13.3    TYPES OF MAILS AND MESSAGES

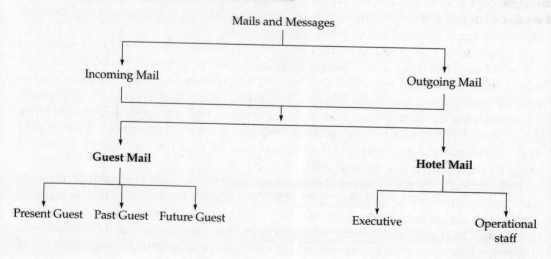

## Hotel Mail (in reference to incoming mails)

Mails of hotel staff may also be called *house mails*. These house mails may be for operational level staff as well as for executive level employees. The mails of operational level staff are usually sent to the personnel office/time office (or security office which is located at the back entrance from where hotel staff enter & exit). In security office, mails are placed in employee rack from where it is collected by operational level staff during their exit hours. All mails of the hotel executives should be directly delivered by bellboys in their respective offices. A brief description how to handle mails of hotel employees is given below:

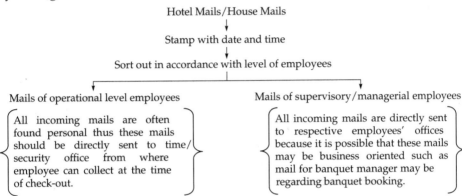

## Guest Mail (in reference to incoming mails)

Under three circumstances, mails/messages are received on behalf of the guest, i.e., mail for a guest who has a reservation (*future guest*) and is expected to check-in, secondly mail for an in-house guest (*present guest*) who is away from his room and finally mail for a guest (*past guest*) who has checked-out. In all these three circumstances, the hotel is held responsible for passing the mail/message, within a reasonable time, to the particular guest. Hotels keep a track of these mail handling instructions with the help of *mail logbook* maintained by mail and message section of front office, in which receiving of mail, etc., are noted down.

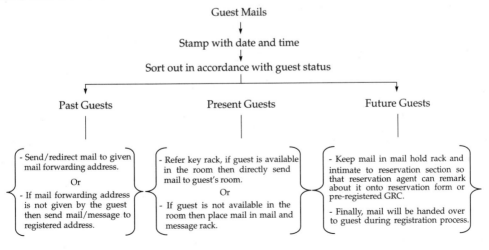

## Procedure to Handle Incoming Mails

Step 1     After receiving the mail, message or package, it should be date and time stamped. If mail or parcel is threadbare then take countersign of the postman.

Step 2     Sort out all mails, enter the detail of guest mail into the *mail logbook* (or *mail signature book*) and then place in the mail & message rack.

Step 3     Now inform the resident guest about his mail and ask permission to deliver it. After getting approval, send bellboy to deliver the mail to the guest room.

**Note**

- In many hotels, properties notify to their resident guests (about received message) by switching *on* the in-room message light of the guest room telephone.
- If mail is marked as urgent then the guest is immediately contacted whether in the room or on his/her contact number to inform him of the same.

## Procedure to Handle Outgoing Mails

Step 1     After receiving the mail or message, check whether it is a hotel mail or guest mail then quantify its weight. On the basis of weight, calculate its carrier cost.

Step 2     Now all the collected mails should be franked via *franking machine*, otherwise affix the postage stamp of correct denomination.

Step 3     Issue the receipt for registered mails or attach the receipt into a special book kept for that purpose. Receipts are an evidence of mail.

**Note**

- A specific time should be set, by mail and message section, as a deadline till which all outgoing mails (of guest plus employees) must be received, especially when it is to be despatched on that day only.
- A regular collection of mails should be made throughout the day; this will ease the pressure (of outgoing mails) that may build up at the end of the day.

| Activity |
|---|
| From the following terms you need to choose correct answers and answer the given questions. |

- Franking machine      - Registered mail      - Future guest      - House mails
- Time stamp      - Time office      - Message forward card

| S. No. | Questions |
|--------|-----------|
| 1 | All incoming as well as outgoing mails of hotel management and its employees are broadly known as........................ |
| 2 | Mails of past guests should be sent to their address to which guest has given in........ .......................... |
| 3 | Mails of operational level employees are usually sent to..................... so that they can collect after their shift. |
| 4 | Generally, confidential papers and other similar documents are preferred to be sent through.................... |
| 5 | Mails of ..................are kept with reservation, thereafter attached with pre-registration card. |
| 6 | ..............................machine is used for the stamping of all outgoing mails with different denominations. |
| 7 | All incoming mails must be............................so that later on guest or management cannot blame for late arrival of mail to them. |

## CONCLUSION

In large hotels, mail and message section may be a separate subdivision of front office department thus often termed *mini post office* whereas in small hotels, it is the responsibility of reception desk or bell desk to undertake the responsibility of mail and message section. After receiving any mail or message, the preliminary task is to stamp with date and time and thereafter, sort out the mails. Mails of resident guests should be directly delivered to their respective rooms. Mails of future guests should be sent to reservation section whereas past guest mails must be sent to given mail forwarding address. If guest has not given any mail forwarding address then send mail to his/her correspondence or business address which is given in guest registration card. Mail/ message of operational level hotel employees like steward, bellboy, cook, etc., should be directly sent to time office/security office whereas mails of supervisory/managerial level employees must be directly delivered to their respective offices as they may be related with hotel booking.

| Terms (with Chapter Exercise) | |
|---|---|
| Franking machine | A machine that marks the date and postage on letters so that the sender does not need to use stamps. It is mainly used when the volume of mails are high because it saves a lot of time. |

| House mails | Hotel mails, i.e., mails of hotel employees, including mails of operational level and executive level employees. |
| --- | --- |
| Registered mails | *Certified mails* that are recorded by the post office when sent and at each point on its route, so as to assure safe delivery. |
| Mail log book | If the registered/ensured envelope or parcel is received in an open stage, record it in the mail logbook and have it countersigned by the postman or delivery man. |
| Mail rack | Rack used to keep the received mails, particularly guest's mails. In large hotels, there is a separate section for handling mails. |
| Key rack | Manual front office equipment that is used to keep guest room keys (metal keys). In this chapter, it is used in lieu of keeping received mail of currently occupying guest because he/she is not present in the room. |
| Mail signature book | This book contains the relevant information in relation to received mails which are collected on behalf of the hotel guests' like guest name, type of mail (such as parcel or letter), date & time of receiving along with remark column. There is one more column for counter signature of guest after handing over the mail. |
| Future guest | Guest who has given his/her room booking but still hasn't arrived. In this chapter, it is used in lieu of attaching mails of such guests with his/her reservation form, afterwards with GRC (before a day of arrival). |
| Past guest | Guest who has already settled his/her account and left the hotel. In this chapter, it is used in lieu of transferring received mail of such guest at *mail forwarding address*. |
| Hotel mails | It refers to *house mails*. |
| GRC | It stands for *guest registration card*. In this chapter, this term is used in lieu of attaching mails of future guest with GRC. |
| Present guest | Guest who is currently staying in the hotel. In this chapter, it is used in lieu of transferring received mail into his/her room. |
| In-house-guest | *Resident guest* whereas the term out-house-guest refers to a *non-resident guest*. |
| Employee mails | It refers to both received mails as well as yet to send mails of hotel employees. |

# 14 CHAPTER

## Handling Situations

⟨ **OBJECTIVES** ⟩

*After reading this chapter, students will able to...*

- identify the various usual and unusual situations;
- explain the general guidelines for handling various usual and unusual situations;
- describe how probable skipper guests and guest with scanty baggage can be identified and what necessary steps are required to be taken;
- express the value of guest complaints and how they should be addressed in the first phase; and
- identify different types of guest complaints.

## 14.1 SITUATION HANDLING

The term *situation handling* refers to tactfully solving diverse kind of state-of-affairs and circumstances. Situation handling is an art as well as science. Because the response of every employee towards any situation is different. Some react typically, others react positively whereas some react negatively. This is because of differences among employees' personality trait and background in terms of age, gender, attitude, education, knowledge, experience, qualification, skill and so forth. Apart from it, employees' demographic, psychographic and geographic background also affects a lot to their situation handling attitude and aptitude. Consequently, it is the art of an individual employee of how much he/she is effective in handling any usual or unusual situation.

Irrespective of usual or unusual situation, every situation needs some meticulous dogma to handle which is inherent after so many years of work experience. Due to the applicability of meticulous dogma a critical situation can never be fully eliminated but it can curtail the degree of its adverse effect. For instance, it is always preferable to collect enough advance payment from *scanty baggage guest* to avoid future revenue loss probability. Similarly, a guest-with-non-guaranteed-reservation request should always give a cut-off-date with cut-off-time to give confirmation in order to avoid future room sale and revenue loss. Remember, if guest does not confirm his/her reservation till the cut-off-date/time, the room will be released for normal sale. Often this cut-off-date is also known as room *release date*.

## Activity

Fill in the below given spaces with appropriate variables that reflect the given parameters of guest bifurcation.

| S. No. | Demographic | Psychographic | Geographic | Socio-cultural | Personality traits | Behavioural | Environmental |
|--------|-------------|---------------|------------|----------------|--------------------|-------------|---------------|
| 1 | ? | Loyalty | ? | ? | ? | ? | Technology |
| 2 | ? | ? | ? | ? | Attitude & Aptitude | ? | ? |
| 3 | ? | ? | ? | Society | ? | ? | ? |
| 4 | Income | ? | ? | ? | ? | Collateral views | ? |
| 5 | ? | ? | Country | ? | ? | ? | ? |

## 14.2   USUAL SITUATION HANDLING

It includes customary/normal situations like walk-in, no-show, skipper, scanty baggage, walking, overbooking and stay-over. These are recurring in nature and have to be dealt efficiently and effectively. Even though one has handled these situations, they may need different treatment depending on the guest's temperament and other probabilities. A situation handler needs to be quick-witted, sympathetic and a good conversationalist. A basic guideline to deal with above-mentioned customary situations are as follows:

**Walk-in Guests** (Chance Guest, In-person or Off-street Guest)

These guests arrive at an accommodation establishment without reservation or prior information. So, rooms and reservation status should be verified before offering a room. Such guests help in increasing room revenue because hotel predominantly offers *rack rate* to these chance guests. It is good to take advance amount (or paid-in advance) from such guests unless they are regular.

If room is not available then front office personnel should "turn away" the guest. *Turn away* means to refuse the room request due to non-availability of rooms. Most of the hotels prepare a "refusal report" that shows the name of those guests who were refused accommodation due to non-availability of rooms. This report aids in room and revenue forecasting. A systematic approach to handle walk-in guests is as follows:

| Step 1 | • Greet and welcome the guest, thereafter check for room reservation status. If guest has no reservation then ask for relevant information (in order to accept or deny the booking request of guest) like length of stay, room preference, number of rooms required and so forth. |
|---|---|
| Step 2 | • On the basis of given information check whether requested room is available or not for the demanded period of stay. If room is available then offer it at rack rate, if room is not available then offer alternate room, even in some situation agent may suggest alternate property too. |
| Step 3 | • If guest agrees for alternate room suggestion then quote the room rate and room facilities. After guest's approval, proceed for registration formalities. |
| Step 4 | • Remember, it is usual to take substantial deposit from walk-in guest or block anticipated amount in his credit card, before allotting room. If guest is not satisfied with your alternate room suggestion, you may suggest alternate property. |

## Skippers (Walk-out guest or Runners)

Skippers are those guests or patrons who depart from an accommodation establishment without paying their bills. Thus, it is a usual practice to take advance from suspicious or unknown guests to avoid such revenue loss. And when this advance is exhausted, fresh advance should be collected. Care should be taken both at the time of offering the room as well as during his stay (especially by the housekeeping department). Room service as well as security department should also be vigilant for any such possibilities. There is no golden rule to deal with these kinds of guests but one may follow these two steps to reduce its probability:

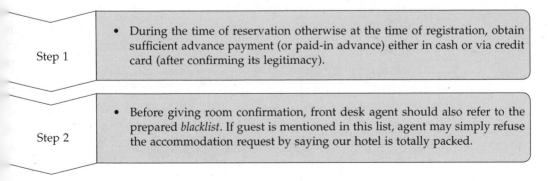

| Step 1 | • During the time of reservation otherwise at the time of registration, obtain sufficient advance payment (or paid-in advance) either in cash or via credit card (after confirming its legitimacy). |
|---|---|
| Step 2 | • Before giving room confirmation, front desk agent should also refer to the prepared *blacklist*. If guest is mentioned in this list, agent may simply refuse the accommodation request by saying our hotel is totally packed. |

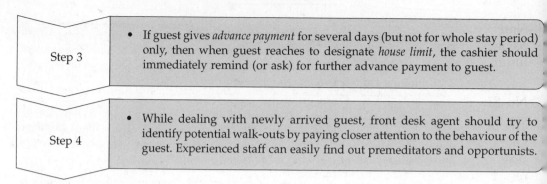

| Step 3 | • If guest gives *advance payment* for several days (but not for whole stay period) only, then when guest reaches to designate *house limit*, the cashier should immediately remind (or ask) for further advance payment to guest. |

| Step 4 | • While dealing with newly arrived guest, front desk agent should try to identify potential walk-outs by paying closer attention to the behaviour of the guest. Experienced staff can easily find out premeditators and opportunists. |

## Scanty Baggage (or light luggage)

It means a guest with light baggage/luggage. These types of guests are sometimes fraudulent. It's normally not a problem with regular guests or corporate guests. But it usually happens with unknown walk-ins or may also be a guest with reservation who looks suspicious. It is a good practice to take sufficient advance over the room rent for the number of days' stay. A guest with scanty baggage would have hand baggage or no baggage at all. Such guests are a hazard as they can slip out of the hotel without paying their bills. There is no way to determine whether a *scanty baggage guest* is walking out of the hotel with intention of not returning. Thus, most management stipulate a policy that scanty baggage guests are required to pay a deposit in advance, as a safeguard against skipping out of the hotel. There is no definite procedure to keep control on guests with scanty baggage. But one may use the following steps to minimize it.

| Step 1 | • After recognizing the scanty baggage, a bellboy should notify the lobby manager and the front desk staff as soon as a guest arrives/checks-in. |

| Step 2 | • Bell staff should stamp scanty baggage on the arrival errand card and front desk agent should also stamp scanty baggage on the registration card. |

| Step 3 | • Now, enter the particulars in the scanty baggage register. Bell staff should completely fill up all the columns in scanty baggage register. |

| Step 4 | • Get the registration card signed by the lobby manager who has a discretionary power to ask for an advance payment from the guest. Finally, get the scanty baggage register signed by the lobby manager. |

**Note.** Front desk agent or cashier should take an advance payment (as well as interim payment during regular intervals) from guests-with-scanty-baggage in order to avoid future loss of revenue.

## No-Show

The term no-show is used for those guests who fail to be present for a previously guaranteed room reservation. It may also happen with previously booked services/seats such as seat reservation in restaurants. To reduce the no-shows, front desk agent/reservation agent should ensure that the information taken during the reservation process should be accurate, in respect to date and time of arrival, mode of arrival, number of room blocks, cancellation charges, cut-off date and advance deposit. If possible contact by telephone or e-mail to check that guest is arriving or has changed his/her plan. A step-by-step process for reducing guest's no-show is given below.

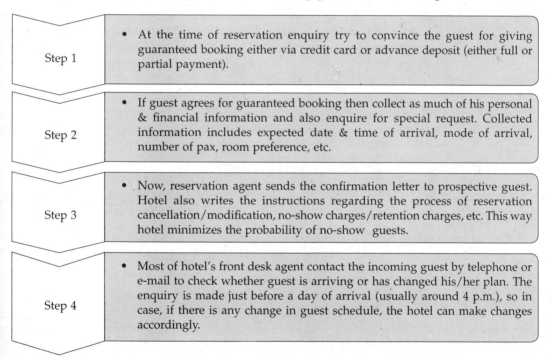

| Step 1 | • At the time of reservation enquiry try to convince the guest for giving guaranteed booking either via credit card or advance deposit (either full or partial payment). |
| --- | --- |
| Step 2 | • If guest agrees for guaranteed booking then collect as much of his personal & financial information and also enquire for special request. Collected information includes expected date & time of arrival, mode of arrival, number of pax, room preference, etc. |
| Step 3 | • Now, reservation agent sends the confirmation letter to prospective guest. Hotel also writes the instructions regarding the process of reservation cancellation/modification, no-show charges/retention charges, etc. This way hotel minimizes the probability of no-show guests. |
| Step 4 | • Most of hotel's front desk agent contact the incoming guest by telephone or e-mail to check whether guest is arriving or has changed his/her plan. The enquiry is made just before a day of arrival (usually around 4 p.m.), so in case, if there is any change in guest schedule, the hotel can make changes accordingly. |

Afterwards, reservation agent should send the confirmation letter to the guest. On the back of confirmation letter, hotels may write down their instructions regarding the process of reservation cancellation, amendments, no-show charges and so forth.

**Walking Guests** (Farm Out, Bounced Reservation or Walking Reservation)

Walking-off guest means relocating the guest into another hotel, even when he/she holds confirmed or guaranteed reservation, as property cannot accommodate him/her because of lack of available vacant rooms. If room status and management policy permits, front desk agent can also opt for upgraded room before diverting guest to alternate hotel of same/better standard. For diverting confirmed and guaranteed reservations, hotel should also arrange a vehicle for guest to reach the alternate hotel. Several other steps could also be taken by the hotel property to avoid losing of these guests in near future. But a basic procedure for dealing with walking guest includes:

| Step 1 | • After welcoming these kinds of guests, first ask the guest to wait for a moment. Meanwhile inform your manager and let the manager decide whether guest can be accomodated or not. |
| --- | --- |
| Step 2 | • If guest cannot be roomed then politely inform the guest about the situation and take permission to relocate him/her in another hotel of same standard. |
| Step 3 | • Now, call other properties to check room availability. If room is available then arrange a taxi to send the guest to an alternate property. You must encourage the guest to return the next day. |
| Step 4 | • While valet attendant arranges taxi for guest, meanwhile, a front desk agent must check for guest's mail & message. If there is any mail, message or parcel then hand it over to guest before he/she leaves. |
| Step 5 | • Now, inform your telephone operator of the alternate property right away and also forward a message to the alternate property. |
| Step 6 | • If guest returns the next day, ask your supervisor to provide an upgraded accommodation (if possible) to the guest or give him/her best available room for the remaining stay. |
| Step 7 | • If possible then tell your manager to welcome the guest. Your manager may also send a letter of apology or a gift basket (complementary) when the guest returns. |

**Note**

✓ Walking-out guest means turning away (or *pledge relocates*) the guest to another property even after holding confirmed room reservation. This problem mainly arises due to wrong overbooking forecasting, full occupancy, rooms suddenly going into out-of-order condition, extension in stay of resident guests, etc.

✓ In case, if guest agrees to stay in alternate hotel of same standard, all extra expenses like any increased room charge at an alternate hotel will be abided by the booked hotel itself, as inconvenience to guest due to hotel's fault.

✓ In majority of cases, the taxi fair and one day/night accommodation charges (of alternate hotel) is paid by the hotel itself on behalf of walking guest.

| **Activity** | | |
|---|---|---|

Differentiate between bounced and bumping reservation:

| S. No. | Bounced Reservation | Bumping Reservation |
|---|---|---|
| 1 | ? | ? |
| 2 | It refers to *farm out reservation* | ? |
| 3 | ? | ? |
| 4 | ? | It refers to *overbooking* |
| 5 | ? | ? |

**Stayover** (guest stay before EDD/ETD) **and Overstay** (guest stay after EDD/ETD) **Guests**

Stayover is room occupancy status which indicates that the guest is not departing the present day. A guest who was expected to check-out but now he/she wishes to stay for some more days is called as *stayover guest*. If the room and reservation status does not permit, it could be handled alike to *walking of guest*. A step-by-step procedure for dealing with *stayover* guest is given below:

| Step 1 | • Front desk agent should confirm (or intimate) with guest about his/her previously given expected date/time of departure, just a day before (or 12 hours prior). |
|---|---|
| Step 2 | • If guest wants to stay for some more days (or at least one more night) then front desk agent must check the room status first to confirm whether room will be available or not. |
| Step 3 | • If room position allows then its fine otherwise next confirmed or guaranteed booking of the same room can relocate into alternate room or in alternate hotel but it must be first confirmed to the prospective guest. |

The term *overstay* is different from stayover. *Overstay* is related with those guests who are staying beyond their declared date of departure. A step-by-step procedure for dealing with overstay guests is given below:

| Step 1 | • Front desk agent should remember that overstay guests may be guest with confirmed reservation, non-confirmed reservation or walk-in. So, deal with all overstay guests in a similar demeanour. |
|---|---|
| Step 2 | • Overstay problem can be minimized by verifying the date of departure to guest during check-in plus also intimate to guest whether hotel allows extension of stay or not. If allows then in what manner, especially when the hotel is at or near to full occupancy and has no provision for such overstays. |
| Step 3 | • If problem still exists then front desk agent should go for room change option or relocation of new arrivals, as situation and guest permit. |

**Note**

✓ If a hotel relocates the resident guest in other room then the entire process will be identical to *room changing process* which is given later in this unit.

✓ On the other side, if there is no room in the hotel (for adjusting overstays) then hotel should try to relocate the resident guest in another hotel (of same standard). The procedure for relocation, here, will be same as in *walking reservation*.

✓ Overstay may prove drastic during room block or group arrival, as it may create huge revenue loss. Overstay is also problematic during suite room or similar bookings, as these arriving guests can be VIP, HWC, etc.

---

**Difference between overstay and stayover**

*Overstay* is a term that is given to a guest who has extended his or her stay at the accommodation establishment after the scheduled date of departure while *stayover* refers to room status term indicating that the guest is not checking-out today and will remain at least one more day. Simply, stayover refers to the situation where guest extends his/her previous stated duration of stay.

---

## Late Cancellation

Late cancellation means a cancellation of a previously given room reservation in an accommodation establishment that occurs after a specified time (i.e., cut-off-time/date). It depends on the circumstances or hotel's policy whether a fee should be charged or not. A step-by-step process for late cancellation is given below:

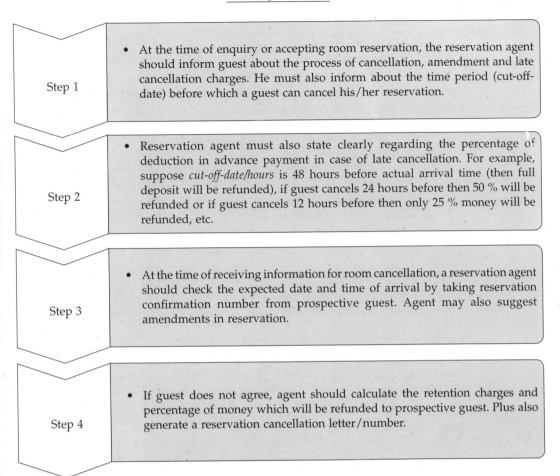

**Step 1**
- At the time of enquiry or accepting room reservation, the reservation agent should inform guest about the process of cancellation, amendment and late cancellation charges. He must also inform about the time period (cut-off-date) before which a guest can cancel his/her reservation.

**Step 2**
- Reservation agent must also state clearly regarding the percentage of deduction in advance payment in case of late cancellation. For example, suppose *cut-off-date/hours* is 48 hours before actual arrival time (then full deposit will be refunded), if guest cancels 24 hours before then 50 % will be refunded or if guest cancels 12 hours before then only 25 % money will be refunded, etc.

**Step 3**
- At the time of receiving information for room cancellation, a reservation agent should check the expected date and time of arrival by taking reservation confirmation number from prospective guest. Agent may also suggest amendments in reservation.

**Step 4**
- If guest does not agree, agent should calculate the retention charges and percentage of money which will be refunded to prospective guest. Plus also generate a reservation cancellation letter/number.

**Note**
- ✓ The time period before which a group or a person can cancel the guaranteed reservation without any deduction (as *advance deposit* received) is called "cut-off date".
- ✓ Cancellation letter/number protects both prospective guest and hotel. Because in the event of any misapprehension from the guest's side, it can prove that hotel releases the reservation upon receiving cancellation request.
- ✓ Similarly, it also relieves the guest of an obligation to pay for any retention charges (in case of timely cancellation). Often, it also demands credit card company to cancel the credit card based guaranteed reservation.
- ✓ Cancellation letter/number also prevents disputing situation which may arise between prospective guest and hotel for no-show billing.

## Late-show

*Late-show* means a guest who arrives in an accommodation property after the designated check-in time. If guest has confirmed or guaranteed reservation (or positive space) or he/she has given the *advance payment* for the hotel room then hotel usually does not offer the blocked room to another guest. A step-by-step process for late show is given below.

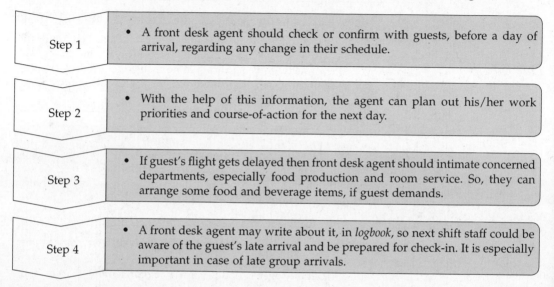

| Step 1 | • A front desk agent should check or confirm with guests, before a day of arrival, regarding any change in their schedule. |
| --- | --- |
| Step 2 | • With the help of this information, the agent can plan out his/her work priorities and course-of-action for the next day. |
| Step 3 | • If guest's flight gets delayed then front desk agent should intimate concerned departments, especially food production and room service. So, they can arrange some food and beverage items, if guest demands. |
| Step 4 | • A front desk agent may write about it, in *logbook*, so next shift staff could be aware of the guest's late arrival and be prepared for check-in. It is especially important in case of late group arrivals. |

**Note**
- ✓ This problem usually arises during the winter season when many flights and trains get late due to fog. But in case of guaranteed booking (with no intimation of cancellation), a hotel is required to hold the room up-to-a-day or more.
- ✓ The information regarding late arrivals should also be given to the front desk and bell desk staff who are responsible for working in night shift/late shift.

## Late Check-Out

Late check-out refers to the situation where guests check out after the hotel's established check-out time (in majority of hotels, the usual check-in & check-out time is same, i.e., 12 noon). Late check-out shall not be considered a guest right, rather a privilege which might be honoured by the front office manager upon room availability. It also depends on the hotel policy whether property charges fee for late check-out or offers at free-of-charge (number of hour/s also affects it). This situation arises in two ways:

### Late Check-Out (Individual Guest)

- ✓ But sometimes guest may request for departing from the room after the designated check-out time. In this situation, first of all front desk agent should check whether room position allows it or not. If situation permits then decide whether a late check-out charge is to be taken from the guests or not. Most of the establishments mention about the check-in and check-out time plus extra charges for late check-out on the back side of *room key cards* as well as in the information brochure kept in the hotel guest rooms.

✓ To avoid confusion between guests and hotel, guests are often asked about their expected check-out time at the time of check-in. At this time, front desk agent informs the guest about the extra charges which could be levied for late check-outs.

## *Late Check-Out (Group Guests)*

✓ In case of group traveller, it is a very typical situation because if hotel takes late check-out charges then it is possible that group does not want to come again due to such petty expenses. In such situations, hotel stores the luggage of all group guests in left luggage room or in concierge section and allot one or more hospitality room/s at no extra charges. This type of alternate solution gives opportunity to hotel to generate revenue by selling room at the right time as well as satisfy the departing group by providing timely alternate room at no extra charges.

✓ As we learnt, the instructions in relation to late check-out has been written on the back side of *key card* as well as stated in information folder kept in the guest room. After having all these alternatives, still a problem can arise in account settlement at the time of check-out and we can't force or blame the guest for it because "guest is always right". So, a receptionist can do the following:

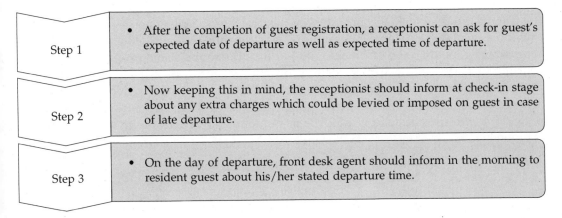

| Step 1 | • After the completion of guest registration, a receptionist can ask for guest's expected date of departure as well as expected time of departure. |
| --- | --- |
| Step 2 | • Now keeping this in mind, the receptionist should inform at check-in stage about any extra charges which could be levied or imposed on guest in case of late departure. |
| Step 3 | • On the day of departure, front desk agent should inform in the morning to resident guest about his/her stated departure time. |

## Early Check-In

It means guest arrives before the given arrival time. In simple terms, hotel allows guest to access a hotel room prior to the normal check-in time (or prior to guest stated time), which is called *early check-in* or *early arrival*. It is important that every staff member of front desk should be aware that there is a guest who is waiting for a room. In case of early check-in, the receptionist should choose any alternate (step-by-step) from the below mentioned options.

| Step 1 | • Receptionist should check for alternate type of available room. If alternate room is not available, receptionist should inform (or apologise) to the guest. He/she also needs to tell estimated time by which room will be ready. |
| Step 2 | • The luggage should be kept in luggage room, so guest will be free from the tension of luggage and can roam anywhere in the hotel. |
| Step 3 | • If guests want to go anywhere then front desk agent asks for the time when they would return to collect their room key or where they may be contacted, as soon as, the room ready. |
| Step 4 | • Give information to the housekeeping department about the presence or early check-in of the guests. |
| Step 5 | • Complete the registration work without taking guest's signature. Signature will be taken after allotting the room. |

**Under-Stay** (Premature departure or early departure)

It means a guest wants to check-out or vacate the occupied hotel room before the previously stated date/time. In simple terms, it means that a booking has been made for a longer time than the guest actually requires the room. Some properties also call it "curtailment". In peak season, usually property prevents the early check-out, as it is a part of high demand yield management tactics. Consequently, early check-out guests are allowed to depart after the payment (either in full or partial, depending on the hotel's policy) for whole requested period. A basic guideline how to deal with under-stay guest is given below.

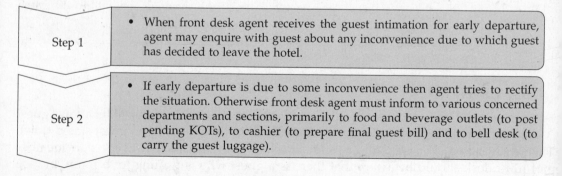

| Step 1 | • When front desk agent receives the guest intimation for early departure, agent may enquire with guest about any inconvenience due to which guest has decided to leave the hotel. |
| Step 2 | • If early departure is due to some inconvenience then agent tries to rectify the situation. Otherwise front desk agent must inform to various concerned departments and sections, primarily to food and beverage outlets (to post pending KOTs), to cashier (to prepare final guest bill) and to bell desk (to carry the guest luggage). |

**Note**

- Most of the hotels allow their guests to depart early without deducting any further charges or full amount. But some hotels take full payment.
- Apart from cashier desk and bell desk, front desk should also inform all other relevant sections/departments. For example, to reservation (to take new booking), to housekeeping (to make room ready), to food & beverage (like room service, restaurant, bar, etc.) outlet for transferring outstanding guest's account balances.

## Late Charges

These are the charges which come into notice after guest departure, as these charges are ascertained after guest's check-out, thus called as *late charges*. There is always the possibility that a guest will leave without settling his/her account or some guests may honestly forget to settle their accounts. Some guests may intentionally leave the hotel without settling their accounts; such guests are known as *skippers*. Regardless of the reason, after-departure charges or balances represent as *unpaid account balances*. Late charges may be a major concern in guest account settlement. A *late charge* is a transaction requiring posting into guest account that does not reach the front office until after the guest has checked-out and closed the account, like, restaurant, telephone & room service charges are examples of potential late charges. Even if late charges are eventually paid, the hotel incurs the additional costs involved in billing the guest.

Sometimes, the extra expenses for labour, postage, stationery and special statement forms may total more than the amount of the late charge. Few hotels (especially large hotels) can afford a large volume of late charges. But in prolongation, these late charges can adversely affect the hotel's profitability. Front desk can take several steps to reduce the occurrence of late charges; such as given below.

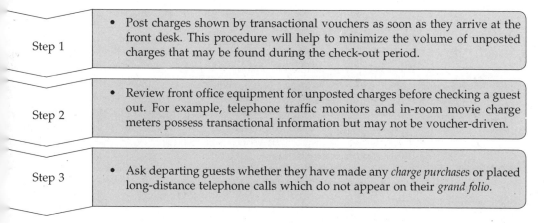

| Step 1 | • Post charges shown by transactional vouchers as soon as they arrive at the front desk. This procedure will help to minimize the volume of unposted charges that may be found during the check-out period. |
| Step 2 | • Review front office equipment for unposted charges before checking a guest out. For example, telephone traffic monitors and in-room movie charge meters possess transactional information but may not be voucher-driven. |
| Step 3 | • Ask departing guests whether they have made any *charge purchases* or placed long-distance telephone calls which do not appear on their *grand folio*. |

The problem of late charges also occurs when hotels provide *hospitality room* to late departures. In case of late check-out, it utterly depends on the hotel policy whether to charge or not. But in majority of cases, hotel tries to avoid these late departure charges even when guest wants to remain for a few additional hours in hotel. Whereas in certain other cases, for instance, in peak-season a hotel may again allow the guest to stay for one or two more hours after designated/standard departure time but on a chargeable basis.

**Note**

✓ When a late charge is processed (after guest departure), a letter is forwarded to the guest together with a copy of the credit card voucher. If the guest has already paid by any means other than the credit card, a payment request letter for the items or services is sent.

✓ This problem (late charges) does not exist in computerized front office/hotel because, here, all charge privileges are transferred into respective guest accounts as soon as they occur.

## Extra Charges (Like VPO's)

Visitors Paid Out is also called as "Cash Advance, Cash Paid Out or Guest Disbursement". It refers to cash payments made on behalf of the guests or the management for any external services rendered to them. Cash Advances/Paid-outs are debit transactions since they increase a folio's outstanding balance. A step-by-step process to handle it is given below.

| Step 1 | • Concierge or information agent must assist guests in selecting tour or destination by recommending various options (on guest demand) with the help of information directory and other informatory material. |
|--------|---|
| Step 2 | • During the conversation, information agent must obtain sufficient information from guest about the requested services. For instance, how much time and money guest wants to spend on a tour or destination. |
| Step 3 | • On the basis of guest's response, arrange the tour or ticket for destination via travel agent or middlemen. In these situations, hotel raises a VPO (visitor's paid out vouchers). |
| Step 4 | • Concierge should obtain guest's signature on VPO then transfer the voucher to cashier, so he can further transfer (or debit) it into the respective guest account or master account. |

**Note**

✓ Extra charges/visitor paid outs do not include only the charges of arrangement of taxi and theatre ticket but also includes other expenses like porter charge, emergency medical expenses, ticket confirmation charges, floral delivery, etc.

✓ The payment on behalf of guest is made from the shift cashier's *cash bank* which is received at the beginning of the shift from the head cashier. Be careful that paid outs are only made in local currency.

## Drunken Guests

A drunken guest may be your resident guest as well as walk-in guest. The basic guideline how to deal with a drunken guest is same in both situations is given below.

| Step 1 | • Remember, during check-in process if guest is found in drunken state then front desk agent should avoid to register him as a guest whether he is a chance guest or guest-with-reservation. |
| Step 2 | • Drunken guest should be taken away from the lobby as early as possible but be careful not to irritate/offend him. |
| Step 3 | • If it is difficult for you to handle a drunken guest (as he is in public area), inform your lobby manager and let him handle the guest. |
| Step 4 | • It is always preferable to take the drunken guest to the back area or directly to his room (if resident guest) and if he does not behave well, the hotel security must be called. |

### Activity

Fill in the below given blank spaces with appropriate answers.

| S. No. | Questions | Answers |
|--------|-----------|---------|
| 1 | The guest will stay for the night. | ? |
| 2 | Front desk cashier returns remaining cash balance from advance payment to checking-out guest, and needs to raise voucher. | ? |
| 3 | The guest has settled his/her account, returned the room keys, and left the hotel. | ? |
| 4 | The guest is allowed to check-out later than the hotel's standard check-out time. | ? |

| | | |
|---|---|---|
| 5 | Guest holds confirmed or guaranteed reservation but property transfers him into alternate hotel. | ? |
| 6 | Room is occupied by guest, but has not been checked into the computer. | ? |
| 7 | A guest checking-in with light luggage. | ? |
| 8 | The room has been locked so that guest cannot re-enter until he/she is allowed by a hotel official. | ? |
| 9 | A guest does not hold any advance booking with hotel but now requests for room. | ? |
| 10 | Room has been vacated and guest has left without settling the bill. | ? |
| 11 | The room is occupied, but the guest is assessed no charge for its use. | ? |
| 12 | A guest holds guaranteed reservation but did not arrive. | ? |
| 13 | A guest room status indicates that guest will not check-out today. | ? |
| 14 | A guest is requesting to leave the hotel room before previously given expected date of departure. | ? |
| 15 | A guest who was expected to check-out today but now wishes to stay for some more days. | ? |

## 14.3 UNUSUAL SITUATION HANDLING

### Lost & Found

The hotel's reputation is at stake when the guest's lost property is not traced and handed over to the guest. Guest's article may be lost/found in two ways: first, when hotel itself found lost article and secondly when guest informs about his/her lost article to front desk. In both cases, when the guest's lost article will be found by any housekeeping staff, they have to immediately inform the front office department. The found article may be in the form of perishable, non-perishable and valuables. There is stipulated time period allowed to guest for the collection of these found articles like 24 hours for the perishable items, 6 months for the non-perishable and valuable items. This duration may also vary as it usually depends on the individual hotel's policy. A step-by-step process for handling lost & found article for room boy or the one who found the article is given below:

| | |
|---|---|
| Step 1 | • The found article should be handed over to the duty manager or assistant front office manager immediately. |
| Step 2 | • A lost & found form has to be filled-in by the person who has found a lost article and a copy of it will be given to him/her for reference. |
| Step 3 | • Assistant front office manager tries to get the contact information of the departed guest (from whose room article was found), e.g., mobile phone numbers, office telephone numbers & email address from business card. |
| Step 4 | • The assistant front office manager requires filling in the *Lost & Found Record Sheet* for record purpose as well as enter the details in the "logbook" for management's information. |
| Step 5 | • The assistant front office manager should check all lost & found items on daily basis to ensure all items match the lost & found record. |

## Activity

Organize a group session and discuss the below given scenario, thereafter collect the views of students in order to rectify them and tell them what & where they are not clear which may make the situation adverse.

Scenario - A guest enquires about a missing article in his room, but the lost and found logbook has no record of it. How would you handle this situation if you are the assistant front office manager?

## Safe Deposit Box

Nowadays, almost every hotel offers safe deposit box (SDB) facility to its resident guests. This facility is usually located nearby the cashier's office or behind the front desk. An individual guest may have the exclusive use of a safe box to deposit any valuables such as airline tickets, jewellery, ornaments, passports, foreign currency and traveller's cheques. A systematic approach to handle safe deposit box provision at front desk includes following steps.

**Safe Deposit Options**

- Safe deposit box
- In-room safe box

| Step 1 | • Front desk should intimate to guest about the availability of SDB provision during the check-in process. After receiving the request for SDB, front desk agent/cashier should arrange then dole out SDB to the requested guest. |
| --- | --- |
| Step 2 | • Now guest will place articles in the SDB. Next he/she locks the SDB and retain its key. Now front desk agent should give instructions in relation to SDB & take guest's signature on SDB card and issue SDB receipt. |
| Step 3 | • Place SDB card with the registration card or in a file box at front desk. Allow only same guest to access the SDB and also record each access on the SDB card and obtain guest's signature on each access. |
| Step 4 | • During each access, make sure that the guest's signature match with the previous signature. Front desk agent should also record the date & time on the SDB card as well as his/her own signature. |
| Step 5 | • If guest wants the SDB facility again, repeat the whole process. During final stage, record the final guest access on the SDB card and handover the SDB to the guest. |
| Step 6 | • Make sure that guest removes all the contents. Lastly, thank the guest for using SDB facility and keep the SDB card for at least six months. |

## Black Listed Guest

Black list (a list of *unwanted guests*) is a record authorized by hotel management of those persons who are not allowed to access/welcomed into the hotel due to various reasons like –

**Source of Blacklist**

- Police record
- Local hotels
- Corporate offices
- Logbook
- Credit manager

- People have a criminal background, thus recorded in police hit list/blacklist.
- People involved in fraud or cheating cases, in accordance to police hit list.
- May be any previous hotel guest who left without paying his bill.
- May be any guest who was blacklisted by other hotels of same area.
- The guest's representing company whose previous account balance was bad debited (due to non-settlement of account).

The list of undesirable people is usually prepared by the top level hotel management people who then circulate this blacklist to the front office and other concerned departments. It is treated as a highly confidential matter, thus never show to any unauthorized person. The blacklist can be compiled by using information from local hotels, corporate offices, logbook, credit manager and so forth. A standard approach to handle unwanted guest at front desk includes following steps:

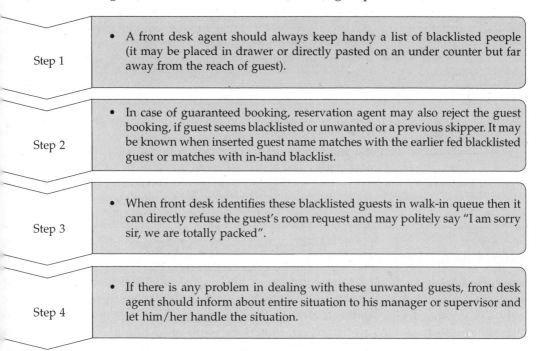

**Step 1**
- A front desk agent should always keep handy a list of blacklisted people (it may be placed in drawer or directly pasted on an under counter but far away from the reach of guest).

**Step 2**
- In case of guaranteed booking, reservation agent may also reject the guest booking, if guest seems blacklisted or unwanted or a previous skipper. It may be known when inserted guest name matches with the earlier fed blacklisted guest or matches with in-hand blacklist.

**Step 3**
- When front desk identifies these blacklisted guests in walk-in queue then it can directly refuse the guest's room request and may politely say "I am sorry sir, we are totally packed".

**Step 4**
- If there is any problem in dealing with these unwanted guests, front desk agent should inform about entire situation to his manager or supervisor and let him/her handle the situation.

### Activity

✓ What are the differences between a hit-listed and a blacklisted guest? Enlist any 5 differences and also suggest probable ways to handle them.

## Paging

Paging is finding out or locating the unseen guest by the hotel personnel. It refers to system whereby a guest identifies himself in such cases where the hotel staff cannot identify him personally. There are two systems of paging: manual and mechanical system of paging. In *manual system*, paging board with marker and portable/attached bell is used, whereas in *mechanical system* electronic pagers and other communication devices like mobiles are used to locate the guest. This paging facility may be adopted to locate a resident guest (the method is known as *in-house paging*) as well as non-resident guest (the method is known as *out-house paging*).

| Steps | In-house paging (concerned with paging/locating of resident guest) | Out-house paging (concerned with paging/locating of non-resident guest) |
|---|---|---|
| Step 1 | A front desk agent may call from reception and inform the resident guest that someone has come to meet him. In certain situations, a page boy may also be sent to search for the guest. Nowadays, hotel uses *location slip* to locate the resident guest. | A page boy should make the page board ready (with guest's name, hotel name or both) and timely (or 15 minutes before) reach the airport or railway station. He must stay at exit point or terminal (at airport) or nearby the given coach number at railway station. |
| Step 2 | Generally, page boy uses a bleep system (and announces the guest's name) to reach out to the guest and when guest is found, page boy informs the guest about the matter. | When page boy meets the guest, he and airport representative must greet and welcome the guest then escort him/her towards the hotel pick-up vehicle (a page boy should carry the guest luggage and airport representative escort him & assist in custom formalities). |

## Room Change Process

If a guest wishes to change room, and his or her arrival has already been entered on the arrival list, the change of room must be treated as a departure from one room and arrival in another room. As the change of room will have no effect on *house-count* (the number of guests staying in the hotel), thus there is no need to make separate arrival & departure list or change in house-count status. If the arrival and departure list has already been circulated, a change of room notification slip (or transfer slip) must be sent to all departments in order to inform them of the room change. A bellboy/room boy can transfer guest luggage (or room change) either in front-of-guest (called as *live move*) or in the absence-of-guest (called as *dead move*). A step-by-step procedure for both situation is given below. But remember, room change differs from *room swap*.

| Steps | Dead Move (Room change process in absence of guest) Teamwork of room boy and bellboy | Live Move (Room change process in presence of guest) Bellboy will perform this task alone |
|---|---|---|
| Step 1 | Before entering, room boy knocks then opens the door lock (with his/her *section key*) and makes sure that guest is not present in the room. | Before entering, bellboy knocks at the door and when guest opens the door, he should greet the guest and take permission to enter the room for luggage assistance. |

| Step 2 | Room boy will systematically pack all guest articles/luggage, if guest has not packed. | If guest requests, bellboy should assist in packing of guest articles/luggage. |
|---|---|---|
| Step 3 | Now bellboy takes out guest luggage and then locks the guest room door. | Bellboy takes out guest luggage and then locks the door. Finally, escort guest towards new allotted room. |
| Step 4 | Bellboy knocks (to ensure there is no one in the room) then opens the door of new allotted room. | Bellboy knocks (to ensure there is no one in the room) then opens the door of new allotted room. |
| Step 5 | Switch *on* the light and place guest luggage at appropriate place (like into rack/cupboard). | Switch *on* the light and allow the guest to enter first into the room. Thereafter, bellboy will enter and place the luggage at appropriate place (into rack/cupboard). |
| Step 6 | Lastly, bellboy will close the door, lock it and return the room key to the reception desk. | Now, bellboy will greet the guest and handover the key of new allotted room. Lastly, return the room key of vacated room to the reception desk. |

### Activity

Differentiate between room change and room swap.

| S. No. | Room Change (it means allotment of new room to resident guest) | Room Swap (it means room interchanges between guests) |
|---|---|---|
| 1 | ? | ? |
| 2 | ? | Room swap depends on guests' mutual understanding |
| 3 | Room change depends on new room availability | ? |
| 4 | ? | ? |

## 14.4 GUEST COMPLAINTS AND THEIR TYPES

"Complaints are opportunities to improve products/services."

Guest complaints are negative feedback about the offered products/services. But this complaint gives an opportunity to the hotel to improve itself as well as serve better to their guests. Therefore, complaints are like quality guarantees which direct management and other supervisory level staff towards the troubled areas.

Guest relation staff are mainly deployed (or empowered) to systematically identify and handle the most frequent guest's problems and complaints. By reviewing a properly kept front desk logbook, management can often identify & address recurring complaints and problems. Another way to identify complaints involves the evaluation of room history record, guest comment cards or questionnaires. Guest questionnaires may be distributed at the front desk, placed conspicuously in the guest room, or mailed to guests following departure. Guests can complain about anything that causes inconvenience to them. Broadly, guest complaints can be classified into four parts:

| Attitudinal Complaints | • Complaints that usually occur due to guest's behaviour. Simply, these complaints arise due to guest's outlook or mindset. For example, guest complaints after overhearing the staffs' personal conversation.<br>• In this kind of situation, a hotel manager should listen and attend to the guest complaint and respond positively. |
|---|---|
| Mechanical Complaints | • Complaint in relation to hotel equipment that are not working properly. These complaints may be in relation to in-room (like door lock, HVAC, telephone & television) as well as public area equipment (like elevators & vending machines).<br>• After receiving complaint, reception desk must immediately inform engineering & maintenance department then enter into work order book. |
| Service Related Complaints | • Complaint in relation to hotel services like poor room service/restaurant service, lack of assistance in luggage handling, missed wake-up call, etc.<br>• These complaints are usually received during peak periods. Thus, front desk agent should carefully listen to guest complaint, thereafter take appropriate action. |
| Unusual Complaints | • Complaints that are difficult to sort out by the front desk staff, like complaint related with absence of swimming pool, night club, building construction, road conditions, weather, lack of public transport, etc.<br>• Here, front office staff is required to carefully listen to guest and try to cope up with the situation (but never demarket the hotel's services and facilities). |

## Complaint Handling Process

The front desk may receive complaints about anything, it may be related to mechanical equipment, food & beverage operations in the hotel, regardless of whether those operations are managed by the hotel. Unless the hotel and the food & beverage operators establish procedures for referring complaints, guests may continue to be upset and the hotel will continue to receive the blame. The hotel & its *revenue outlets* should maintain close communication and develop procedures in order to satisfactorily resolve guests' complaints. A standard guideline to resolve guest complaint includes the following steps.

**Step 1**
- Avoid conflicts with the guest, as it may spoil hotel's goodwill as well as loss of customer that ultimately affect the hotel's revenue. Thus, never argue with guests.

**Step 2**
- Listen carefully (without interrupting him/her) to the guest complaint in order to understand why guest is angry or unsatisfied. Remember, 50% of guest's problem can be sorted out by careful listening.

**Step 3**
- During conversation, show regret towards the guest as it creates a positive impact on the guest's mindset (because it shows that you are serious and taking favour of guest).

**Step 4**
- Do not justify who is wrong and who is right because it may create a critical situation. You have to keep yourself calm and remember the famous caption *guest is always right* even when you know he is wrong.

**Step 5**
- Clarify the matter by asking questions in order to get an idea of what actually happened with the guest due to which he/she is hurt/upset.

**Step 6**
- Sort out the matter as soon as possible and make sure that you agree with the guest and take appropriate action as quickly as possible.

**Step 7**
- Even if the complaint was resolved by someone else, you should contact the guest and ensure that the problem was satisfactorily solved or not.

**Step 8**
- Enter every complaint (with detail of situation) into *logbook*. Thereafter, report the entire event, the actions taken and the resultant conclusion of the incident to the management or reporting authority.

## CONCLUSION

Hotel industry is a more *service oriented* than *product oriented* industry. Therefore, it is impossible to imagine that critical and difficult to handle situations will not emerge. Some of these situations are customary thus termed *usual situations* whereas some rarely occur, termed *unusual situations*. Securing paid-in advance from walk-ins & scanty baggage guests and allow (or not to allow) overstay and so forth situations are a few perfect examples of usual situations whereas dealing with blacklisted, drunken and under-stay (during peak season) guests are some excellent examples of unusual situations. These situations add value to experience of indulged employees too. Apart from situations, guest complaints are usual in hospitality industry. Guest complaints should always be accepted positively because it supports in recuperating the product and service, consequently modified/upgraded product/service can meet or exceed the expectation level of guests. Remember, the organization which claims that they do not have any guest complaints, these organizations have weak CRS therefore complaints are not reaching to the top level.

| Terms (with Chapter Exercise) | |
|---|---|
| Early arrivals | It is related with all those guests who arrive into hotel before their designated arrival time. |
| Skipper | ? |
| ? | It refers to the resident guest does not expect to check-out today (as stated earlier) and will remain at least one more night. |
| Overstay | It indicates to a guest who has extended his/her stay of duration or is not checking out on the previously stated date. |
| ? | *Charge purchase* made by guest who has already settled his/her grand folio but afterwards consume additional services on charge basis. |
| Unpaid account | ? |
| No-show | ? |
| ? | It refers to *walking/farm out* reservation in which guest is directed to alternate property/ accommodation due to non-availability of room even when he/she holds guaranteed reservation. |
| Release date | It refers to the cut-off-date when all temporarily reserved rooms are released for normal sale. |

| Pledge relocates | It means rooms for guests housed at another property, but paid for by the hotel, due to walking reservation (or as guaranteed reservations are powered by advance deposit) of that guest. |
|---|---|
| Skipper | Guests who check-out from the property without settling his/her account. |
| ? | Guest who has no-luggage or light luggage. |
| Blacklist | A list of those guests who are not allowed to register in a hotel; it may be due to criminal police record, a skipper, etc. |
| ? | A list of companies to which (or to employees of these companies) hotel should not entertain due to previous bad experience in terms of settlement of account, delay in payment, etc. |
| ? | A situation whereby hotel diverts its guest to alternate property due to non-availability of accommodation. Simply, this term is used for a guest who is transferred to alternate lodging. |
| Walk-in | It is also known as *chance guest*. Walk-in refers to all those guests who approach the property without any prior reservation. |
| Live move | ? |
| ? | Room shifting in the absence of guest. |
| Late check-out | ? |
| ? | It is also known as *early departure/un-expected departure* and concerned with all those guests who depart from hotel before their designated check-out time. |
| Location slip | A *slip* on which guest intimates that at which specific location he/she can be contacted when he/she is not available in the room. It is also known as *locator message*. |
| Outstanding balance | It is also known as *unpaid account balance* which represents late charges or after-departure charges or balance (or balance remaining after closing account or amount yet to pay). |

# 15 CHAPTER

## Guest Registration

**OBJECTIVES**

*After reading this chapter, students will be able to...*

- describe the process of registration for different guests; and
- explain different methods of registration and their pros and cons.

## 15.1 GUEST REGISTRATION

The term *guest registration* is a legal agreement between a guest and hotel in support of boarding and lodging for a specified period of time. Generally, the registration process differs for the different categories of guests like registration of guest-with-reservation is different from chance guest; group guest registration is different from transient guest registration. Similarly, registration of alien guest (or foreigners) is different from Indian guest. In the same way, VIP guests and group guests are mostly pre-registered whereas it is not possible with a chance guest.

The registration procedure takes maximum time during a *chance guest* (or walk-in guest) because hotel cannot do the pre-registration of him. Inevitably, the front desk agent will have to find out his requirement before assigning and allocating the room to match his need, expectation and room preference. In case of walk-in guests, the registration records are created during the time of check-in at front desk; not by the reservation agent as in case of a guest-with-reservation. Remember, the guest-with-reservation simplifies the registration procedure (as he/she is often pre-registered) but it doesn't mean walk-in client makes registration critical, else it increases the registration process. A step-by-step standard process for registration is given below.

| | |
|---|---|
| Stage 1 | • First of all front desk agent should refer, as a part of opening duty, to *movement list* received from reservation section and pay special attention for VIP/CIP/MIP/SPATT guest. Perform/ensure pre-registration (possible with only guest-with-reservation and also perform *pre-key process* for group arrival). |
| Stage 2 | • Greet and welcome guests (if guest is known then address him/her with title/name otherwise with sir/madam) and confirm about reservation status (confirm whether guest holds advance room reservation or walk-in then proceed further). |
| Stage 3 | • Reconfirm all available financial and personal information (via pre-registration card) with guest-with-reservation. Whereas start from the initial stage of room enquiry with walk-in guest and when guest agrees then proceed towards room assignment (on the basis of guest's room preference). |
| Stage 4 | • Perform guest registration (it can be in-room, at front desk, at self-registration terminal or at satellite check-in counter). Now convert *non-guest account* into *guest account* for guest-with-reservation and open respective guest-account (*incidental account* for individuals and *master account* for group) for walk-in. |
| Stage 5 | • Confirm mode of payment (it means whether guest wants to settle his account via cash payment, charge card, traveller's cheque, travel agent voucher, deferred payment or through any other negotiable instrument) then obtain *paid-in-advance* or tag/adjust received advance deposit to guest account. |
| Stage 6 | • Next issue room key and call bell desk for *rooming* the guest (handover room key either directly to respective guest, to group leader, to bellboy, or to guest relation agent, depending on designation/status of guest). |
| Stage 7 | • Bellboy/s escort guests towards their allotted rooms (bellboy escorts regular guests whereas guest relation agent escorts VIPs, CIPs, and often SPATT guests). |
| Stage 8 | • Finally, front desk agent informs its own sections (like telephone operator, cashier, reservation, concierge, etc.) and other departments (like room service, housekeeping, etc.), via arrival notification slip (ANS) in manual front office, about guest arrival. |

## 15.2   CONCEPT OF REGISTRATION

The registration process begins when the guest arrives at the front desk of hotel and completed when he has signed the registration card or guest arrival register and has been assigned a room. The flow of transactions in the registration process can be divided into following stages:

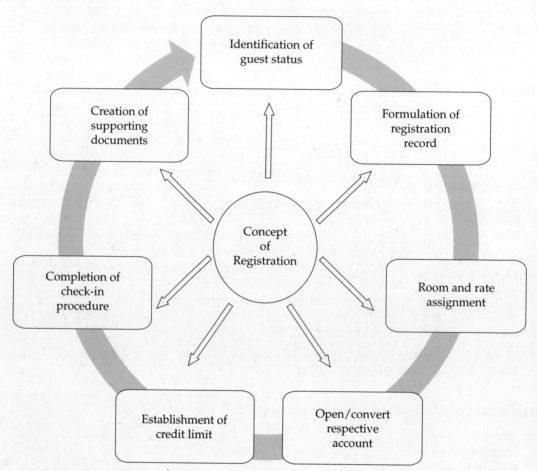

## Identification of Guest Status

✓ The identification of transient guest status is very important. A hotel can directly process the registration of a guest-with-reservation whereas chance guest has to start from the initial stage of room enquiry.

✓ In case of guest-with-reservation, the front desk agent either refers to reservation rack for pre-registration transaction information or requests the transient guest to complete a registration form. The guest-with-reservation generally is of two types: FITs and GITs.

# Formulation of Registration Record

✓ Once the guest has signed the registration card the process of formulation of registration record is complete. The hotel on the basis of this card develops the related hotel documents. This is a permanent record and is used for check-in process.

✓ Further, this card is used to create a guest history record for future planning and sales promotion. The completion of this form is a legal requirement and this should be stored for a period of one year as a legal requirement.

# Room & Rate Assignment

✓ The details regarding the type of room and rate help the front desk agent in deciding what kind of room is to be assigned to the guest. Once the room type is decided the front desk agent would try to pick a room accordingly particularly as far as rate is concerned.

✓ Although the reservation record will indicate as to what type of room the guest requires yet there may be various rooms in the hotels of the same type. For example, single room facing swimming pool, away from stairs, with studio bed, near to public areas, etc. And also with different room tariffs.

# Open/Convert Respective Account

✓ Front desk agent converts *non-guest account* into *guest account* when dealing with guest-with-reservation. Whereas he directly opens *guest account* for walk-in guest and requests him to deposit paid-in-advance with cashier.

✓ Front desk cashier posts/tags/adjusts received advance deposit of guest-with-reservation into his converted guest account.

# Establishment of Credit Limit (or *house limit*)

✓ The determination of guest's credibility and how he will settle his bill is the next important aspect. The guest's credibility is determined on the basis how he wants to settle his/her account. Account settlement can be done either by the use of charge card (like credit card), advance cash payment or deferred payment (like BTC).

✓ Thereafter, establish house limit. The term *house limit* indicates to charge purchase facility for the guest. The house limit primarily depends on guest's previous account settlement history and deposited advance payment with hotel. If hotel does not assign house limit to any guest then *no-post status* will reflect in opened account in PMS.

# How is Guest's Credibility Established?

### Charge card (or credit card)

✓ The front desk agent would ask the guest to produce his credit card at the time of arrival; it would pass through a small PDQ/EDC machine to obtain authorization to charge the guest account. Machine generates either *pre-authorization code* (or acceptance code) or *denial code* (or rejection code).

✓ Pre-authorization code helps in blocking the amount on guest account which assists in establishing the *house limit* for the registered guest. This house limit offers *charge privilege/credit privilege* facility to guest for the purpose of *charge purchase*. Many times credit card companies indicate the *floor limit* of their credit card which aids hotel to determine *house limit* to guest.

### Deferred payment (or BTC)

✓ Generally, hotels have legal agreement/contract with companies (like MNCs & Airlines) for the purpose of obtaining permanent room sale for a stipulated period of time. Here rooms are offered at contractual rates (or company rates) for the payment of which company is generally responsible; such kind of payment by third party (like BTC) is known as *deferred payment*.

✓ While registering company guests, front desk demands letter of authorization (LoA) then cross refer it against approved company list to determine guest legitimacy and company's floor limit. Guest account of company clients usually splits into two parts. *Incidental folio* for guest whereas *company folio* for representing company as per LoA. Guest has to pay for incidental folio, simultaneously needs to certify company folio.

### Advance cash payment (PIA)

✓ It usually happens with walk-in guest and guest-with-confirmed-reservation Both have to pay advance amount (termed as paid-in-advance) to cover room and other related forecasted expenses. This PIA provides base to determine house limit to extend *credit privileges* to guest.

✓ When paid-in-advance amount is exhausted, it will reflect in night auditor's *high balance report*. The basic purpose of this report is to intimate front desk that guest has reached the allotted house limit, so inform him for *interim payment*.

## Completion of Registration Process

✓ It also fulfils the legal requirements that each guest must be registered. Guest signatures are important as it signifies his/her consent to abide by the rules and regulations of the hotel and at the same time it is a proof of his/her stay in the hotel from a certain date to certain date. Nowadays, in large hotels guest registration card (GRC) is used instead of a *hotel register*.

---

**Where guest registers in fully automatic system**

Front desk directed (these can be simple computer systems as well as software powered systems)

✓ It is also called computerized check-in/registration where front desk agent requires to open the booking particulars on the computer screen when guests arrive and perform the registration and related procedures. The booking profile can also be modified as per the requirement of the guest.

Guest directed (these are powered by self-guided software packages like PMS)

✓ It is also called automated or self check-in/registration. This machine is almost identical to ATM machine. It is mainly used in large hotels where guest turnover rate is very high. The machine activates through *credit card* and can handle both types of guests, i.e., guest-with-reservation as well as chance guests. The machine is installed either in hotel foyer or at airport terminal.

---

## 15.4 PROCEDURE FOR CHECK-IN

This topic is related with the manner of registering a guest in the hotel. Generally, hotels have two broad approaches for guest registration, first registration via front desk agent and second guest can self-register himself/herself. Most of hotels use the former approach whereas newly built techno savvy luxury hotels offer self-registration kiosks/units. These self-registration kiosks can be on-site as well as off-site. A step-by-step procedure for both is given below.

| | Front desk check-in<br>(via front agent or guest<br>relation agent) | Self check-in/Self-registration<br>(it may be on-site or off-site self<br>registration kiosk) |
|---|---|---|
| Step 1 | Front desk confirms reservation status with guest. If *chance guest* then check room status. If status allows then provide room related information. Next, proceed towards registration (as for guest-with-reservation) | Guest can use self check-in machine to check room status (chance guest) as well as to register (guest with reservation) himself. Self check-in machine can be installed at hotel lobby or off-site like at airport terminal. |
| Step 2 | - Guest-with-reservation will directly verify his/her pre-registered card.<br>- If chance guest gets room then front desk completes the registration formalities, else refuse or may *farm out* the guest. | - Guest-with-reservation will directly verify fed information and may update to it, if required.<br>- If chance guest gets room then he/she proceeds towards self-registration otherwise search for another property.<br>- Successful registration demands paid-in-advance. |

| | | |
|---|---|---|
| Step 3 | - After registration, front desk agent will issue room key and *key card* to registered guest.<br>- Guest requires showing *key card* every time when he/she takes it back from the front desk. | - After registration, self-check-in machine will issue room key to registered guest.<br>- In case of off-site registration, guest is required to show his/her registration slip to guest relation agent. Registration form will be generated and signed here only then deposited with agent. |
| Step 4 | In manual system, front desk agent generates ANS to inform relevant sections/departments about guest arrival. Whereas in fully automated system, software itself gets updated at each terminal, as and when new arrival is registered successfully. | |
| Step 5 | - Bellboy will carry guest luggage and escort guest towards allotted room.<br>- During the way, bellboy needs to provide information about hotel and locality. | - Bellboy will carry guest luggage and escort guest towards allotted room.<br>- During the way, bellboy needs to provide information about hotel and locality. |
| Step 6 | Lastly, bellboy will open the room door, place guest luggage into cupboard and introduce guest with room functions. | Lastly, bellboy will open the room door, place guest luggage in cupboard and introduce guest with room functions. Finally, handover room keys to guest. |
| Step 7 | In manual front office system, bellboy is required to deposit *arrival errand card* with bell captain after the completion of guest rooming. | |

## dvantages & Disadvantages of Above-Mentioned Registration Methods

| System | Advantages | Disadvantages |
|---|---|---|
| Bound book | ✓ All records are available at one place or in one book only.<br>✓ The separate filing of records is not required in this method.<br>✓ It is economic in nature in comparison to registration card and loose leaf book.<br>✓ It has minimum wastage of time and paper. | ✓ Due to frequent use, book becomes loose and looks dirty.<br>✓ Guests' provided information cannot be kept confidential.<br>✓ Not possible for the registration of VIP, CIP, etc., guests.<br>✓ If books get misplaced, all the data and record will also be lost at one time. |
| Loose leaf register | ✓ Up to some extent, it helps keep the guest stay incognito or confidential.<br>✓ The loss of one sheet will mean loss of only one day information.<br>✓ After registration sheets are put under the counter, so it gives neat and clean look to the reception counter. | ✓ A single sheet can easily be misplaced or lost due to lose book.<br>✓ It has more chances of wastage of sheet on certain days when there are only few check-ins or arrivals.<br>✓ Only one guest can register at one time, so during rush hours, other guests have to wait for their turn. |

| | | |
|---|---|---|
| ...istration ...l | ✓ It helps m aintain complete privacy of the gue st's personal details.<br>✓ During ru ısh hours, many guests can be regı istered at the same time.<br>✓ Low rate of paper wastage and guests can also be pre-registered.<br>✓ Registratioı ı cards are easy to store and more systematic. It can be arranged ei ither alphabetically or in order to c late of arrival. | ✓ The individual guest registration card is expensive in nature.<br>✓ There are more chances of loss or misplacement of cards, if reception desk agent does not pay proper attention.<br>✓ It requires additional space or slot or rack to keep these registration cards. |
| ...nt desk ...cted | ✓ Front desk a gent can try to do up-selling by s uggesting upgraded room.<br>✓ Front desk a gent in coordination with guest relation desk can provide addit tional assistance by answering gue st queries.<br>✓ If situation de mands then guest registration for malities can also be performed in g uest rooms, like in ca se of SPATT a nd VIP/CIP guest. | ✓ Front desk agent in coordination with housekeeping department may also try to earn unau thorized money by showing same occu ipancy.<br>✓ Front desk agent may try to offer discount unnecessarily when agent does not succeed in selling room in first approach.<br>✓ Generally, the in-room registration of guests may be found time consuming as well as demands extra manpower which enhances direct operational cost over the company. |
| ...uest ...rected | ✓ It is a perfect option for *timid guests,* especially when they do not hold any adv ance reservation with hotel.<br>✓ Timid g uest-with-reservation can also opt for upgraded room during self-registı ation process.<br>✓ In case of hotels with high guest traffic, gue st is not required to stand in queue.<br>✓ Property can reduce its manpower costs which enhance its differential profit margin (DPM). | ✓ Some guests may not feel comfortable (or may get confused) in self-registration as machine can display some industry jargons.<br>✓ If guest is not found comfortable in handling self-registration machine then he/she may also move/divert into another property.<br>✓ If machine is jammed or does not function properly or electricity-fail-out occurs then guest turnover will be affected. |

# Activities Involved in Different Methods of Guest Registration

| Registration activities | Non-Automated | Semi-Automated | Fully-Automated (PMS powered) |
|---|---|---|---|
| What is it? | It refers to the manual system (or front-desk-directed) of registration. | It refers to a combination of system (equipment & machines) and manpower. | It is composed of computer equipment & electronic devices or simply PMS. |
| Where it refers? | It refers in small hotels (due to low business volume) | It refers in medium hotels (due to moderate business volume) | It refers in large hotels (due to high business volume) |
| Expenses involved | Less expensive | Moderately expensive | More expensive |
| How is registration performed? | The guest is provided with a hotel register (primarily either loose leaf or bound book). | Here also the guest can be provided with a hotel register or individual guest registration card. | Registration is an integral part of reservation, thus registration details are obtained from reservation file. |
| Where is registration generated? | In hotel register/loose leaf ledger/bound book. | Mainly in pre-printed guest registration card (GRC) but hotel register may also be used. | In PMS, thereafter agent generates hard copy and soft copy of registration. |
| How are documents produced? | The documents are produced manually. | Use of office equipment like typewriter to produce documents. | The soft copies (as well as hard copy) of documents are produced. |
| How is information distributed? | Documents (like ANS) are distributed manually to all relevant sections and departments. | Documents are distributed either manually or sometimes mechanically through the usage of pneumatic tubes. | Soft copies of documents are distributed on-line to various sections & departments or it may get updated automatically when added/removed. |
| Report/s generation | Required reports are prepared manually in order to submit to management for review and room/revenue forecasting. | Required reports are prepared by the use of typewriter, national cash register, electronic cash register, etc. | Various reports are generated automatically when interfaced with reservation and registration module. |

| Feature/s of system | It is a slow process with more chances of mistakes & errors. | It is a moderate process with limited chances of mistakes & errors but content clarity will be more in this method than manual system; but less as compared to fully automated. | It is a speedy process with minimal chances of mistakes and errors plus content will be absolutely clear and accurate. |
|---|---|---|---|
| Examples of equipment | Manually prepared formats with pen and office stationery. | Pre-printed formats with equipment like typewriter, pneumatic tubes, NCR, etc. | Hardware like VDU and various expensive & comprehensive software packages. |
| Who will register the guest? | Mainly front desk agent will register but GRE may also be involved like in VIPs registration. | Mainly front desk agent will register but GRE may also be involved like in VIPs' registration. | Front desk agent as well as guest can self register like in self-check-in machine that may be installed either in lobby or at airport terminal. |

## Standard Conversation between Guest and Front Desk Agent

**Situation 1** (*Guest-with-reservation*): Mr. Black works as a front desk agent in a hotel. He is dealing with Mr. White who holds advance room reservation (for twin room) for 3 days through advance deposit of Rs 50,000. Mr. Black reminds about the allotted price of twin room, i.e., Rs. 25,000 per night, to Mr. White. Mr. White agrees with the given information and tells Mr. Black that he wants to use credit card to settle his account. Mr. Black allows him and accepts the given credit card. Standard conversation phrases between Mr. Black and Mr. White are given below:

| Guest | Mr. White |
|---|---|
| F.D. Agent | Mr. Black |
| Room Type/No. | Twin room/306 |
| Room location | Second floor |
| Payment mode | Credit card |
| Extra services | Welcome booklet |
| Booking period | 3 days |
| Date of arrival | 10/Jan/2017 |
| Date of departure | 13/Jan/2017 |

| Mr. Black | Good Morning sir, how may I assist you? |
|---|---|
| Mr. White | Good morning sir, I have already made reservation in your hotel. |
| Mr. Black | May I know your name and reservation confirmation number/letter, please? |
| Mr. White | I am Mr. White. And sorry, I forgot the confirmation number/letter that you had given me. |

| | |
|---|---|
| Mr. Black | Just a moment Mr. White. Let me check our movement list (or expected arrival list). |
| Mr. White | Ok |
| Mr. Black | Sorry for keeping you waiting. Yes, there is a twin room booked for Mr. White for 3 days. |
| Mr. White | Yes that's me. |
| Mr. Black | Ok sir. May I have your passport, Mr. White? |
| Mr. White | Oh please! |
| Mr. Black | Thanks a lot, would you please fill in the registration form? |
| Mr. White | Of course, why not .. |
| Mr. Black | Here is the pen Mr. White. |
| Mr. White | Oh, thank you. I think I have finished. Is it alright? |
| Mr. Black | By the way, may I know your next destination so that we can forward your reservation to the Holiday Inn there. |
| Mr. White | Mumbai, India. |
| Mr. Black | Thank you very much Mr. White, may I ask you, how will you be paying the bill? |
| Mr. White | I want to use my credit card. Is it fine? |
| Mr. Black | Yes sir. And to express your check-out, may I have your credit card to have an imprint please, Mr. White? |
| Mr. White | Yes please. |
| Mr. Black | Thank you, here is your credit card and you will only need to sign your name when you check-out. Sir, this is your welcome booklet. It will provide you useful information. |
| Mr. White | Oh thank you! |
| Mr. Black | Here is the key to room and your *key card*. You may drop your key at the front desk/ information counter when you go out and collect it afterwards. The key card should be shown each time you collect your room key. Your room number is 306. It is a twin room and located on 2nd floor. Your room rate is Rs. 25,000 per night including GST (% will be as per prevailing rate) and your expected check-out date is 13/Jan/2017. You have already deposited Rs. 50,000. Mr. Brown is my colleague who will take care of your luggage and show you your room. |
| Mr. White | Oh thanks a lot. |
| Mr. Black | My name is Black, please call me any time, I'll be glad to be of service. Enjoy your stay with us. |
| Mr. White | Thank you so much. |
| Mr. Black | You are welcome sir. |

**ion 2** (*Chance guest*): Mr. Mango works as a front desk agent. He is dealing with
~range who directly comes to the hotel for a room. Mr. Mango obtains room
~ence and duration of stay from Mr.
~e then refers to room status position
~ shows that room is available for
~Now Mr. Mango offers deluxe twin
~ (as per room preference) to Mr.
~e at a special rate of Rs. 19,500 per
~ Mr. Orange tells he would like to
~his account via cash. Mr. Mango
~ and allots room number 206 to Mr.
~e. Mr. Orange also requests to keep
~ggage in safe custody for 6 hours
~departure. Standard conversation
~es between Mr. Mango and Mr.
~e are given below:

| Guest | Mr. Orange |
|---|---|
| F.D. Agent | Mr. Mango |
| Room Type | Deluxe twin room |
| Room No. | 206 |
| Room location | Second floor |
| Payment mode | Cash |
| Extra services | Left luggage |
| Booking period | 3 days |
| Date of arrival | 10/Jan/2017 |
| Date of departure | 13/Jan/2017 |

~Iango    Good afternoon sir, how may I assist you?

~Orange   Good afternoon, can I get a room in your hotel please?

~Iango    Have you made an advance booking, sir?

~Orange   Sorry, I haven't done any booking.

~Iango    What kind of room do you want sir: a double room or : twin room? This
          is our *room tariff card*.

~·ange    I want a twin room, please.

~ango     For how many nights will you require the room?

~ra inge   For 3 nights.

~Iar ngo   Just a moment, Mr. Orange. Let me check. Yes we ha ve a standard twin
          room for Rs. 15, 000 per night as well as a deluxe tw in at special rate of
          Rs. 19,500 per night.

~ran ge    I'll take the deluxe one. Thank you, could you pleas se send someone to
          get my luggage out of the taxi?

~Iang o    Of course (Mr. Mango will instruct the bell desk via i telephorwe), please
          get the gentleman's luggage out of the taxi.

~Orang e    Thank you so much.

~Mango     Its my pleasure sir, may I have your passport, please ?

~Orange    Oh, sure.

~Mango     Thank you so much. Will you please fill in your name and addre ·ss in the
          registration card and then sign at the bottom of the card. Would   you like
          a pen, Mr. Orange?

| | |
|---|---|
| Mr. Orange | Yes, please. I think I have finished. Is it alright? |
| Mr. Mango | Yes, thanks you very much. What time do you plan to check-out on 13 January, Mr. Orange? |
| Mr. Orange | Well, I can be out of my room at 10 a.m. but I would like to leave my luggage here until 4 p.m. |
| Mr. Mango | Yes that will be fine, Mr. Orange. The bellboy will store your luggage for you. |
| Mr. Orange | Oh I am very thankful to you. |
| Mr. Mango | Mr. Orange, may I ask you, how you wish to settle your account? |
| Mr. Orange | I want to pay cash. Is it ok? |
| Mr. Mango | That is so nice of you, Mr. Orange. Here is your room key and key card. We've given you room number 206. The bellboy will take your bags to your room. Is there anything else you would like, sir? |
| Mr. Orange | No, thank you very much. |
| Mr. Mango | Mr. Orange, you may drop your key at the front desk/information counter when you go out and collect it when you come back. The *key card* should be shown each time you collect your room key. Your room is a deluxe twin and it is located on 2$^{nd}$ floor. Your room rate is Rs.19, 500 per night including GST (% will be as per prevailing rate) and your expected check-out date is 13/Jan/2017. Mr. Water Melon is my colleague who will take care of your luggage and show you your room. |
| Mr. Orange | Oh thanks a lot! |
| Mr. Mango | My name is Mango, please call me any time, I'll be glad to be of service. Enjoy your stay with us. |
| Mr. Orange | Thank you so much. |
| Mr. Orange | You are welcome sir. |

## Standard Phrases While Performing Registration Task

| | |
|---|---|
| - When you want to know guest's first name | - May I have your first name, please? |
| - When you want to know guest's last name | - Could you possibly spell your last name, sir? |
| - For obtaining passport | - May I have your passport please? |
| - For getting guest's signature on GRC | - Your initials, please… |
| - For additonal services | - Is there anything else I can do for you, madam? |

| | |
|---|---|
| - Happy wishing when walking guest to another hotel | - Have a pleasant stay in the 123 hotel, sir. |
| - When guest agrees with your concern and supports you | - Thank you for being so understanding. |
| - For happy wishing after allotting room in your hotel | - Have a pleasant stay with us, madam. |
| - When guest wants undue favour like to accept invalid credit card | - I am terribly sorry sir, I cannot accept it. |
| - When you are not able to understand your guest | - I am sorry sir; I don't get your point. |
| - During seeing off the guest | - Thank you for coming. |
| - During check-out, for getting repeat business | - We all look forward to serve you again, sir. |

## Activity

Mr. Avocado, a chance guest, will stay in your hotel for one night only and will be fully responsible for all charges incurred. As a front desk agent, how would you explain to the guest that you have to collect one night room rate (INR 20,000) + GST (% will be as per prevailing rate) + an extra INR 10,000 for hotel signing/charge privileges from him as the deposit for check-in?

## CONCLUSION

Guest registration is a legal formality between the guest & hotel. It is a written agreement bound to both hotel as well as guest on specified accommodation oriented terms and conditions. For example, guest is allowed to stay for the requested duration against cash or any negotiable instrument. After the completion of this time period, it depends on hotel whether it will allow *overstay* or not; guest can never force a hotel for overstay. The way of registration may differ from property to property, like most hotels use guest registration card whereas others may use *bound book* or *loose leaf ledger*. Luxury hotels offer *self-registration-counter* or kiosk which may be equipped either on-premises or off-premises. The registration mode may be manual, semi-manual or fully automatic. Loose leaf ledger and bound book are examples of manual system; pre-printed guest registration formats are semi-manual instruments whereas electronic registration file/ printout is a fully automatic tool. Each registration way and system has its own pros and cons over another. Remember, registration of transient guest is always different from group registration. Group can be registered at *satellite check-in* counter as well as group leader may also be handed over with key envelopes (this system is known as *pre-key system*) to distribute to its group members while on the way towards hotel.

| Terms (with Chapter Exercise) | |
|---|---|
| ? | It is a system/software package that is used to manage/run entir operation on single window software or common software platfo |
| Authorization code | It is on-line verification of credit card that shows requested transac been accepted by the credit card company. Machine generates acc code that is known as *authorization code*. |
| ? | It is a guest-directed-registration system whereby guest can dir registered with his/her hotel. It is also known as self-check-in and may be installed at airport terminal too (then it is known a *registration kiosk*). |
| "C" form | ? |
| Pre-key system | It is also known as *key-pack system* in which rooms are pre-assi rooms are allotted before guest arrival) and their keys are p an envelope along with welcome letter. It occurs in group arri allotted room number and guest name is also mentioned on th envelope. |
| Denial code | It is on-line verification of credit card that shows requested tra has not been accepted by the credit card company. Machine g rejection code that is known as *denial code*. |
| House limit | It is concerned with hotel established guest's credit limit. In sim it shows *charge privilege limit* or up to what amount a resident *charge purchase* from the hotel. |
| ? | It is guest registration before actual arrival in the hotel. For many hotels offer self-guest-registration through self-check-in at airport terminal. |
| Registration card | ? |
| ? | In India, this term refers to the *hotel register* whereas in Americ to the publication of AH&LA which contains the name of all its members. |
| SPATT | ? |
| ? | It is a kind of record keeping system whereby individual pap are punched and used to maintain records. |

| Bound book | It is a kind of record keeping system whereby a bound book is used to maintain records. |
|---|---|
| ANS | It is *arrival notification slip* found in manual front office. It is a kind of slip in respect of intimating that guest has been registered into a respective room, thereafter its copies sent into relevant sections and departments. So that they can offer required services and supplies. |
| Guest history card | This card contains relevant information (such as name, address, contact number, room number, his/her likes and dislikes and so forth) about departed guests. |
| Floor limit | It is the maximum credit limit given by the credit card companies to the customers on their issued credit cards/charge cards. |

# 16 CHAPTER

# Guest Cycle

**OBJECTIVES**

*After reading this chapter, students will be able to…*

- define guest cycle and outline its stages;
- discuss the activities that occur during the different stages of guest cycle;
- identify major activities performed between hotel and guests from different marke segments; and
- describe the guest cycle under different operating modes.

## 16.1 GUEST CYCLE

The *guest cycle* is made up of the different stages and activities from which a gues has to go through. This *cycle* comprises several stages, pre-arrival, arrival, occupancy departure and after departure. The hotel business is competitive in nature. Therefore, i should have a strong system to handle guests efficiently and quickly during all stage of the guest cycle. Nowadays, more and more hotels are using various systems to kee their guests happy and also remain in touch with the past guests particularly for th purpose of generating repeat business.

The different stages of guest cycle contain different sets of activities and the operatin mode of these sets of activities varies with standardization of the front office system. I manual system, all activities are performed manually whereas in semi-manual syster front office staff uses pre-printed formats and limited/available electronic periphera to perform the assigned tasks. In fully automatic front office system, hotel uses propert management system (or PMS) to perform all activities.

Similarly, while dealing with different market segments hotels follow different sets ( procedures. For example, registration of walk-in and regular guest is usually performe by front desk agent whereas registration of VIP, CIP & SPATT guests is performed i their allotted rooms by guest relation executive or agent. Likewise, the registration ( group segment is performed at *satellite-check-in counter* (or *remote check-in counter*) aga by guest relation executive or agent.

The guest cycle is profoundly identical to *moment of truth* (MoT) concept, as MoT al: comprises series of activities that materialize between prospective guest and hotel fro

he preliminary stage of reservation enquiry/room enquiry and remains till departure
age.

| Activity |
|---|
| Enlist the activities that occur during the different stages of guest cycle which directly/ indirectly affect the moment of truth experience of guest. |

| Stages | Moment of Truth |
|---|---|
| Pre-arrival | |
| Arrival | |
| Occupancy | |
| Departure | |
| Post-departure | |

## 6.2   STAGES OF GUEST CYCLE

**Pre-Arrival Stage:** It is the stage before actual guest arrival in the hotel. In this stage, the guest chooses a hotel on the basis of certain factors like previous experience, advertisement, recommendation from friends, travel agents, relatives, through business associates, hotel's location, reputation and so forth. From hotel's perspective, the following activities are performed by the front office staff before the guest arrival.

### Attend Room Queries and Make Reservation

- ✓ Reservation section provides necessary information like room type, room rate, available services and facilities. The reservation request may be obtained through telephone, e-mail, telex, fax, letter as well as in-person.
- ✓ *Make reservation* in accordance with guest's room preference and ensure whether guest is giving guaranteed, non-guaranteed or confirmed reservation. Accordingly, intimate to guest whether booking is waitlisted/guaranteed.

### Open Account, Adjust Advance Deposit and Send Confirmation

- ✓ Open non-guest account for future transient arrival who gives guaranteed reservation in lieu to posting advance deposit. For group arrivals, *master account* is opened in which charges will be transferred into group folio/master folio.
- ✓ Send reservation confirmation letter that contains confirmation number only for guaranteed bookings. It may be sent either through e-mail or post.

## 16.3 STAGES OF GUEST CYCLE (ALONG WITH VARIOUS ACTIVITIES THAT ASSOCIATE WITH EACH STAGE)

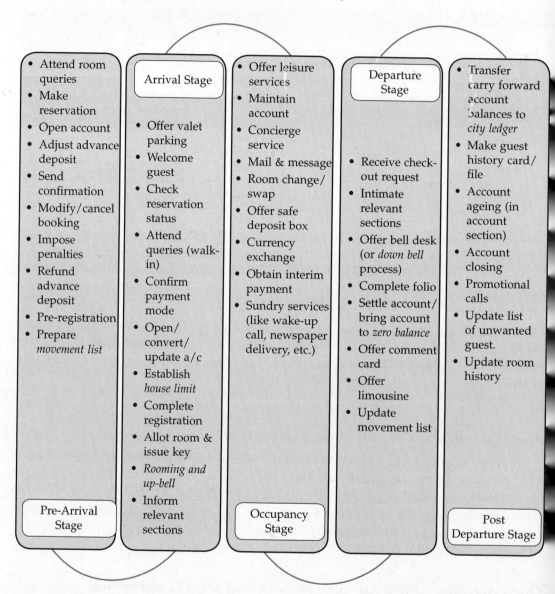

**Pre-Arrival Stage**

- Attend room queries
- Make reservation
- Open account
- Adjust advance deposit
- Send confirmation
- Modify/cancel booking
- Impose penalties
- Refund advance deposit
- Pre-registration
- Prepare *movement list*

**Arrival Stage**

- Offer valet parking
- Welcome guest
- Check reservation status
- Attend queries (walk-in)
- Confirm payment mode
- Open/convert/update a/c
- Establish *house limit*
- Complete registration
- Allot room & issue key
- *Rooming and up-bell*
- Inform relevant sections

**Occupancy Stage**

- Offer leisure services
- Maintain account
- Concierge service
- Mail & message
- Room change/swap
- Offer safe deposit box
- Currency exchange
- Obtain interim payment
- Sundry services (like wake-up call, newspaper delivery, etc.)

**Departure Stage**

- Receive check-out request
- Intimate relevant sections
- Offer bell desk (or *down bell* process)
- Complete folio
- Settle account/bring account to *zero balance*
- Offer comment card
- Offer limousine
- Update movement list

**Post Departure Stage**

- Transfer carry forward account balances to *city ledger*
- Make guest history card/file
- Account ageing (in account section)
- Account closing
- Promotional calls
- Update list of unwanted guest.
- Update room history

## Modify/Cancel Booking, Impose Retention Charges & Refund Outstanding Amount

✓ Reservation section may also receive a request to modify/cancel earlier received room reservation. Agent should politely handle request, confirm guest legitimacy, suggest to modify in-place of cancellation (if room status allows).

✓ Otherwise, refer cut-off-date/time, impose retention charges and refund outstanding balance accordingly. Finally, generate cancellation letter as many credit card companies demand it to cancel the card based guaranteed booking.

## Pre-registration and Preparation of Movement List (both not possible for Chance Guest)

✓ Pre-registration is performed before actual guest arrival. It is only possible for guest-with-guaranteed reservation, either transient or group guest. In case of group reservation, often *pre-key process* is also incorporated as a part of pre-registration activity.

✓ Subsequently, reservation section prepares movement list on daily basis for next day's due-ins and due-outs and sends to relevant divisions. The term *due-in* indicates expected arrivals (or number of rooms going to be occupied) whereas *due-out* indicates expected departures (or expected number of rooms going to be vacant).

**Arrival Stage:** On arrival, the guest fills in a registration form/card giving his name, address, passport number and other relevant personal and financial details plus provide apposite identification instrument to support registration. The front desk transfers this information to an accounting card, called *folio* which is used to record the guest charges made during his stay. The following activities are usually conducted during the arrival stage.

## Offer Valet Parking and Welcome Guest

✓ Valet parking attendant allows guests to leave their vehicles at the front door of hotel entrance and have them parked by a valet attendant. After parking, attendant raises a *car docket* and hands it over to guest along with car key.

✓ As first impression is base for further impressions, thus offer pleasant and warm welcome to guest. The first person who welcomes guest is valet attendant then *commissionaire*. Commissionaire primarily offers door service thereafter intimates to bell desk to offer bell service.

## Confirm Reservation and Handle Queries (of Chance Guest)

✓ Now front desk agent first confirms whether guest holds advance reservation or is straightforwardly a walk-in guest. Next, agent demands reservation confirmation letter/number from guest-with-reservation then proceeds to registration formalities.

✓ On the other side, walk-in guest will be provided with all room related information and on the basis of response and room availability front desk agent proceeds further.

## Confirm Payment Mode and Complete Registration

✓ Now front desk agent confirms with both walk-in as well as guest-with-reservation guests about their mode of payment. Next obtains paid-in-advance from walk-in and adjust/tag advance deposit of guest-with-reservation. Guest may use cash, charge card, traveller's cheques or any other negotiable instrument for settling account.

✓ Guest-with-guaranteed reservation needs to verify the pre-registered card then certify guest registration card with signature. *Alien guests* register on C form whereas *rooming list* is used for group guest registration. After room and rate acceptance, Indian chance guests directly register on F form or hotel register.

## Open/Convert Account and Establish House Limit

✓ Straightforwardly, open *guest-account* for chance guest, whereas convert *non-guest account* into *guest-account* for guest-with-reservation. And open *house-account* or *management account* for complimentary guests. Front desk agent should further enquire for *incidental account* and *master/company account* with group/company guests.

✓ Finally, assign *house-limit* to guests on their incidental accounts and extend *charge privilege/charge purchase* facility. If agent does not grant house-limit then indicate *no-post* status in PMS.

## Allot Room and Issue Key

✓ Allot room in accordance with guest's room preference else room change request may emerge during occupancy stage. Guest-with-guaranteed-reservations get room before their arrival whereas chance guests get room during check-in.

✓ Afterwards, issue room key to guest or bellboy. Conventionally, metal keys were issued but nowadays electronic key is offered. But in most modern techno savvy hotels RFID technology is used. Metal keys were accompanied with *key card* and electronic key offered in *key jacket*. For group room allotment, *pre-key-pack* system is used.

## Rooming, Up-bell Activity and Informing Related Sections

✓ The term *rooming* means to make aware a guest with room and its interiors. In simple terms, it means escorting guests to their allotted room. Meanwhile, bellboy provides guest with available hotel services & facilities. *Up-bell activity* is related with luggage handling during arrival stage. Bellboy performs both the tasks.

✓ Finally, front desk agent informs all relevant departments and sections about guest arrival through *arrival notification slip* or it will automatically reflect/update in each department, in PMS powered hotels.

**Occupancy Stage:** This stage refers to the period when guest uses hotel's room. This can be for the purpose of accommodation, conference or any other use which may not allow the hotel staff to allocate the room to another guest. During the occupancy stage, the hotel guest consumes certain hotel services, products & facilities against payment as well as sometimes on credit. Some of them are given below.

## Leisure and Concierge Services

- ✓ It is the responsibility of travel desk to provide leisure services to resident guests like arranging vehicle for sightseeing, mountain/adventure trekking and so forth activities. This travel desk can be a part of hotel organization structure or space may be rented out to an outside agency.
- ✓ Concierge service is a value addition to the luxury of any hotel. It is responsible for providing information in relation to in-house activities/services/facilities as well as out-house (locality) information. For instance, happy hours, babysitting, table reservation in restaurant & information about nearby drug store, bank, airport terminal, etc.

## Maintaining Account

- ✓ Generally, guest account is affected as and when guest purchases anything from any *point of sale* outlet plus when guest pays. Remember, either guest incurs expense or not but its room rent & applicable taxes are generated every night which night auditor charges to respective guest account.
- ✓ Expenses and *charge privileges* are transferred on debit side of guest account (either in *room/incidental account* or *master/company account*). An expense paid by hotel on behalf of guest claims through *guest's disbursement voucher* (or VPO) and its entry is posted on debit side of guest account. Guest payment is transferred on credit side of guest account.

## Mail & Message

- ✓ Hotel may receive mail & message for their current staying guests as well as for future guests and past guests. Mails of future guests are attached with reservation form then transferred with movement list to front desk (to attach with GRC) whereas mails of past guests are sent to their *mail forwarding address*. Mails of present guests are sent to their rooms or may be kept at front desk so guest can be notified when he/she returns to collect room key.
- ✓ Mails of hotel employees are sent to their respective offices. Like management employees mails are directly sent to their respective offices whereas operational level employees' mails are sent to time office/security office.

## Room Change and Room Swap

✓ When guest is not satisfied with the room then he/she may request for room change. The guest's request for room change may be due to improper functioning of bathroom, air conditioner, television or any other facility. Room change request can take place either via *live move* (in presence of guest) or *dead move* (in absence of guest).

✓ *Room swap* is concerned with interchanging of room among/between the resident guests. Room change as well as room swap also affects room rate, thus earlier performed guest registration is required to be updated/modified.

## Safe Deposit and Currency Exchange

✓ To keep guests' valuables safe and secure, hotel offers *safe deposit box facility* (at front desk). But nowadays, hotels offer *in-room safe facility* to resident guests.

✓ Currency exchange facility is offered to *alien guest* for which hotel cashier is responsible. He is required to keep himself/herself updated with daily exchange rates. Hotel requires an appropriate licence/s when dealing in foreign exchange. Cashier must issue an *encashment certificate* against every foreign exchange transaction.

## Night Audit, Obtaining Interim Payment and Sundry Services

✓ Night auditing is concerned with *cross referencing* of all prepared accounts on daily basis by night auditor. Afterwards, night auditor prepares necessary reports and sends to management and departmental heads for review and forecasting. He also sends report of guest with *high balance* to front office manager.

✓ High balance report indicates to those guests who flourish with allotted maximum credit limit (or who reached their allotted house limit). Front desk intimates to such guests for interim payment so as to prevent probable skippers.

✓ There are enormous sundry services and facilities offered by hotels to their resident guests like placing *wake-up* call, delivering newspaper in guest rooms, connecting international calls, offering business services and so forth.

**Departure Stage-** It is the fourth stage of guest cycle in which guest departs from the hotel after settlement of account. Generally, front desk confirms departure status from received movement list as well as registration record also intimates about it. Often it has been observed that stayover guests may also request for overstay, whether property allows it or not depends on room position and hotel's policy. The following activities are usually performed in hotel, during the departure stage.

## Receiving Check-out Requests and Intimating to Related Sections

✓ Front desk agent may receive guest's check-out request via room telephone or in-person. The check-out can be *front-desk-directed* (either computerized check-out or express check-out) or *guest-directed* (either video check-out or self-check-out). It varies with standard of hotels.

✓ After receiving check-out request, in manual system front desk agent sends *departure notification slip* to all relevant sections and departments. So, all outstanding charges can be timely transferred to respective guest's accounts. But in PMS powered organization, all charge transactions are instantly transferred into respective accounts when they occur.

## ell Desk Service (or *down bell activity*) and Complete Grand Bill

✓ Now front desk agent gives instruction to bell captain to send bellboy to provide luggage assistance to departing guest. Until bill settlement is not complete, the luggage will be kept in the custody of bell desk. Ensure that guest must return room keys (may return either to bellboy or front desk) as well as keys of safe deposit box, if it has been issued.

✓ Front desk cashier should make ready to *grand folio* of checking-out guest before he/she reaches to the counter. Once guest reaches, hands over grand folio and allows him/her to check bill details. If guest wants any assistance then assists him/her in clarifying to posted entries.

## ccount Settlement and Bringing to Zero Balance

✓ Once guest checks his bill and agrees with all posted charges then cashier will precede to account settlement and payment. Guest can pay either in cash, through charge card, deferred payment or use any other negotiable instrument to settle down his account. In case of group, usually group leader is responsible to settle *master account* (or master folio) whereas *incidental account* (or incidental folio) is settled by individual guests.

✓ In settlement, cashier brings guest account to *zero balance (or zeroing out)*. If guest pays cash then guest account will automatically close down, whereas in deferred payment guest account converts into non-guest account and carry forward balance transfers into *city ledger*.

## Offer Comment Card, Limousine Service and Update Movement List

✓ Comment card (or feedback form) is used to obtain guest's satisfaction level with offered services, amenities and facilities. Comment card must have sufficient close-ended questions with limited open-ended questions. Nowadays, instead of comment card *guest satisfaction index* (GSI) is prepared for comprehensive review.

✓ Next, cashier intimates the bell desk to release the guest luggage and transfer it into a vehicle. Luxury hotels offer limousine for transient and mini-bus for group departure. The transportation fare to be charged or not, depends upon management policy.

✓ Finally, front desk should update the room status position with every check-in, check-out, overstay and under-stay transactions. Otherwise, hotel may face problems like *double rooming, sleeper* and so forth situations which lead to revenue and customer loss.

**After departure stage:** It is the stage when guest has already left the hotel premises. Now, the front office staff need to perform or complete the following activities.

## Transferring Account Balances

✓ Front desk cashier transfers carry forward balance of converted non-guest accounts into city ledger (including carry forward balance of *house accounts*) along with their respective signed bills/folios plus supporting vouchers and documents to accounts section. Now, the bill collection responsibility is also transferred from the front desk cashier to accounts section.

## Account Aging and Account Closing

✓ The term *account aging* is related with account receivable system. Every hotel divides this account aging system into three broad categories i.e. *current, delinquent* and *overdue account*. Each category represents different set of time periods for outstanding account settlement.

✓ If payment is not received even after maximum allotted overdue period then hotel adjusts (or *bad debt*) this amount as a revenue loss and *writes-off* eventually. Adjustable revenue loss can be written-off from the same year's revenue but non-adjustable revenue loss splits among number of years and write-off eventually.

## Preparing/Update Guest History and Room History

✓ In manual system, guest history is prepared for every checked-out guest who came first time and is updated with check-out of every regular guest. In automated system, this guest history is prepared/updated automatically with every check-out transaction. It also provides a powerful databank for strategic marketing, room and revenue forecasting.

✓ Similarly, room history is also prepared for every new room that has been checked-out whereas it gets updated with every regular guest's checked-out transaction. It supports in tracking room record in terms of revenue generation, room complaints, turnover rate and so forth.

## Updating List of Unwanted Guests and Placing Promotional Calls

✓ The term *unwanted guest* is used for blacklisted guest. After guest departure stage, front desk provides information about *skippers*. On the basis of this, prepared blacklist of unwanted guests gets updated.

✓ The term *promotional call* is used for every call that is placed to departed guest in order to make personalized business relations and to intimate the available offers for the purpose of getting business again. These promotional calls are placed with the help of guest history card/file.

✓ Promotional calls support hotel marketing, intimating guests about new offers and discounts. These calls are also placed on guest's anniversary, birthday and other similar occasions.

---

### Activity

Fill in the appropriate stages with the correct alphabet (A-E), each alphabet denotes the five different stages of the guest cycle.

| A | Pre-arrival stage | B | Arrival stage |
|---|---|---|---|
| C | Occupancy stage | D | Departure stage |
| E | After departure stage | | |

| S. No. | Guest Activities | Stage |
|---|---|---|
| 1 | Placing wake-up call | |
| 2 | Raising location slip | |
| 3 | Preparing movement list | |
| 4 | Currency exchange | |
| 5 | Booking of movie tickets | |
| 6 | Transferring guest account balance into city ledger | |
| 7 | Brining account to zero balance | |
| 8 | Entertaining room change request | |
| 9 | Generation of high balance report | |
| 10 | Preparing rooming list and completing key pack activity | |
| 11 | Generation of guest history file | |

| 12 | Video check-out | |
|----|-----------------|--|
| 13 | Self check-in at remote location | |
| 14 | Request for safe deposit box facility | |
| 15 | Removing non-lettable rooms from room inventory | |
| 16 | Maintaining guest accounts | |
| 17 | Assigning house limit or stating charge privilege | |
| 18 | Confirming room availability to chance guest | |
| 19 | Generation of reservation confirmation number/letter | |
| 20 | Considering a delinquent account settlement | |

## 16.4  GUEST CYCLE ACTIVITIES UNDER DIFFERENT OPERATING MODES

The term *operating mode* is concerned with the method of working, i.e., manual, semi-automatic or fully automatic in different sections of front office department. The operating mode is the foundation for work speed, accuracy, readability, correctness of the written documents/vouchers/forms. In general, electronic/computer generated forms/reports or any other material is always considered more readable and accurate than manually prepared forms/reports. Handwritten material (either fully written or partially filled on pre-printed formats) are often found more erroneous than computerized ones. A brief description of different operating modes and their uses & effects during different stages of guest cycle is given below.

| Guest cycle activities | Manual system (Manually prepared formats) | Semi-automatic system (Pre-printed formats) | Fully automatic system (Electronic files of PMS driven) |
|------------------------|-------------------------------------------|---------------------------------------------|----------------------------------------------------------|
| What is system? | Front office uses paper work like record-keeping relied on handwritten forms & cards. | Front office uses some electro-mechanical equipment which rely upon both handwritten and machine-produced forms. | Front office uses software programs like PMS which can easily integrate with various standalone software also. |

| Pre-arrival | Check room status with density control chart, advance letting chart, booking journal. | Check room status with electronic room status board which updates manually. | Refer reservation module to check reservation status. |
| --- | --- | --- | --- |
| Arrival | Manually open account/GWB (guest weekly bill) and perform registration in registration diary/ bound book or loose leaf ledger. | Open account via EBM (Electronic Billing Machine) or NCR (National Cash Register) & perform registration in GRC (Guest Registration Card). | Open/update account on guest account module and perform registration in registration module or guest can self-register. |
| Occupancy | Generate bill manually and send to front desk cashier to enter into guest's *room account/ GWB*. | Fill pre-printed formats (for *audit trail*) and send to front desk cashier so that entry can be transferred into EBM/ NCR. | Transfer charges on-line (through established terminals) into respective guest accounts. |
| Departure | Manually complete and generate guest bill/ GWB. | Perform calculation, generate outstanding balance on EBM/NCR and accept payment. | Generate hard copy of guest bill and may also send soft copy to guest's e-mail account. |
| After departure | Prepare summary of guest experience with property. | Fill/update room history and guest history with the help of GRC. | Room history and guest history file is created/ updated automatically. |

## Activity

Differentiate between the following:

| POS | PMS |
| --- | --- |
| (it means point of sale terminals which reflects revenue outlets) | (it means property management system which reflects single window/platform software for entire property) |
| | |
| | |
| | |
| | |
| | |

## Guest Cycle Activities Under Different Target Segments

| Guest cycle activities | Transient Guests | | Group Guests | Alien Guests (with or without advance reservation) | VIP Guests |
|---|---|---|---|---|---|
| | Guest with Guaranteed Reservation | Chance Guest | | | |
| Pre-arrival stage (pre-registration of guest) | Perform pre-registration & open non-guest account | No activities | Perform pre-key process and key-pack | May (if guest with reservation) or may not pre-register (if chance guest) | Perform pre-registration & open house-account or non-guest account |
| Welcoming & greeting | Commissionaire thereafter front desk agent | Commissionaire thereafter front desk agent | GRA (may offer garlands/flower bouquets/ aarti, tilak) | Airport representative | GRE (may offer garlands or flower bouquets) |
| Registration | Update pre-registration thereafter take guest signature | On-the-spot registration | Group registration at satellite-registration-counter. Fill C form (common/individual) for foreign group. | Registration and genera-tion of C form (on-line) by front desk agent | Registration by GRE in his/her allotted room |
| Open account | Convert non-guest account into guest-account (plus update, if any additional payment received) | Open guest-account, payment may be cash or charge card base (usually these are PIA) | Open master account for group leader and incidental account for its members | Open guest-account (accept foreign currency and return balance in Indian currency with encashment certificate) | Convert non-guest account into guest-account or open house account (as often these are complementary) |
| Assigned house limit (or charge privilege) | Usually offered (if guest pays more than room charge) or depends on guest's past record | As these are PIA guest, thus not offered (if PIA is more than room charges then may be offered charge privilege) | Usually no charge privilege for incidental account. | Usually these are PIA guest, thus not offered (if PIA is more than room charge then offer charge privilege) | Offer house-limit |

*Contd.*

| Guest cycle activities | Transient Guests | | Group Guests | Alien Guests (with or without advance reservation) | VIP Guests |
|---|---|---|---|---|---|
| | Guest with Guaranteed Reservation | Chance Guest | | | |
| Occupancy stage activities | Transfer charges into room/guest account (if assigned house-limit), otherwise take cash payment | Mostly guest requires to pay cash (if not sufficient advance received or not sufficient amount in credit card block) | Transfer room charges of group members into master account and incidental charges into respective group members account | Transfer charges into room/guest account (if assigned house-limit), otherwise take cash payment | Transfer into room account + Offer butler (or personal server) |
| Departure stage activities (check in-balance and out-balance) | Zero out to account (if guest pays cash), otherwise transfer into city ledger to bring account to zero balance | Zero out guest account through full settlement of account (either through cash payment or charge card) | Zero out master account (by sending balance to city ledger) and incidental account must settle immediately | Settle guest account immediately (issue encashment certificate either guest pays via foreign currency or traveller's cheque of foreign origin) | Zero out by transferring balance amount into city ledger (to write off when room is offered as complementary) |
| After departure stage activities | Send bills for clearance (to CCC), if incentive guest (send bill to his company) and give promotional calls | Promotional calls for the purpose of repeat business | Offer promotional calls and account section sends bill to group leader for payment | Place promotional calls | Account section sends bill for clearance when guest is not offered complementary room. |

## Activity

After the completion of this chapter, try this activity by filling in your answers, using the guest cycle provided below.

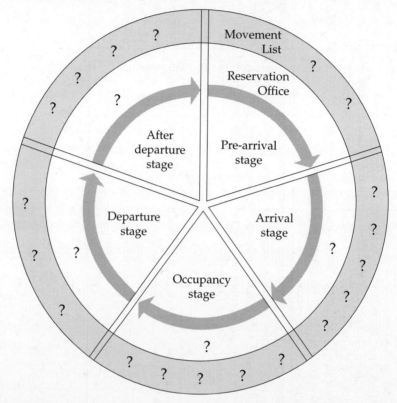

## 16.5   TASK FORCE IN DIFFERENT STAGES OF GUEST CYCLE AT FRONT DESK

### Task Force before Guest Arrival

Task 1    Prepare reception area for service and check all necessary equipment prior to use.

Task 2    Check and review daily arrival details, prior to guest arrival.

Task 3    Allocate rooms in accordance with guest requirements and hotel policy.

Task 4    Follow up uncertain arrivals or reservations in accordance with hotel procedures.

Task 5    Compile and distribute accurate arrivals lists to relevant personnel/departments.

Task 6    Inform colleagues and other departments about special situations or requests in a timely manner.

### Ideal phrases to be used while agreeing on anything

Phrases for all situation    "Certainly sir" or "Definitely sir" or "That's true" or "Absolutely, sir" or "Exactly, sir" or "Why not?"

### Ideal phrases to be used when you do not understand guest's language

Use as per situation    "Could you repeat please?" or "I am sorry, sir. I don't get your point." or "Could you speak more slowly?" or "I'm sorry, I don't quite understand. Should I get the manager?" or "Pardon sir / madam?"

**Note.** The above-mentioned phrases can be used on telephonic conversation with guest as well as while directly interacting with guests at front desk.

## *Task Force during Guest Arrival*

Task 1    Welcome guests warmly and courteously.

Task 2    Confirm details of reservation with guests.

Task 3    Follow hotel procedures correctly for the registration of guest with or without reservation, and complete registration within acceptable timeframes and in accordance with hotel security requirements.

Task 4    Follow correct accounting procedures in accordance with hotel practices.

Task 5    Explain clearly to guest about the relevant details such as room key/electronic card, guest mail, messages and safety deposit facility arrangements.

Task 6    Follow correct hotel procedures where rooms are not immediately available or overbooking has occurred, in order to minimize guest inconvenience.

Task 7    Monitor arrivals and check actual arrivals against expected arrivals, report any deviations in accordance with hotel procedures.

### Ideal phrases for addressing guests

| *To address any* | | *To address any* | |
|---|---|---|---|
| Gentleman (both married & unmarried) | Mr. | Woman (only married) | Mrs. |
| Woman (only unmarried) | Miss | Woman (both married & unmarried) | Ms. |

**Note-** Sometimes we are not sure whether the lady guest is married or not then it is a good approach to address her with Ms. But sometimes few lady guests become irritated at your addressing with "Ms". So, if possible it is better to ask "How should I address you, Sir/Madam?"

---

**Ideal phrases for welcoming guests**

- Welcome to XYZ hotel or XYZ restaurant
- Thanks for your coming Mr. Y

- We are glad to see you again, Mr. X.
- We are honoured to get the chance to serve you, sir / madam.

---

## Task Force During Guest Occupancy

Task 1     Inform guest, if there is any call, mail, message or someone waiting for him.

Task 2     Offer additional guest services like pager, foreign exchange, safe deposit box, etc.

Task 3     Fulfil guest request, if possible, for instance, guest may request you to arrange tickets for a movie.

Task 4     Promptly handle all guest complaints and intimate to management.

Task 5     Keep handy *contact information material,* as guest may demand to contact outside resources.

Task 6     Give direction to guest towards other locations, like guest may want to know where the nearest bank, shopping centre, medical store, theatre or cinema is.

Task 7     Language translator – many alien guests are not comfortable in either national language of visiting destination as well as English; suppose a French guest may speak only in French. The hotel should provide a language translator. Often concierge performs this activity.

Task 8     Offer supportive services and facilities to guest like phone calls, fax, Internet & other business centre services. In addition, hotel employee should know what the hotel charges for these supportive services.

---

| **Ideal phrases for asking questions and handling enquiries** | |
|---|---|
| For asking any question | Could I ask you, a question, sir? |
| For enquiring guest's needs | Do you need anything else, sir? |
| For knowing guest's name | May I know your first name? or, may I have your name, please? |
| For taking guest's sign | Your initials, please? or, may I have your signature here, sir? |
| For knowing guest's room number | May I know your room number? |
| For assisting guest | May I assist you? or, may I help you with that, madam? |
| For offering extra services | Is there anything else I can do for you, madam? |

## Ideal phrases for apologizing

| | |
|---|---|
| For apologizing | "I am terribly sorry, sir." or "I'm awfully sorry." |
| For asking anything | "Excuse me" or "excuse me for interrupting." |
| For repeating order | "I beg your pardon, sir." |
| For going to deal with other | "I'm sorry, a guest is waiting for me." |
| To put guest on waiting | "I'm sorry to have kept you waiting, sir." |
| For not excepting | "I'm sorry. The house is fully booked." |

## Ideal answer to apologizing

| | |
|---|---|
| For replying against courtesy words | "It's Ok" or that's all right" or "It's my pleasure" or "you are welcome sir/madam" |
| Other courtesy replying phrases | "No problem" or "no problem at all" or "it doesn't matter" or "it's nothing" |

## *Task Force during Guest Departure*

Task 1  Review departure lists, checking for accuracy.

Task 2  Seek information on departing guests from other departments in a timely manner to facilitate preparation of account.

Task 3  Generate guest accounts and check for accuracy.

Task 4  Explain account clearly and courteously to the guest, and accurately process the account.

Task 5  Recover keys/electronic cards from guests and process correctly.

Task 6  Action against guest requests for assistance with departure courteously, or refer requests to the appropriate department for follow-up.

Task 7  Process express checkouts in accordance with hotel procedures where appropriate.

Task 8  Follow correct procedures for group check-out and process accounts in accordance with hotel procedures.

| Ideal phrases during farewell | |
|---|---|
| When seeing off a guest | "Good bye, and have a nice trip" or "have a nice day" or "see you" or "thank you for coming" |
| When excusing yourself | "Sorry, I have to go. Nice talking with you" or "we all look forward to serving you again," or "hope you enjoyed staying with us" |

# CONCLUSION

Guest cycle is composed of five different stages, i.e., pre-arrival, arrival, occupancy, departure and after-departure stage. Each of these stages further comprise immense set of activities which play a decisive role in meeting guests' expectation level and enhancing guest satisfaction. The ease, efficiency and effectiveness of these guest cycle activities largely depends on the operating mode of front office. Remember electronic front office is predominantly more productive and effective than manual front office, simultaneously also reduces error, work pressure (as once data has to be filled then it will be automatically retrieved every time) and time. Front office staff can utilize this time in offering best services to guests. In fully automatic system guests can register as well as check-out themselves. There are few exceptions or additional work required to be performed as and when guest cycle interacts with different targeted market segments. For instance, VIP guests are welcomed with garlands/bouquet and their registration is performed in their rooms, group guests are welcomed with welcome drink and they are usually registered at satellite counter. On the other side, sports team is welcomed with *aarti* and *tilak*. Lastly, remember guest cycle plays a crucial role in fabricating positive *moment of truth*.

| Terms (with Chapter Exercise) | |
|---|---|
| Guest directed check-out | A check-out option where guest can self check-out, for example, express check-out, video check-out or self check-out terminal. |
| ? | It refers to the stages from which a guest has to pass; it includes before actual arrival, arrival, stay, departure and after departure stages. |
| In-room safe | ? |
| Front-desk directed | A check-out option where front desk itself performs all check-out activities for departing guest. |
| Charge privilege | ? |
| Promotional call | It refers to all those sales calls placed to guest in order to generate business (or future business) or to sell hotel products or services. |

| Check-in | The guest cycle stage in which guests arrive at the hotel thereafter fills-in registration information, and arranges a paid-in-advance or adjust/tag advance deposit. |
|---|---|
| ? | The guest cycle stage in which a guest is given an accurate statement of charges, pays the bill, and is given a receipt. |
| GSI | It stands for guest service/satisfaction index which is prepared for comprehensive review. In simple terms, it is a modified version of guest feedback form. |
| Audit trail | Also known as *audit log*. It is a chronological sequence of audit records, each of which contains evidence directly pertaining to account and resulting from the execution of a hotel process or system function. It serves as a documented history of transactions. |
| High balance | ? |
| Ledger | Secondary book of recording transactions. In simple terms, ledger is a grouping of accounts and also known as *tab*. |
| Waitlisted reservation | Type of guest's room reservation acceptance in which hotel neither gives guaranteed reservation nor cancels it. Hotel places reservation request in queue (or in waiting list), simultaneously also intimates about it to the guest. It is a kind of overbooking. |
| Remote check-in | ? |
| Cross referencing | ? |
| No-post status | It occurs in fully automatic front office (or it is a term used in property management software) which indicates that guest does not have *charge privilege* facility, thus guest has to pay instantly after consuming hotel services or facilities. |
| Rooming list | List of guest names provided by a group leader to the hotel to inform names of group members occupying the block of room booking. |
| Dead move | ? |
| Room account | Individual guest account. |
| Key-pack | ? |
| Zero out | Brining guest account to zero balance or close down to guest account. |

| Bad debt | It is a term of accounting industry which indicates that the earned money (or account receivable amount) has been lost for forever. It is a kind of monetary business loss which will later on be written-off from profit amount. |
|---|---|
| Master account | It is also known as *master folio* or *group folio* which is opened to record the charges of group members. It is the sole responsibility of group leader to settle master account/folio. |
| ? | A pre-arrival stage activity whereby only guest-with-guaranteed reservation is registered before his/her actual arrival. |
| PMS | It stands for property management system/software which manages the operation of entire organization on single window/platform software. |
| POS | It stands for point of sale and refers to all those outlets which sale products like restaurant, coffee shop, gift shop, etc. If these outlets use electronic devices for recording & processing transactions then it is known as EPOS. |
| Satellite check-in | *Remote-check-in system* which is used for the successful registration of group arrival. |
| Occupancy stage | Stage of guest cycle in which a registered guest stays in a hotel. All types of account maintenance predominantly occur during the stage of occupancy like room rent posting on daily basis. |

# PART B

## Housekeeping Operations

# 17 CHAPTER

# Role of Housekeeping

⊂ **OBJECTIVES** ⊃

*After reading this chapter, students will be able to...*

- define housekeeping and draw the layout of department;
- discuss the activities executed by housekeeping department and its sections;
- identify and enlighten the housekeeping equipment, guest room amenities and supplies; and
- describe the structural foundation of housekeeping department.

## 17.1  HOUSEKEEPING

*Housekeeping* is the task of doing housework like area cleaning, laundering and changing bedsheets. In hotels, housekeeping department is responsible for in-house as well as out-house cleaning. The term *in-house* is used for internal areas of hotel building whereas the term *out-house* is used for external areas of hotel building. Cleaning of hotel lobby, restaurant, cloakroom, guest room, corridor, lounge, garden, swimming pool and other public areas are a few examples of in-house and out-house activities. Apart from it, housekeeping is also responsible for the cleanliness of staff uniform and sometimes for guest clothes also.

Nowadays, many hotels outsource most of their housekeeping errands, especially public area cleaning and laundry work. Housekeeping department is headed by Executive Housekeeper along with assistance of supervisors in different sections. The operational work of housekeeping department is performed by attendants. Remember, this department supports in creating first positive impressions of hotel into the minds of guests by providing clean, hygienic and pleasant environment. Besides, it also performs some other significant tasks like assisting in guest safety and security. In any hotel, the role of housekeeping is to:

- provide clean, comfortable, hygienic and pleasant atmosphere;
- support guest privacy; safety and security (of hotel and guest);
- play a crucial role in developing internal decor of the hotel;

- work as a back-up department for front office in room selling;
- offer laundry services and support to pest control team; and
- support in room maintenance and deal with key control.

| Activity | | |
|---|---|---|
| Complete the below given table by filling in the columns with appropriate in-house and out-house areas. | | |

| S. No. | In-House Areas | Out-House Areas |
|---|---|---|
| 1 | ? | ? |
| 2 | Guest Rooms | ? |
| 3 | ? | Vehicle Parking Area |
| 4 | ? | ? |
| 5 | ? | ? |

## 17.2 FUNCTIONS OF HOUSEKEEPING DEPARTMENT

The heading is concerned with work profile of entire housekeeping department which starts from the initial stage of cleaning and remains till the decoration of the public areas and guest rooms (both in-house as well as out-house areas). Its functions vary from property to property and each set-of-activities is performed by different sections. The housekeeping department of a large-sized hotel comprises laundry, uniform and linen room, housekeeping control desk, guest floors, public areas, health club, and floral & plant arrangement. An epigrammatic work profile of housekeeping department is depicted below:

## 17.3 HOUSEKEEPING EQUIPMENT

*Housekeeping equipment* refers to the various gadgets, appliances and utensils used by the different sections of housekeeping department for maintaining hygiene and sanitation of hotel organization. Here, equipment for on-premises laundry operation are also included. These equipment are categorized into three broad groups, i.e., manual, semi-manual and automatic. A list of several important housekeeping equipment, under manual, semi-manual and automatic housekeeping operations, is given ahead.

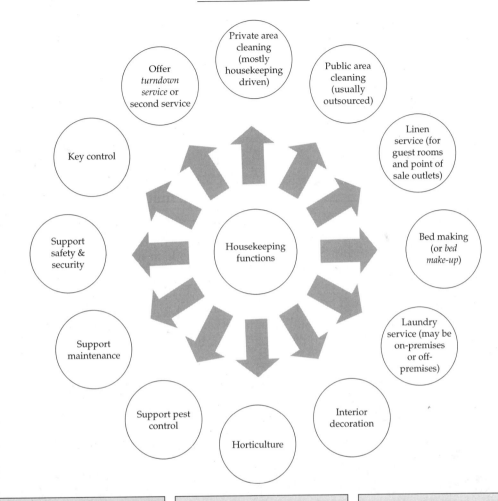

| Manual Equipment | Semi-Manual Equipment | Mechanical Equipment |
| --- | --- | --- |
| <ul><li>Brooms (indoor & outdoor)</li><li>Brushes (scrubbing, soft and hard brush plus brush for upholstery, toilet and so forth purposes)</li><li>Mops (dry and wet mops)</li><li>Cloth (like *rags*, wash, polish, duster, etc.)</li><li>Polish applicators (natural lamb's wool, synthetic wool and solid wax pressurized) and its trays</li></ul> | <ul><li>Trolleys (*chambermaid*, linen, *mop wringer, Janitor's*)</li><li>Spray bottles</li><li>Carpet sweeper</li><li>Commercial washer and dryer</li><li>Iron with board</li><li>*Carpet beaters*</li><li>Ladders</li><li>Airing racks</li><li>Fit pumps</li></ul> | <ul><li>Property management system</li><li>Vacuum cleaner (only dry, and dry & wet vacuum cleaner)</li><li>Floor scrubbing and polishing machine (it is used for *scrubbing, buffing,* polishing, *burnishing* and *scarifying*)</li><li>*Spray cleaning* machine (with different coloured pads, i.e., beige, green and black pads)</li></ul> |

| Manual Equipment | Semi-Manual Equipment | Mechanical Equipment |
|---|---|---|
| • Laundry sacks<br>• Dustpan + bins (dustbins, sanibins)<br>• Basins and bowls<br>• Mattress and overlays<br>• Containers and buckets<br>• Linen (bed, bathroom & dining)<br>• Cleaning agents<br>• Personal protective equipment<br>• Melamine foam<br>• Squeegees<br>• Scrubber (like deck scrubber)<br>• *Hand caddies*<br>• Abrasive pads<br>• *Door hangers* | • Choke removers<br>• *Box sweepers* | • Carpet cleaning machine (steam extraction, cylindrical brush dry foam, rotary brush, wet shampoo)<br>• Automatic soap dispenser<br>• Wet extraction machine (hot water and solvent extraction machine)<br>• Scrubber drier sweepers (power, pedestrian, petrol & gas and self-propelled driven sweepers)<br>• *Calendering machine*<br>• High pressure washers<br>• Scarifying machine (heavy duty scrubber polishers & self-propelled scarifies) |

## Activity

There are different types of clothes used in housekeeping department. Each of these clothes is used for different kinds of operation which affects its durability. In order to increase its durability, proper care and cleaning is required. Fill in the below given spaces with appropriate answers:

| Cloth | Description | Types |
|---|---|---|
| Rags | | |
| Wash cloth | | |
| Duster | | |
| Polish cloth | | |
| Wiping cloth | | |

## 17.4 GUEST ROOM AMENITIES AND SUPPLIES

It is always desirable that hotel management should fulfil every level-headed request made by the guests. This can include many different services and actions. For example, if a guests needs a babysitter especially when they go out, housekeeping may help arrange a babysitter for them. If guests need firmer mattress, an iron, a new set of bathroom amenities, or a rollaway bed, the housekeeping will help them get, set up, or take away. Therefore, housekeeping facilitates to make guest's stay more comfortable and pleasant. Hotels place different standards of *amenities, supplies, expendables* and *loan items* into guest rooms. All these offered items may vary from property to property due to differences in their standards and *level of services*. A brief description of each is given below.

# Supplies

✓ The term *supplies* or *guest supplies* refers to various items that guest requires as a part of the hotel stay like toilet tissues, hangers, etc. Simply, guest supplies are essentials provided for the guest's needs and convenience. Nowadays, many hotels also provide *pillow menu* in their guest rooms.

✓ The standard of these guest supplies will vary as per standard of property. Supplies can be classified into two broad groups, i.e., *guest room supplies* and *bathroom supplies*. The housekeeping department is responsible for storing, distributing, controlling and maintaining adequate inventory levels of both—room supplies and bathroom supplies.

# Amenities

✓ The term *amenities* or *guest amenities* refers to various services or items offered (such as *in-room safe*) into guest rooms or directly to guests for their expediency and comfort usually at no additional costs, although the cost of these items is often hidden in the room rate. Simply, amenities are non-essentials but enhance the guest's experience.

✓ These amenities enhance the standard and extravagance of the organization. Subsequently, the eminence of these amenities will vary with level of organization. Often these amenities, especially in case of VIP guests, are offered as complementary stuff (or *complementary amenities*).

# Guest Expendables

✓ The term *expendable* or *guest expendables* refers to all those items that are expected to be used up or taken by the guests such as soaps, laundry bags, etc. Whether guest has used expendable items or not its replenishment is obligatory when the room is made ready for the next arrival.

✓ Remember, these guest expendables cannot be classified as opulence even at many budget class properties. All guest expendable items are stocked and supplied by the housekeeping department but sometimes these may also be supplied by any other hotel department.

## Guest Loan Items

✓ The term *loan item* or *guest loan item* refers to all those loan items that are not maintained in the guest rooms but are made available when guests request on a receipted loan basis.

✓ The guest loan items are usually available in main linen room and delivered to guest rooms when requested. Remember, guest loan receipt should specify when the item may be picked up, so that guest is aware that it is not available free of cost and is not allowed to be carried along with them.

| Activity | | | | |
|---|---|---|---|---|
| Fill in the columns with appropriate guest supplies, amenities, expendables and loan items. | | | | |
| S. No. | Guest Supplies | Guest Amenities | Guest Expendables | Guest Loan Items |
| 1 | ? | ? | ? | Adapter |
| 2 | Bath Towels | ? | ? | ? |
| 3 | ? | Mini Bar | ? | ? |
| 4 | ? | ? | ? | ? |
| 5 | ? | ? | Soaps | ? |
| 6 | ? | ? | ? | ? |
| 7 | ? | ? | ? | ? |
| 8 | ? | ? | ? | ? |
| 9 | ? | ? | ? | ? |
| 10 | ? | ? | ? | ? |
| 11 | ? | ? | ? | ? |
| 12 | ? | ? | ? | ? |

## 17.5  LAYOUT OF HOUSEKEEPING DEPARTMENT

*Housekeeping layout* refers to location planning of different subdivisions of housekeeping department. When planning the layout of housekeeping department, the designer must consider certain factors such as architectural features; sequential work flow; hygiene; safety; directly and indirectly related activity areas; and wherever possible ensure that the plan does not impede the smooth flow of operations. An ideal layout of housekeeping department is shown below:

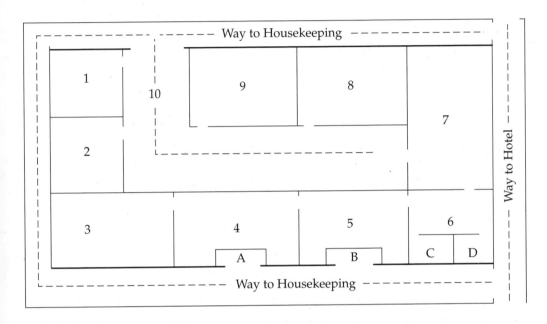

1 (Executive Housekeeper's Cabin), 2 (Control Desk), 3 (Linen Store Room), 4 {Linen & Uniform Room (for/with Hotel Linen and Uniform)}, 5 (Guest Linen), 6 (Tailor's/Sewing Room and Ironing Room), 7 (Laundry Section), 8 (Housekeeping Store Room), 9 (Flower Room), 10 (Way or route to enter and exit housekeeping department), A (Staff's uniform and hotel's linen exchange counter), B (Guest linen and other clothes exchange counter), C (Tailor's or Sewing Room), D (Ironing Room).

## 17.6  STRUCTURAL FOUNDATION OF HOUSEKEEPING DEPARTMENT

*Structural foundation* refers to underneath constitution of housekeeping department. In simple terms, it is related with different sections/subdivisions of housekeeping department which may be located in-house or out-house of the organization. An ideal structural foundation of housekeeping department (and its in-house and out-house areas) along with brief description is given ahead.

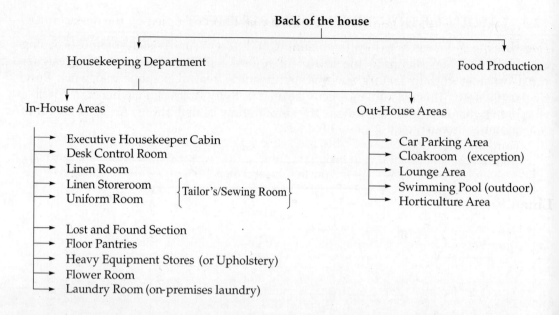

## 17.6.1   In-House Areas

*In-house* area refers to the internal areas of hotel building including guest rooms and interior public areas. For instance, guest rooms, food and beverage outlets, gift shops, lobby, guest waiting area, corridors and predominantly all front-of-house outlets/ sections from the perspective of guests. The responsibility of cleaning these areas is always on the shoulder of housekeeping department. Hotels may out-source the cleaning contract of their out-house areas but cleaning of in-house areas is never preferred to be out-sourced due to security reasons. A brief description of major in-house areas of housekeeping division is given below.

### Executive Housekeeper's Cabin/Office

- This is the main administration centre of the entire housekeeping department. This cabin/office is used by the key position of housekeeping department, i.e., executive housekeeper. It must be an independent cabin to provide the executive housekeeper a silent atmosphere to plan out the work and hold meetings.

- It should be a glass panelled office so as to give a view of what is happening outside the office. The location of this office must be in such a place so that executive housekeeper can see the cleanliness, maintenance and aesthetic appearance of most of the hotel areas, especially public areas like corridors, lobby, etc.

### Desk Control Room

- This is the main communication centre of housekeeping department from where all information is sent out and received by the concerned department. This control room is also known as *housekeeping desk* and occupied by the desk control supervisor who is responsible for coordinating with the front desk for information on departure rooms and handing over to cleaned rooms.

- Therefore, the desk control room should have a desk with a telephone and a computer system. It should have a large notice board for the staff schedules and day-to-day instructions. It is also considered a command centre where all staff members report for duty and check-out at the duty end. This desk also receives complaints on maintenance from housekeeping supervisors and transmits it to the maintenance department. It would be next to the executive housekeeper's office.

## Linen Room

- This section/room is headed by the linen room supervisor and used by the linen room attendants. It is the place where linen is stored for issue and receipt purpose. Therefore, the linen room should have a counter across which the exchange of linen can take place.

- The room should be next to the laundry (if on-premises) usually with an interconnected door between the rooms, so that the supply of linen to and from laundry is quick and smooth. But if it is off-premises then it must be near the service entrance.

### Tailor's/Sewing Room

- The working platform of tailer may avaliable either in linen room or uniform room.
- In-house tailor is responsible for mending and stitching torned linens/ uniforms.
- The tailor room is avoided if the mending and stitching works are contracted out.

- The linen room preferably should always be built near the service elevator for easy transportation to various units. Simultaneously, it must be away from the food handling units, especially kitchen, to avoid a fire hazard as well as prevent linen from absorbing food odours, smoke, soot and dampness.

- In large hotels, *linen storeroom* is also avaliable. This room stores the stock of new linen and cloth materials for stitching various hotel linen (such as bed and bathroom linen) and uniforms. This room should be cool and dry with ample shelves, generally 6 inches above the ground level.

## Uniform Room

- This section/room of housekeeping department is headed by the uniform room supervisor and used by the uniform room attendants. It is the place where hotel employees exchange their soiled uniform with clean ones. This room stocks the uniform in current use and sends collected soiled uniforms to the laundry section.

- Uniform room is often found as a part of linen room otherwise it is nearby or adjacent to the linen room and laundry section. It is due to the interdependcy and interrelation of involved departments' activities.
- Therefore, this room must have enough hanging spaces/facilities as many uniforms are best maintained when hung. Additional space is also required for keeping enough stock of uniforms for urgent use. Larger hotels may have enough space for an independent uniform store in addition to a linen store.

---

### Combined linen and uniform room

Often there is a common section/room meant for linen and uniform, especially in medium and small sized hotels (it also depends on the volume of linen and uniforms in circulation). Stock of new linen and uniforms may also be kept here. This stock is only used when the current uniforms and linen in circulation falls short due to damage, loss or any other reason. Common linen and uniform room also reduces the cost of employees as well as requirement of extra space. Staff who are working here will be responsible for both linen and uniform room functions like:

Tailors    Tailors or seamstresses are responsible for mending and stitching the linen, uniforms as well as upholstery.

Helpers    Responsible for the physical work of transporting, counting and then bundling the uniforms and linens

---

## Lost and Found Section

- A hotel's lost-and-found section is operated by the housekeeping department; maintains found items. This section is a small secure space with a cupboard to store all guest articles that are lost and may be claimed later.
- It is necessary to have strict control of this section. This section is controlled by linen room supervisor (hotel may also appoint seperate supervisor for this room) and only authorized people are allowed to enter such as executive housekeeper or assistant and linen room attendants. At the end of the each day, the linen room supervisor will ensure that this section should be properly closed and locked.
- This section/room is generally a part/subdivision of linen room (but may also be located in desk control room). The items found during day shift are directly sent to the linen room office whereas items found during swing or grave shift are sent to the front desk.
- The linen room supervisor will transmit any left item at the front desk to the linen room for proper storage and logging. The entry of each found item will be logged (with date) into the *log book* then placed in a bag, and this bag is marked with the logbook serial number. Next place this bag at the given shelves/cupboard, using a sequential numbering system so as to track the bag easily.

## Floor Pantries

- Each guest floor must have a pantry to keep a stock (plus *par stock*) of linen, guest supplies and cleaning supplies for that particular floor. It works as a housekeeping department's nerve centre for the respective floor in order to acheive effective and efficient result. Floor pantry is also known as *maid's service room*.
- The floor pantry should keep linen for that floor in circulation. Often *linen chute* is also found in floor pantry room from which all soiled linen collected from guest rooms of that floor is directly transmitted to the linen room. Floor pantry should be located near the service elevators and have shelves to stock all linen and other supplies.

## Supply Store and Heavy Equipment Store

- The *supply store* is a room that is used to store routine supplies such as routine guest room and bathroom supplies, cleaning supplies and so forth. This room is usually located nearby the housekeeping control room (or desk control room) and must contain enough shelves to store daily supplies. The room should be clean, dry and lockable.
- *Heavy equipment store* is a room that is used to store bulky items such as vacuum cleaners, shampoo machines, etc. This room is usually located in back area where movement of guests is very low. As this store room contains heavy and expensive housekeeping equipment, therefore it must be a clean, dry and cool. Simultaneously, it can also be locked when not in use.

## Laundry Section

- Laundry section is responsible for the cleaning of guest clothes and hotel linen. In hotels, laundry may be done in-house as well as out-sourced. Out-sourced laundry is also known as *out-house laundry*. The in-house laundry should be adjacent to the uniform and linen room in an attempt to avoid the unnecessary movement of staff. In addition, this also reduces the time needed for transferring clean linen and receiving soiled linen to/from in-house laundry.
- The laundry section usually performs the following two important tasks for guest as well as for hotel—wash and dry, thereafter supply well pressed guest clothes, uniforms and linen. In-house laundry section is headed by laundry manager who is responsible for the overall operation of laundry and for the supervision of laundry staff.

## 17.6.2 Out-House Areas

Out-house refers to external areas of hotel building. Out-house areas are also known as exterior *public areas* of the main hotel building which nowadays are often found to be on-contract to outside cleaning agencies. Therefore, hotel management does not need to hire additional manpower and resources. But in an attempt to obtain excellent cleanliness of public area, hotel may appoint public area supervisor.

## Parking Area

- It is an area which is used to park vehicles and the person who is responsible for parking guest vehicle is known as *valet parking attendant*. He is the first person who greets the hotel arrivals. In many cases, they are also the last employees who see off the departing guest. Thus, this job requires that an attendant must have a gracious and hospitable demeanour.
- The access to parking area is usually located immediately close to the entry gate. In many downtown hotels, the hotels may not have adequate space for parking provision for their guests. So, these hotels choose *outsource* (or outhouse) parking.
- When parking space is available on the premises, the cleaning of parking area is usually outsourced and public area supervisor is held responsible for ensuring and inspecting this area.

## Cloakroom

- It is a room that is used to hang guest's coats, hats, umbrellas, etc., for temporary hours. This room is used by both resident as well as non-resident guests. In luxury hotels, the décor of the cloakroom may also be luxurious—may have carpet, good wall covering, flattering lighting, etc.
- Cloakrooms (of both ladies & gents) are usually located near the entrance door and each room clearly states either men or women. Ladies cloakroom is also known as *powder room* (or *vanity room*) that usually has several separate water closets, wash basins with vanity units, large size mirrors with lighting.
- The cloakroom should always be neat and clean. The person responsible for the cleaning of this room is called *cloak room attendant*. The gents' cloakroom directly comes under the supervision of bell desk whereas the ladies cloak room goes under the supervision of housekeeping department.
- Cloakroom attendants are also responsible for maintaining the supply of hand towels, soaps and other toilet supplies for guest's use. He should also take care of the cleanliness and upkeep of the place.

## Lounge Area

- Lounges are provided for guests who wish to spend time in places other than their bedrooms and where they may be served with drinks (like tea, coffee, milk shakes, juices, etc.) in relaxed atmosphere.

- In city and transient hotels, the lounge may be an extension of the foyer whereas in resort hotels this area is separate from the main building. There may be a separate room set aside for television viewing, reading, writing, games and so forth activities.
- The furnishing of the lounge room should be comfortable & restful and furniture should be available for the guest to converse in small groups.
- In-house lounges should have pleasant lighting. The lighting of the lounge room should give an inviting appearance and *chandeliers* or lampshades should make them look attractive.

# Swimming Pool

- Swimming is perhaps the most popular of all recreational sports among guests. Many properties particularly resorts cater to this interest by providing swimming facilities. It can be located inside (indoor) or outside (outdoor) the hotel building. Its design is as varied as hotel operation and ranges from very basic to very elaborate setting such as some pool areas also include *whirlpool* and *saunas*.
- Most of the pool operating functions (especially cleaning) come under the responsibility of executive housekeeper (or public area supervisor) along with other team members, i.e., pool supervisor, lifeguard and pool attendants. Apart from cleaning, the rest of the tasks of swimming pool area are supported by engineering and maintenance department.
- For the purpose of changing clothes, properties also offer *cabana room* near the swimming pool. The cleaning of this room is also the responsibility of housekeeping department.

# Horticulture/Flower Room

- *Horticulture* room deals with the art, science, technology and business of plant cultivation. Often this section is also known as *floriculture*, especially when it only deals with the cultivation of flowering plants. Many medium and small sized hotels usually out-source their horticulture part whereas large sized hotels maintain this area in their establishments.
- Large hotels have professionally trained horticulturists who maintain the gardens and supply flowers for interior decoration. Flowers are mainly used in banqueting functions, guest rooms, restaurants, lobbies, offices, guest relation desk and so forth places. They are also supplied on the demand of guest. In hotels, the flower room is responsible for all flower arrangement and their placement in the hotel.
- Thus, flower room should be air-conditioned in order to keep the flowers fresh. The room should have work table, sink with water supply and all necessary tools and equipment required for flower arrangement. It can be a section of horticulture or operate independently.

## CONCLUSION

Housekeeping department is responsible for cleaning and maintaining hygiene and sanitation in the organization. For effective cleaning, it is also important that housekeeping must have proper equipment; these equipment may be manual, semi-manual or electronically operated. It is the front office department which generates and rooms *guest traffic* whereas housekeeping department is responsible for keeping the guest in the premises and generate repeat business by offering hygienic and pleasant atmosphere. Housekeeping department comprises different subdivisions like linen & uniform room, laundry section, control desk, horticulture, etc. Some of these subdivisions like horticulture and laundry section are often found to be outsourced. This classification may also be made on the basis of location of subdivision like whether subdivision is in-house or out-house. Apart from cleaning, housekeeping department is also accountable for placing/replacing amenities and supplies into guest rooms and bathrooms. It also offers loan items, expendables and other items to resident guests.

| **Terms** (with Chapter Exercise) | |
|---|---|
| Mop wringer trolley | ? |
| ? | It is a room facing/nearby swimming pool; it can never be used as a sleeping room. |
| Horticulture | It is the work of plant cultivation, especially used for decoration purposes. |
| Chamber Maid | It is a maid to clean and look after bedrooms. This position is also known as *room attendant.* |
| Guest supplies | All the necessary items that guest requires as part of the hotel stay like toilet tissues, hangers and so forth. |
| Chandeliers | ? |
| Guest loan item | All the items that are not maintained in the guest rooms but are available for guest on request through a receipted loan basis. |
| Guest amenities | All those items which give (free of charge) extra comfort to the guests or allow them to get something done, for example, a toothbrush, shoeshine cloth and so forth. |
| Pillow menu | It is a list of pillows available with housekeeping department, usually free of charge. It allows guest to make an alternate pillow choice. |
| Whirlpool | ? |
| Log book | Record at the housekeeping control desk in which all calls, requests and other significant information are entered by the housekeeping clerk for the next shift. |

| | |
|---|---|
| Box sweepers | Carpet sweepers for sweeping up dust and debris from soft surfaces as well as rugs and carpets. It consists of a friction brush. |
| Laundry | One of the major sections of housekeeping department which is primarily responsible to clean guest clothes, hotel clothes and linens. |
| Sacrifying | ? |
| Calendering machine | Often it is also known as *flatwork ironer* or *roller iron* used for pressing flat cloth materials like bed sheets, pillowcases, table clothes and so forth. |
| Linen chute | It is one way shaft which passes from every floor and opens into the laundry section. In satellite-linen-system, all collected soiled linen from guest rooms is passed through it. |
| Floriculture | Cultivation of variety of flowers, especially used for decoration. |
| Carpet beaters | Equipment used for beating carpets (or cleaning carpets by removing dust and dirt through beating process). Position carpet with their naps down then beat with rattan beaters instead of wire beaters. |
| Burnishing machine | Floor cleaning machine that places less weight on floor but allows speed up to 1000 rpm. Speed produces friction which creates high glossiness. |
| ? | A trolley used for public area cleaning or for special cleaning projects for guest rooms. It is used for carting and storing cleaning supplies. |
| Soiled | Covered or stained with dirt or other impurities. |
| Vanity room | Also known as *powder room*. |
| Upholstery | Stuffing fabric and other materials used in upholstering. |
| Wash | Water and soap or detergents to clean clothes and linens. |
| Rags | Discarded cloth pieces obtained from sewing room and used for applying polish and general cleaning. These are disposed of when heavily soiled. |
| Guest essentials | Items essential for the guest rooms and are not normally used up or taken away by the guests. |

# 18 CHAPTER

# Housekeeping Department: Organizational Structure

<hr>

$\langle$ **OBJECTIVES** $\rangle$

*After reading this chapter, students will be able to...*

- create an organizational structure of small, medium and large hotels' housekeeping department;
- identify the major and supportive positions in housekeeping; and
- describe the duties and responsibilities of staff involved in housekeeping operations.

## 18.1   ORGANIZATIONAL STRUCTURE

Almost all hotels generate maximum revenue from room division by selling intangible things especially rooms and extra services. Therefore, it is not wrong to say that intangible things (i.e., services) produce more revenue in comparison to tangible things (i.e., products). Hotels sell services, not products which guests can touch or feel, but the physical part of the hotel is what the guest sees and touches when they stay in the room. Each hotel has rooms, equipment, electronic gadgets and machines that all need to be taken care of. And for guest, room is the most important component among all of these. Therefore, a clean & comfortable room is an elementary requisite of guests in hotels.

The housekeeping department is responsible for the care and cleaning of various in-house and out-house areas. Often out-house areas are outsourced but in-house areas, especially guest rooms and public in-house areas are never outsourced. So it makes the job of housekeeping department one of the most important in the hotel, whether it is a large five-star hotel, chain property or a small two-star guesthouse. In order to perform all housekeeping tasks efficiently and effectively, hotel needs to design structured *span of control* system, so that authority and responsibility among all involved members can flow properly. This systematic relation among all involved members and *chain* or *flow of command* is collectively known as *organizational structure* or *hierarchy*.

## 18.2 ORGANIZATIONAL STRUCTURE OF HOUSEKEEPING DEPARTMENT

Housekeeping department depicts the formal relation among different positions of the department and its subdivisions. It establishes the flow of authority/chain of command which flows from top to bottom. It is important to have *job evaluation* of every position as it provides job details (or duties and responsibilities), especially *job description* and *job specification*. Remember, organizational structure always varies with the size of establishment, for instance, organizational structure of large hotels always carries more job positions than medium and small establishments. An ideal organizational structure of housekeeping department of small, medium and large hotels is shown below.

## Organizational Chart for a Small Economy/Limited-Service Hotel

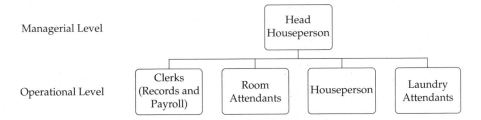

## Organizational Chart for a Medium Service Hotel

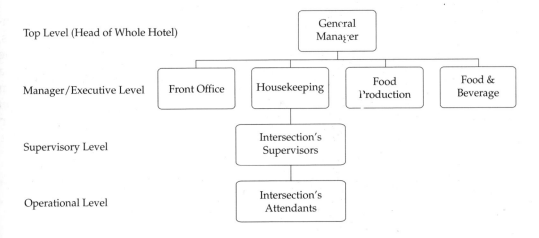

# Organizational Structure of a Large/Chain Hotel

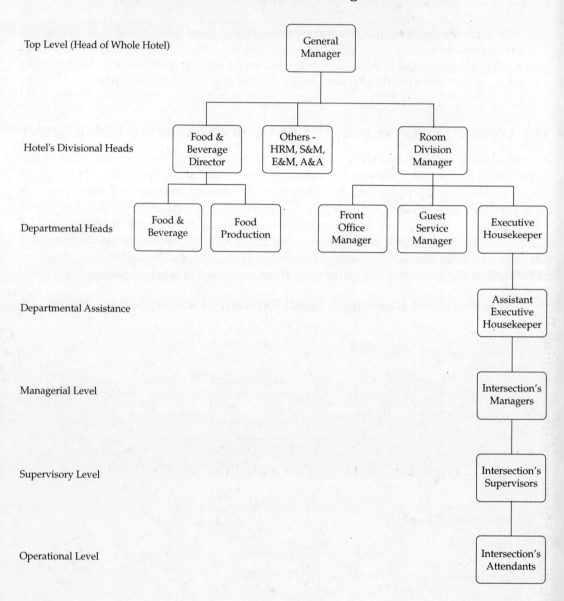

# Sample of Organizational Structure of Housekeeping Department

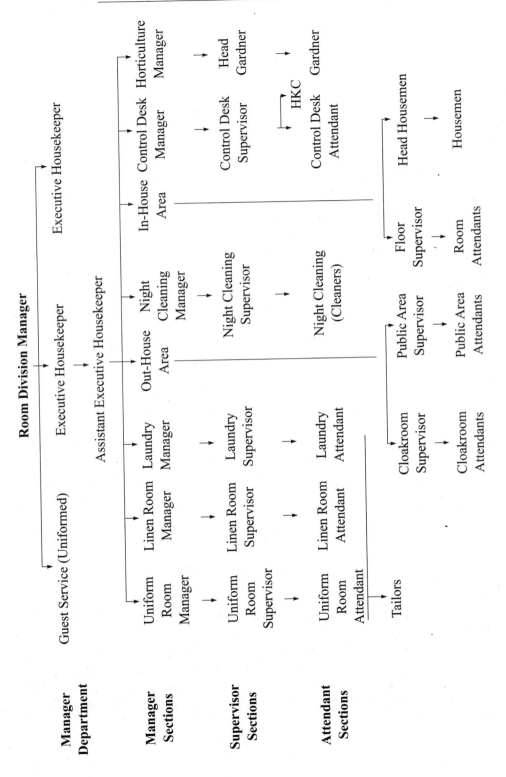

**Room Division Manager**

| Manager Department | Guest Service (Uniformed) | Executive Housekeeper |
|---|---|---|

Assistant Executive Housekeeper

| Manager Sections | Uniform Room Manager | Linen Room Manager | Laundry Manager | Out-House Area — Night Cleaning Manager | In-House Area — Control Desk Manager | Horticulture Manager |
|---|---|---|---|---|---|---|

| Supervisor Sections | Uniform Room Supervisor | Linen Room Supervisor | Laundry Supervisor | Night Cleaning Supervisor | Control Desk Supervisor | Head Gardner |
|---|---|---|---|---|---|---|

| Attendant Sections | Uniform Room Attendant | Linen Room Attendant | Laundry Attendant | Night Cleaning (Cleaners) | Control Desk Attendant / HKC | Gardner |
|---|---|---|---|---|---|---|

Tailors

Cloakroom Supervisor — Public Area Supervisor — Floor Supervisor — Head Housemen

Cloakroom Attendants — Public Area Attendants — Room Attendants — Housemen

## 18.3  DUTIES AND RESPONSIBILITIES OF HOUSEKEEPING STAFF

### Executive Housekeeper

He/she is the chief of housekeeping department, thus accountable for the planning (like budget), organizing, staffing (interview, selection and placement), directing (supervision and coordination), controlling and evaluating performance of all housekeeping employees. He assists in developing, delegating and monitoring departmental goals, objectives and programmes to ensure timely completion.

**Job Specification of EHK**

| | | |
|---|---|---|
| Reports to | - | RDM or GM |
| Qualification | - | HM/HS Graduate |
| Experience | - | 5 to 7 years |
| Expertise | - | Management |
| Proficiency | - | Conceptual skills |
| Supervise | - | All HK sections |

His/her major duties and responsibilities include to prepare standard operating procedure for all sections of housekeeping department then ensure that they are properly implemented; to instruct and advise staff on changes in policies/procedures/working standards; to conduct meetings with the housekeepers and other staff separately (to exchange ideas and solve problems); to supervise inspection performed by assistants of all areas in a frequent and steady schedule; he may also inspect personally; monitor departmental activities; identify correct supplier (for the purchase of housekeeping materials) thereafter recommend to purchase department; to develop lost and found procedure and make policy for the disbursement of unclaimed articles to employees. He works in close coordination with General Manager and head of other departments. Finally, he prepares a report required by the management.

### Assistant Executive Housekeeper

He works under the supervision of executive housekeeper and is responsible for the management and proper utilization of resources given by the Executive Housekeeper to achieve the stated common objectives of cleanliness, maintenance and attractiveness in the given shift. The assistant executive housekeeper may be one for each shift in a large hotel; may be the housekeeper of a small hotel or the only deputy to the Executive Housekeeper of a medium-sized

**Job Specification of AEHK**

| | | |
|---|---|---|
| Reports to | - | E. Housekeeper |
| Qualification | - | HM/HS Graduate |
| Experience | - | 3 to 5 years |
| Expertise | - | Management |
| Proficiency | - | Managerial skills |
| Supervise | - | All HK sections |

hotel. He passes notices, instructions and memos given by the executive housekeeper to supervisors, consequently informs the executive housekeeper about the grievances of employees, if any. He works as a middleman between executive housekeeper and supervisors of different housekeeping sections and in the absence of executive housekeeper he resumes all his duties and responsibilities.

## Housekeeping Supervisors

The term *housekeeping supervisor* does not refer to a single position; but to the supervisors of different sections of housekeeping department like floor, uniform, linen, public, night, etc. The duties and responsibilities of all these positions are almost identical. It includes deployment, supervision, control and training of staff working under them; performing surprise inspections; organizing induction (for newly appointed staff) and training programme for staffs; intimating

| Job Specification of HKS | | |
|---|---|---|
| Reports to | - | A. E. Housekeeper |
| Qualification | - | HM Graduate |
| Experience | - | 2-3 years |
| Expertise | - | Supervision |
| Proficiency | - | Interpersonal skill |
| Supervise | - | Desk activities |

employees about their welfare schemes and other benefits; planning duty roaster; deciding weekly off and also making necessary arrangement for these weekly offs keeping inventories and records of equipment and supplies; record all lost and found items; resolving room status discrepancies; liaisoning with the maintenance department and so forth. Supervisors also check the standard of cleanliness of all areas of the hotel and instruct attendants about their routine activities. Finally, supervisors report to assistant executive housekeeper about staff's working performance.

### Activity

Write down the major duties and responsibilities of respective area supervisors in given spaces who work under different sections of housekeeping department. Few points have been given for your reference:

| Supervisors | Major duties and responsibilities |
|---|---|
| Uniform room | • To provide clean and serviceable uniforms to the employees of the hotel.<br>• To keep inventory control on all uniforms and prepare the budget for them.<br><br>? |
| Desk control | • To provide and receive information concerning housekeeping (to guest and staff).<br>• To coordinate with the front office for information on departure rooms and handing over cleaned rooms.<br><br>? |
| Linen room | • To purchase, store, issue and maintain linen.<br><br>? |
| Night shift | • To handle all aspects of housekeeping during night shift including desk control operation, issue of linen or uniform in an emergency.<br>• To take all necessary decisions in relation to guest rooms, public areas, linen and uniform room. |

| Floor | • To ensure cleanliness, maintenance and attractiveness of allotted guest floor/s.<br>• To inspect guest rooms, corridors, staircases, pantries, elevators, etc., of allotted floor/s. |
|---|---|
| Public area | • To ensure cleanliness, maintenance and attractiveness of all public areas (like restaurant, bars, banquets, gardens, health club, swimming pool, main entrance, parking area, etc.)<br>? |

## Housekeeping Coordinator (HKC)

In order to enhance guest satisfaction, most hotels have an additional position of *housekeeping coordinator* or *order taker*. He/she is a clerk or secretary in the housekeeping department who provides administrative assistance to the housekeeping department. He/she is principally responsible for responding to guest requests and complaints over telephone. He/she answers the phone call promptly and coordinates the fulfilments of guest requests as well as also logs all calls

**Job Specification of HKC**

| | | |
|---|---|---|
| Reports to | - | EH or AEH |
| Qualification | - | HM Diploma |
| Experience | - | 1 to 2 years |
| Expertise | - | Administration |
| Proficiency | - | Human relation |
| Take care | - | Guest orders |

in the *message book* and follow-up to make sure tasks are accomplished quickly. His/her other duties and responsibilities include tracking attendance and working through payroll procedures, assisting work assignments for room attendants and supervisors; reporting room status to proper departments; maintaining accurate records of room status; securing keys in accordance with hotel policies; and taking lost and found calls.

## Housekeeping Attendants/Runners *(in general, duties and responsibilities are almost identical for all attendants /runners whether working in guest rooms or public areas)*

They work under the direct supervision of respective area's housekeeping supervisor and are principally responsible to clean the allotted guest rooms (V/D rooms) or public areas (such as floor/lobby, corridor, restaurant, bar, coffee shop, etc.) on routine basis; to clean the common toilets and bathrooms on a routine basis; remove and dispose of the refuse and rubbish items at the designated area; give report of missing or broken property to the executive

**Specification**

| | | |
|---|---|---|
| Reports to | - | H. Supervisors |
| Qualification | - | High School |
| Experience | - | 6 to 12 months |
| Expertise | - | Operational work |
| Proficiency | - | Practical skill |
| Supervise | - | Houseman |

housekeeper; follow the laid down SOPs, given instructions and cleaning procedures; take care of cleaning materials and ensure economical use of supplies given from the department; attend daily briefings and de-briefing session; hand over the lost and found articles to the supervisor; maintain a polite, dignified and helpful attitude towards the guest and so forth.

## Activity

Write down the major duties and responsibilities of housekeeping attendants in given spaces who work under the different sections of housekeeping department.

| Attendants | Major duties and responsibilities |
|---|---|
| Guest room | To do the actual cleaning of assigned guest rooms and bathrooms |
| Cloakroom | ? |
| Linen room | To exchange linen and fill-in related records as necessary |
| Lost & found room | |
| Linen store room | To ensure that the stock maintained should be great enough to replenish the whole hotel at a time.<br>Also ensure that stocked linen should only be used when required; do not use unnecessarily. |
| Public area | ? |
| Uniform room | To issue uniforms, simultaneously receive the soiled ones from the staff<br>To transfer the soiled linen into laundry section |
| Houseman (of medium/small size hotel) | ? |
| Head houseman (of medium/small hotel) | Supervise the areas assigned by the executive housekeeper and often (in medium sized hotel) works on behalf of Public Area Supervisor, especially during night shift. |

## Horticulturist

He is overall incharge of horticulture section thus predominantly responsible for complete supervision and smooth functioning of in-house garden. He maintains in-house garden/s and supply flowers for interior decoration. He also assists housekeeper in preparing different styles of flower arrangement. His major duties and responsibilities include preparation of budget for horticulture section; making duty roster of gardeners & deciding their weekly *offs*; market survey for best supplier to purchase garden commodities; purchasing in-door and out-door plants; providing training assistance to head gardener and gardners and so forth. Nowadays, this section is mostly contracted out to external agency.

| Specification | |
|---|---|
| Reports to | - EHK |
| Qualification | - Graduate in HC |
| Experience | - 6 years in nursery |
| Expertise | - Botanical aspects |
| Proficiency | - Horticulture |
| Supervise | - Head gardener/s |

## Head Gardener

He works under the supervision of *horticulturist* and is responsible for the supervision of gardeners and keeping them updated each season. He must have proficency about seasonality of plants (both indoor as well as outdoor) and their maintenance. His major duties and responsibilities include ensuring proper landscaping of garden, seeds planted at desired place, lawn maintenance training (like how to handle garden tools and machines) and briefing gardeners; preparing duty roster of gardeners and determining their weekly *offs*; selecting reliable supplier and procuring seeds at reasonable price; procuring, controlling and supervising the usage of manure and fertilizers; maintaining nursery and ensuring steady supply of saplings for planting.

| Specification | |
|---|---|
| Reports to | - Horticulturist |
| Qualification | - High School |
| Experience | - 4 years in nursery |
| Expertise | - Nursery + Green H. |
| Proficiency | - Gardening |
| Supervise | - Gardeners |

## Gardener/s

He works under the supervision of head gardener and is principally responsible for the actual digging, planting, pruning, watering, etc., of plants in garden on a day-to-day basis. His work profile include attend daily *briefing*; planting seeds and saplings; applying manure and fertilizer; taking care of newly cultivated plants; spraying insecticides, pesticides and fungicides; cutting, trimming pruning hedges, bushes, flowers. He is responsible for providing flowers, garlands, wreaths, bouquets as and when required by hotel. He also maintains the nursery, indoor and outdoor plants. It is always favourable that gardener should be techno savvy, creative in arranging the plants and knowledgeable enough to utilize garden tools correctly.

| Specification | |
|---|---|
| Reports to | - Head Gardener |
| Qualification | - High School |
| Experience | - 2 years in nursery |
| Expertise | - Plantation |
| Proficiency | - Gardening |
| Supervise | - None |

all the above given boxes, the synonym/term HM-Graduate is used for Hotel anagement Graduate, HS-Graduate for Home Science Graduate; EH for Executive ousekeeper; AEH for Assistant Executive Housekeeper; and HKC for Housekeeping ordinator.

## Activity

Vrite down the major duties and responsibilities of the given supportive housekeeping taff who work under the different sections of housekeeping department.

| Supportive staff | Major duties and responsibilities |
|---|---|
| unner | • Attends all guest rooms for collecting or delivering guest laundry |
| lat checkers | ? |
| ailor | • Mends all damaged linen and uniform; refurnish all damaged upholstery and stitch new linen and uniforms. |
| alet | ? |

## ONCLUSION

he dimension of housekeeping department varies with the size of organization, r instance large hotels always comprise more operational positions than small and edium hotels. Consequently, number of supervisory/managerial positions also vary. order to accurately understand all these dimensions and their formal relations, the rrect term is *organizational structure*. Organizational structure is a kind of chain of mmands which shows the formal relation among all levels of involved employees. emember, organizational structure also shows that every senior is junior of his/her nior and every junior is senior of his/her junior. For achieving maximum output, ery job position should be thoroughly broken down into *job description* and *job ecification*, thereafter selection should be made in accordance with the *job evaluation*. nd for creating sound internal control system, *span of control* should be followed and r motivating employees, delegation of authority should also be carried out. This apter also depicts the job profile of different housekeeping employees.

| Terms (with Chapter Exercise) | |
|---|---|
| ob description | Work profile of any job. |
| Runner | An attendant/employee who is assigned the duty of collecting and delivering laundry/items to guest rooms and sometimes on floors. |

| ? | A person responsible for laundry service of house guests. |
|---|---|
| Briefing | A meeting session before the starting of any shift. |
| De-briefing | ? |
| ? | It is concerned with doing housework, like room making, bed making and laundering in the hotel. This work is usually done under an Executiv Housekeeper. |
| Job specification | Specific demands and requirements of any job position for which specifi prerequisite qualification and experience is mostly required. |
| Room attendant | A hotel employee responsible for cleaning guest rooms. |
| ? | Person responsible for the cleanliness, maintenance and service in publi area like guest washrooms. |
| Public area attendant | A hotel employee responsible for cleaning public spaces, back-of-the house areas, meeting and banquet rooms. |
| Message book | Record of incoming calls including the callers, their purpose, and th time of the call. Also known as *message log*. |
| Attendance | ? |
| Chambermaid | A hotel employee who takes care of the housework in the guest rooms. |
| HK Coordinator | A housekeeping position, i.e., a clerk or secretary in the housekeepin; department. |
| ? | A specified time period in which different groups of people work togethe for the organization. |
| Work assignment | A pre-printed list of the rooms that need to be cleaned. This list also state the room condition at the end of the shift. |
| Intangible things | Non-touchable things which can only be felt but never touched. |
| Tangible things | ? |
| Span of control | The extent (or width) of number of employees who work under on supervisor. It is a kind of employee control system. |

| Organizational structure | Formal structural foundation of chain of commands in which authority and responsibility flows from top to bottom whereas employee grievances and suggestions flow from bottom to top. |
|---|---|
| Chain of commands | ? |
| Payroll | Monthly calculation of how much salary each employee draws. |

# 19 CHAPTER

# Linen and Uniform Room

OBJECTIVES

*After reading this chapter, students will be able to...*

- describe what is linen and uniform;
- draw the layout of linen and uniform room;
- express the functions of linen and uniform room;
- identify the bedroom, bathroom and restaurant linen; and
- explain the stages involved in linen exchange cycle.

## 19.1 LINEN

In commercial sense, the term *linen* is used in cloth manufacturing field and defined as a stem fibre extracted from the cultivated plant called *flax*. Thereafter, this stem fibre is spun into yarn and woven into fabric. Nowadays, the term *linen* is mostly found in hospitality industry and it can be classified on the basis of their uses in this industry, like bed linen, bathroom linen and dining room linen. A place where all these linen are exchanged on daily basis is known as *linen room* and where the extra stock is kept for emergency or future use is known as *linen store room*. Linen room is headed by linen room supervisor who is responsible for taking work from linen room attendants. A brief explanation on types of linen is given below.

### Guest Room Linen

✓ The term *guest room linen* refers to all the linen that is required to be placed in guest rooms and replaced on daily basis in occupied rooms. These can be further classified as *bedroom linen* and *bathroom linen*.

### Dining Room Linen

✓ The term *dining room linen* refers to all the linen required in different food and beverage outlets. These can be further classified as *table linen* and *back area work linen*.

## Activity

Complete the table by filling in the columns with appropriate bedroom linen, bathroom linen, table linen and back area work linen.

| S. No. | Bedroom linens | Bathroom linens | Table linens | Back area work linens |
|--------|----------------|-----------------|--------------|------------------------|
| 1 | Bed sheet | Bath sheet | Table cloth | Wiping cloth |
| 2 | | | | |
| 3 | | | | |
| 4 | | | | |
| 5 | | | | |

## 19.2  UNIFORM

It refers to the uniform worn by various hotel personnel, for example, waiter's uniform and service jackets, chef's uniform (like apron, cap, coat, pant and neck scarf), housekeeping uniforms and protective clothes, uniform of guest service department (like doorman, bell desk, valet attendant, etc.), cleaner's overalls and so forth. In large hotels, apart from *linen room* there is a separate room for the management of employees' uniforms, i.e., *uniform room*. The employees' uniforms are exchanged over the counter on daily basis.

## Activity

Match the uniform with the respective employee.

| S. No. | Employees | Uniform |
|--------|-----------|---------|
| 1 | Managers or stewards | Aprons |
| 2 | Doorman | Service jackets |
| 3 | Silver polishers | Blouse & sarees |
| 4 | Kitchen stewards, horticulturist | Dungarees |
| 5 | Cooks, drivers and parking attendants | Bush shirts |

| 6 | Restaurant stewards | Black bows |
| 7 | Receptionist & restaurant hostess | Ties |
| 8 | Cooks and utility employees | Gumboots |
| 9 | Managers and other front line employees | Woollen overcoat |
| 10 | Engineering technicians and houseman | Rubber slippers |
| 11 | Mainly for security employees | Caps |
| 12 | Health club, laundry and pool area employees | Turbans and turras |

## 19.3   FUNCTIONS OF LINEN ROOM

The *linen room* is one of the major sections of housekeeping department which supports the whole organization by supplying neat and clean linen, uniform and clothes on daily basis, and simultaneously also supports the guest laundry/ dry cleaning. Linen room can be centralized as well as decentralized. *Centralized linen* room acts as a storage point and distribution centre for clean linen whereas the term *decentralized linen room* (satellite-linen-room system or floor pantries) is related with floor pantry. Centralized linen room should have a counter, across which the exchange of linen can take place. The room should preferably be adjoining the laundry so as to supply linen to and from the laundry. Apart from linen exchange (or issues & receipts), this room is also responsible for repair/mending, renewal, maintenance of proper inventory and stock records of all linen items.

**Linen room equipment**

- Linen room counter
- Storage hamper
- Mobile convertible shelving
- Sewing machine
- Hopper
- Glass washer
- Wicker baskets
- Canvas bags
- Vinyl hampers
- Step ladder
- Iron with ironing board
- Telephone
- Computer systems
- Working tables

Linen room also plays a crucial role in laundry operation because if the laundry is *on-premises* then hotel linen is first collected in the linen room then sent to the laundry (but guest clothes can be directly sent to laundry section). On the other side, if the laundry is *off-premises* then both the guest clothes and hotel linen are collected in the linen room then sent to off-premises for laundering. Thereafter, the laundered linen including guest laundry is collected at the linen room from where it is sent to the guest rooms and other service points. The standard working hours of linen room is from morning 7 a.m. to evening 7 p.m. and during night duty, manager or night attendant supplies the items from emergency store and leaves a note with details of what has been removed. A list of various functions of linen room is briefly given below:

## Collection of Linen

✓ It is collecting of linen from various respective areas of hotel. It involves further process of *sorting, marking* and *counting* of linen for the purpose of effective control of linen. The linen collection is facilitated either through *chutes* (or *linen chutes*), canvas bags, wicker baskets or trolleys.

✓ After collection, linen is sorted out primarily to make counting easy as well as it also streamlines the laundry task. Thereafter, it eases in tallying the exchange of linen between linen room and laundry personnel.

✓ Similar items are placed in one bag. And remember badly soiled linen must be marked and wrapped separately as these require special attention.

## Routing of Linen

✓ It refers to transferring of collected, sorted and counted soiled linen to the respective areas for the purpose of cleaning, i.e., in-house laundry or to out-house laundry. *Wicker baskets* are used to route linen into in-house laundry whereas *vinyl hampers* and *canvas bags* are used to transfer linen into out-house laundry.

✓ In case of out-house laundry, a list of soiled linen is prepared in two copies: first copy is attached with soiled linen bag (or handed over to linen pick-up boy) and sent to out-house laundry whereas second copy is kept in linen room.

## Repairs/Alterations

✓ It refers to the mending of linen which must be carried out before laundering but soiled and wet articles are always preferred to be mended after laundering. Linen are mended by stitching or darning. Bed linen, table linen and towels are often found with small holes and cuts. These are repaired by darning machine. Alteration of uniforms for correct fit is usual.

✓ Condemned lines are not preferred for repairs, as these are converted into useful items called *cut-downs/makeovers*. These cut downs can be used for many cleaning purposes and lastly can be converted into *rags*.

## Monogramming (for new linen)

✓ It is the labelling or marking of linen with property name or department or individual employee name, depending on the requirement. Marking forms may include marking with pens/marker, heat seal machine, iron on or sew-on labels, embroidered or woven.

✓ Monogramming is usually positioned on right hand side of the article whereas marking may be done on any article except aprons, dusters, waiter's jacket, etc.

## Exchange Linen

✓ Linen is exchanged on the basic principle of "One Clean for One Soiled" rule In in-house laundry, linen is exchanged at designated hours whereas out-house laundry delivers linen once or twice a day. In general, the hotel linen is usually delivered in the morning hours whereas guest linen is delivered during evening hours by in-house laundry.

✓ But with out-house laundry, the frequency/schedule of delivery is must to be considered which mainly depends on the amount of linen being sent and the distance of out-house laundry from the hotel premises. In large hotels, the exchange of linen usually takes place on daily basis.

✓ In case of out-house laundry, extreme care is required at the time of exchange of linen because both routing of soiled linen and receiving of clean linen take place simultaneously. Thus, the basket or bag of soiled and cleaned linen should be kept separate.

| Exchange Linen | |
|---|---|
| Soiled linen | It can be either directly exchanged over the linen room counter or first listed then bundled and brought to the linen room at designated time or directly despatched through linen chute to the linen room or frequently collected from the floor linen rooms/maid's service room. |

## Linen Inventory Management

✓ It refers to the storage and rotation of linen (linen must rotate on FIFO principle As linen is one of the most expensive and daily-use article of any hotel, therefor its systematic storage is vital. The linen shelves should be marked for each type of linen to store.

✓ Linen room must have sufficient cupboards and shelves for each specific item an should be lined with *baize cloth*. The shelves must be built on convenient heigh and firmly fixed.

✓ Linen room should be warm and the shelves should be positioned in such a way so that it allows circulation of air. Linen must be stored on paper lined shelve and kept free from dust.

| Linen Rotation | |
|---|---|
| Regularly used | Linen such as restaurant tablecloth, tea towels, glass clothes, etc. To make counting and proper rotation easier, the linen should be stacked with the folds outwards and small linen like napkins are placed in bundles of 10 or 12 and secured with rubber band. |
| Irregularly used | Linen like banqueting tablecloth, curtains, extra blankets, condemned sheets (or linen covers), etc. These should be wrapped and placed at separate shelves in cupboards. |

## Linen Inventory Control

- ✓ It is stocktaking of linen for the purpose of linen available in linen room. Regular stocktaking supports in maintaining *minimum and maximum stock level*, determining *re-ordering point* and also intimates about any loss of any article.
- ✓ *Physical inventory* control method is more effective than *perpetual inventory control* system. Stocktaking should be done at regular intervals or at any frequency in-between, if required. In order to prevent linen discrepancies it is better to perform stocktaking on daily basis, especially when all stock items are taken on the same day.

## Distribution

- ✓ It refers to the division and delivering of laundered linen to the respective areas. Hotel linen is distributed from the linen room counter whereas guest linen is directly delivered to their rooms. Some hotels use other systems of distribution such as *topping-up* or a fixed issue based on forecasted occupancy.
- ✓ When linen is issued from the linen room counter, linen room supervisor designates the specific timings for issue of linen. But before issuing, it is vital to inspect the wash quality of linens, thereafter, at the time of linen exchange, linen attendant must ensure the amount of laundered linen tallies with the amount of soiled linen.

## Maintaining Guest Loan Items

- ✓ It refers to guest supplies usually not found in guest rooms but may be available on guest request. These items are of everyday use but guest may have forgotten to bring. For instance, hair dryer, adaptor, extension cod, transformer, razor, iron with its board and so forth.
- ✓ These items are usually stored in linen room and when guest requests it is delivered to the guest room along with receipt form. The receipt form indicates the time period guest loan items are provided to guests. They also vary from property to property depending on the market segment/s the hotel attempts to reach and satisfy.

## Security

✓ The security of linen room is utmost in order to prevent misuse and pilferage and to guard against fire breakouts. Therefore, only authorized employees must be allowed to enter the linen room, except during emergencies for other staff. For instance, duty manager may have authority to remove items from the linen room but he should leave a note with details of what has been taken.

✓ Linen room supervisor should strictly make the linen room a non-smoking area. Linen room must be kept locked during the non-working hours and its key handed over to a designated employee, in accordance with the hotel's rules and regulations.

## 19.4   LAYOUT OF LINEN ROOM AND LINEN EXCHANGE CYCLE

Linen exchange cycle is concerned with sequential flow of linen room activities whereas the term layout is related with the position of different subdivisions, equipment, working counters, cupboards/racks and passageway of linen room.

### Linen Exchange Cycle

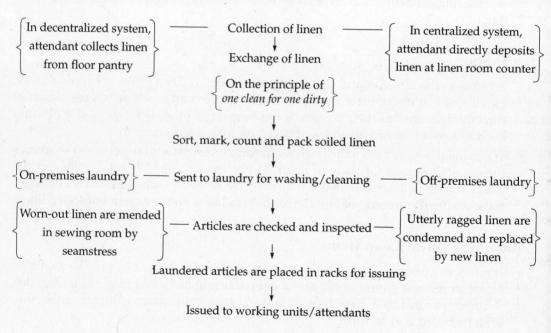

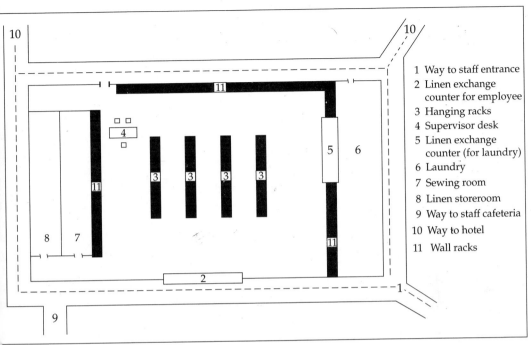

1 Way to staff entrance
2 Linen exchange counter for employee
3 Hanging racks
4 Supervisor desk
5 Linen exchange counter (for laundry)
6 Laundry
7 Sewing room
8 Linen storeroom
9 Way to staff cafeteria
10 Way to hotel
11 Wall racks

**Layout of Linen Room**

## 19.5 BED-LINEN OR BEDROOM LINEN

The term *bed-linen* or *bedroom linen* refers to various individual pieces of linen required for bed making or for laying over the bed. For instance, pillowcases, bedsheets, blankets, bedspreads, pillows, comforters, coverlets, mattress protector, shams, quilts, dust ruffles and so forth. These bed linen can be made from linen, *flannelette*, cotton or synthetic fibres. These can also be made from the combination of various materials, for instance blend of linen/cotton fibres or polyester/cotton fibres. Most hotels prefer white bed linen whereas other hotels also use colourful bed linen. A brief description of major bed linen is given below.

## Bedsheet

✓ First, bedsheets should be long enough so that they can properly tuck around the bed. Bedsheets should have soft & smooth texture and made from non-crease fabrics. Simultaneously, it should be absorbent and free from static.

✓ The bedsheets must have equal hem. Bedsheets are often called sheets but they are different from *re-sheeting*. At the time of purchasing bedsheets, purchaser should take care that they should be easily launderable and should not fade in colour under repeated washes.

| Particulars | Bed Sizes (in inches) | Bed Sheet Size (in inches) | Mattresses (in inches) | Quilts (in inches) |
|---|---|---|---|---|
| Single | 34 × 75 | 66 × 104 | 34 × 75 | 55 × 83 |
| Twin | 38 × 75 | 78 × 104 | 38 × 75 | 60 × 83 |
| Double | 54 × 75 | 81 × 104 | 54 × 75 | 71 × 83 |
| Queen | 60 × 80 | 90 × 110 | 60 × 80 | 83 × 83 |
| King | 78 × 80 | 108 × 110 | 78 × 80 | 95 × 83 |

## Pillows and Pillowcases

✓ Nowadays, the traditional concept of *bolster* has been replaced by two pillows. Pillows are placed on the top of the bed (near the headboard) for the purpose of head rest. It can be filled with various fillings which may be either soft or hard. But good pillow should be resilient, evenly filled (with no lumps) and light in weight. The pillow cover and its filling should be fire retardant and *ticking* should be water and stain proof.

✓ Pillowcases are used to cover pillows that augment the life of pillows, simultaneously also prevent it from stains. These are made from the same fabrics used for making bed linen. Pillows may be filled with either *goose down, synthetic fibres, expandable polyurethane, kapok* or *small feathers*.

| Pillow | Size (in inches) | Pillow cases | Size (in inches) |
|---|---|---|---|
| Standard size | 20 × 22 | Standard size | 20 × 26 |
| King size | 20 × 36 | King size | 20 × 40 |

## Blanket

✓ It refers to a bed linen that works as an insulator, as its basic motive is to make the body warm by keeping the body heat *in* and cold air *out* of blanket. A good blanket is light in weight, soft, smooth and resilient. Nowadays blankets are replaced by *duvet covers*. Duvets are made up of filling that is stitched in a fabric case with a changeable cover.

✓ Comfortable and durable blankets are made from woollen that is often blended with synthetic fibre (or acrylic) whereas its less expensive alternatives are made from nylon

**Standard Size (in inches)**

Blanket Single
70 × 100

Blanket Double
90 × 100

M. P. Single
64 × 96

M.P. Double
90 × 96

fibre. Nowadays, the concept of electric blanket replaces the traditional standard blankets. The way a blanket is woven also plays a crucial role in making it best for warmth.

## Mattress Protector (M.P.)

✓ *Mattress* is a vital part of bedding usually available in rectangular shape pad of heavy cloth and filled with soft material or an arrangement of coiled springs. There are usually five types of mattresses, i.e., innerspring, foam, latex, water and air-inflated. These mattresses may be (or its surface) medium, firm, extra firm or super-firm.

✓ Mattress protector is a kind of waterproof or water-resistant pad used to protect the mattress from stains and spills. It is also known as *mattress pads, mattress topper, overlays* or *bed toppers*. But remember, mattress pad is different from pillow-top mattress.

## Bedspread

✓ Bedspreads are a kind of *coverlet* used to cover laid bed linens. There are two main styles of bedspreads, i.e., throw spreads and tailored spreads. A bed spread may reach the floor, covering the mattress, frame and box spring. If coverlets are used, a *dust raffle* is added to cover the box springs and the frame.

✓ The dust raffle is usually cleaned when the bedspread is cleaned. In a formal bed making, the bed is also decorated with *shams* whereas *quilted comforters* are preferred to be put on bed in an informal bed making. These are also known as *bedcover* or *counterpane*. Around 8 metres of fabric are required to stitch a single bedspread.

| Styles of Bedspreads | |
|---|---|
| Tailored spreads | Fit the corners of the mattress snugly |
| Throw spreads | Bulge at the corners and near the foot of the bed |

## Draperies

✓ It refers to various curtains used to cover the open space like windows and between the rooms (on doors). For instance, sheer curtains/net curtains/glass curtains combined with heavy draperies are usual in a guest room. Nowadays, draperies are available in various styles, materials and shades.

✓ The fabric used for curtains and draperies should be fire resistant, soil & wear resistant, wrinkle resistant, resistant to molds, mildews and sun damage. In hotels, generally white, vinyl-lined fabric is preferred for draperies because of its durability. A heavy fabric is used for public areas whereas a lighter one for guest rooms. Remember, dry-cleaning is a perfect method for draperies as it holds the shape better than laundered fabrics.

## Cushion Covers

✓ Cushions are kind of pillows mostly placed on sofas or couch. In order to enhance the durability of cushions, cushion covers are used. Cushion covers should be launderable and non-crease. It is also vital that they should be dirt resistant and should not accumulate dust and sagging.

✓ The fabric used to make cushion covers should be non-slip without being rough and must be free from static so that it does not cling to the guest's clothes. It should also easily not lose colour or lint.

## Upholstery

✓ It is a soft padded textile covering that is fixed to furniture such as chairs and sofas/couches. The fabrics used to make upholstery should not stretch after they have been fitted. Used fabric should be firm with a close weave.

✓ Generally, blends of natural and synthetic fibres are used to make upholstery. Upholstery can be woven (textile) into knit (*stretch knits* and *double knits*) as well as into a jacquard texture and fine texture. The upholstery should be durable, match with decor, resistant to snagging, abrasion, soil and fading. It should also be non-flammable, pest-proof, easily cleaned and have stretch recovery characteristics.

| Soft Furnishings | | |
|---|---|---|
| The term *soft furnishing* is used for all those articles/materials made up of clothes and used to decorate a room. These are also known as *software items* but both are different. Remember all soft furnishings must match the decor of the room. A list of furnishings is given below; use Y if it is a soft furnishing or N if it is not. Also give appropriate description. | | |
| **Soft furnishings** | **Brief description** | **Y or N** |
| Blankets | | |
| Bedspreads | | |
| Curtains | | |
| Bathroom linen | | |
| Draperies | | |
| Bed linen | | |

## 19.6   BATHROOM LINEN

It refers to the linen required in guest room's bathroom, for example, range of towels, bathrobe, wash cloth and so forth. Bathmat is also a part of bathroom supplies. Hotel must be very careful while selecting bath-linen, especially towels. Towels should be gentle on the skin, with a high degree of absorbency and lint-free. Remember, colour of bath-linen should also be chosen carefully like colourful towels may be selected for public areas like swimming pool whereas white towels are always preferred for guest room's bathroom. The different colour scheme may be used for the purpose of identification. A brief description of various bath-linen is given below.

## Bath Sheet

✓ It is an extra large bath towel, usually 35 inches long and 60 inches wide, which gives the power of extra absorbency and substantial size. Bath sheets, due to size and volume, add considerable luxury to the bathroom amenities.

✓ Due to size and absorbency, bath sheets are expensive. The textile or fabric used to make bath sheets is same as for bath towels. In hotels, generally white bath sheets made from cotton, rayon, bamboo, non-woven fibres or made with similar type of other materials are preferred.

## Bath Towel

✓ It may be of linen or cotton but bath towels are invariably made from Turkish towelling, i.e., *terry weave* which has a looped pile on both sides. The durability and strength of towel depends upon the foundation material used to make a towel. A blend of polyester/cotton foundation cloth is often used for strength but it reduces the absorbency power of towel.

✓ A good bath towel should have close pile (for greater absorbency), short thread (as long threads may pull), strong selvedge and firmly stitched corners of the hem. In hotels, white towels are mostly preferred as they reflect the hygiene, simultaneously due to regular laundering with alkaline the colourful towels may fade.

| Activity | | |
|---|---|---|
| Differentiate between bath sheet and bath towel. | | |
| **Base/Criteria** | **Bath sheets** | **Bath towels** |
| Dimension | ? | Small in dimension |
| Cost | Expensive than bath towels | ? |

| Preferred by | ? | Mainly small properties and in homes, people with small height |
|---|---|---|
| Match with decor | Not easily matched with other in-room fabrics | ? |
| Further use | Due to extra surface area and absorption, it can be used at poolside or beaches. | ? |

## Shower Curtain

✓ Actually, it is not bathroom linen; it is a kind of curtain used to divide the bath area (bathtub) with remaining bathroom space. The best type of shower curtain for any hotel operation is a curtain made of 260 denier nylon with a white or pastel coloured *vinyl liner*.

✓ 260 denier nylon is better than plastic curtain because it is easy to maintain, it does not show soap stains, plus it is mildew resistant and it does not stiff or get brittle over time.

## Hand Towel

✓ It may be of linen or cotton. Traditionally, *huckaback* towel is offered in guest rooms as it has a close, fancy weave and mostly made of linen. But nowadays, Turkish hand towels are offered in most star hotels.

✓ These hand towels are also provided in the cloakrooms but here the size of hand towels are small. In many places, however, disposable paper towels are offered. Although nowadays, hand towel has been replaced by electric hand drier.

## Face Towel/Wash Cloth

✓ It may be of linen or cotton. In general, one face towel with one bath towel is provided for each guest. The term face towel itself indicates that this towel is used to clean/wipe the face.

✓ Wash cloth can also be called *face cloth* or face towel because it is also used to clean face and hands. It is slightly different from face towel as it is moderately small.

**Standard Sizes (in inches)**

Bath Towel
20 × 40
or
22 × 44
or
24 × 50
or
27 × 50

Bath Sheet
36 × 70

Hand Towel
16 × 26
or
16 × 30

Face Towel
16 × 27

Bath Mat
18 × 24
or
20 × 30

Wash Cloth
12 × 12

## Bathmat

✓ It is a *rug* or *doormat* usually placed at the outer surface of the bathroom. It can be made up of cork, rubber or any other suitable material like *candlewick*. But it needs to be very absorbent and non-slippery. Therefore, it is mostly made from Turkish towelling or candlewick.

✓ Bathmats made from Turkish towelling or candlewick are required to be laundered frequently, thus considered more hygienic than rubber and cork bathmats. Nowadays, disposable bathmats are also available in markets but luxury hotels do not prefer them.

## Bidet Towel

✓ It is a small bathroom towel that is usually hung on rod, hook or toilet paper stand near the toilet and bidet. Its dimension can vary from washcloth to hand towel. These towels should be softer and environment-friendly and work as an alternative to toilet paper after bidet use in order to wipe away any excess moisture.

✓ In hotels, mostly white cotton linen bidet towels are used which can be washed in hot water and bleached when necessary. Bidet towels can also be made from smooth or waffled linen or plush cotton of varying thickness. Remember, wash these towels regularly to ensure proper hygiene.

## Bath Robe

✓ It is a dressing gown or housecoat (for ladies) which is usually made from towelling or other absorbent fabric such as cotton, silk, microfibre, wool and nylon. It can be weaved into different forms like flannel, velour and waffle.

✓ It is usually worn before or after taking bath and usually serves both as a towel and an informal garment. A dressing gown is loose, front-open, closed with a fabric belt. In hotels, it is usually found in white colour and in towelling linen.

## 19.7 RESTAURANT LINEN

It is an array of linen used in restaurants and other food and beverage outlets. For example, range of table linen, waiter's cloth, wiping cloth, dusters, runners, frills, baize cloth, slip cloth, tea cozy cover, placemats and so forth. Often the term *napery* is also used for restaurant linen. But it indicates various table linen used to furnish dining tables which may be in restaurant, banquets or in any other food & beverage outlet. A brief description of several major kinds of restaurant linen is given ahead.

## Tablecloth

✓ It indicates to a table cover used to cover the table top and should hang 12 inches from the edge of the table. Table linen is available in various shapes (like round, square and rectangular), sizes and designs.

✓ The selected tablecloth should have the ability to retain colour, shape, easy to launder and should match the decor and lustre of good finish. The fabric should have ability to sustain the power of stain removal agents. Therefore, linen is better than cotton but it is more expensive.

### Size of Tablecloths

| Table of | Square Table (in inches) | | Round Table (in inches) | | Rectangular Table (in inches) | |
|---|---|---|---|---|---|---|
| | Tablecloth | Table | Tablecloth (in diameter) | Table (in diameter) | Tablecloth | Table |
| 2 pax | 54 × 54 | 28 × 28 | 64 | 36 | 54 × 72 | 30 × 48 |
| 4 pax | 64 × 64 | 36 × 36 or 40 × 40 | 81 | 60 to 66 | 54 × 96 | 30 × 72 |
| 6 pax | 72 × 72 | 48 × 48 or 52 × 52 | 90 | 72 to 76 | 64 × 96 | 40 × 72 |
| 8 pax | - | - | - | - | 64 × 120 | 40 × 96 |

## Baize Cloth

✓ It is a table *base cloth* that is placed under the tablecloth or directly on the table surface in order to protect the table top. It protects the tablecloth from tear & wear (also absorbs the sound). Nowadays, it is pre-fixed on the table by the manufacturer. It also prevents the tablecloth from slipping.

✓ But in restaurants where baize clothes are not directly attached on the table top, cloth such as *multans* may be used. Due to its purpose of use, it is also known as *silencer cloth*.

# Frills

- ✓ It is a coloured and lustrous cloth usually made from satin or rayon which may be plain or patterned. While purchasing frill, purchaser should consider the length and width of the fabric required. Width must also correspond with the height of the table.

- ✓ There are various styles (or design) of setting frills when draping around the table. The pleats may be stitched or done when draping the table. Frills are mostly used in banqueting functions and also known as *jupone*.

| Standard Sizes of Napkin (in inches) |
| :---: |
| Cocktail |
| 6 × 6 or 8 × 8 |
| Breakfast |
| 10 × 10 or 12 × 12 |
| Lunch |
| 18 × 18 or 20 × 20 |
| Dinner |
| 20 × 20 or 22 × 22 |

# Napkins

- ✓ Cloth napkin or linen napkin is costlier than paper napkin (as linen cloth involves high initial cost and repetitive laundry cost). But it imparts standard in formal table setting; a cloth napkin can be placed in the centre of the cover or on side plate.

- ✓ Many hotels offer different napkins for different occasions like breakfast, lunch, dinner and cocktail napkins. The sizes of these napkins may vary with organization to organization.

# Tray/Salver Cloth and Waiter's Cloth

- ✓ Tray is different from salver. Tray (used in room service) is rectangular in shape while salver (used in restaurant) is round in shape. Consequently, tray cloth will be rectangular whereas salver cloth will be round in shape. Both are used to cover the surface of tray/salver in order to protect it from food particles and other spillage.

- ✓ Waiter's cloth is also known as *serviette*. It is like a napkin and used to carry hot dishes (like hot portion bowl, hot platters, etc.) while providing table service to the guest. It protects the palm and hand from getting burnt.

# Runners and Placemats

- ✓ *Runners* are used to give the formal décor (like in meetings) to the restaurant table. Table runners are a long strip of cloth, laid across the table lengthwise to add warmth and beauty. Table runners bring about a splash of colour to the table allowing to see a beautiful table instead of hiding it beneath a tablecloth.

✓ *Placemats* are mainly used in those food and beverage outlets where table linen is not placed on the tables. It looks attractive on the table at the same time also reduces the laundry expenses of table cloths.

## Wiping Cloth and Dusters

✓ *Wiping cloth* is used to wipe the various restaurant cutlery, crockery, glassware and hollowware. It soaks the water at the time of wiping. Therefore, "Damask" is most preferable wiping cloth. However, most of the properties use their discarded or rejected linen for this purpose.

✓ Duster is used for dusting the furniture, railings, doorknobs, wall mounted fixtures and fittings, television, writing table and so forth in-room equipment & furniture.

**Standard Sizes (In inches)**

Tray Cloth
16 × 27

Placemats
10 × 15

Waiter's Cloth
18 × 27

Runners
17 × Varying Length

Slip Cloth
36 × 36

Buffet Cloth
72 × 144

---

### Channel of Linen Management

The term *channel* refers to the mode or system of issuing and receiving clean and soiled linen. Generally, there are two channels for handling linen, centralized and decentralized linen room channel. A brief description of each is given below.

#### Centralized linen room channel

✓ Issue and collection of clean linen as well as soiled linen from the same point or place. The term centralized linen room indicates that all tasks (such as work distribution, staff scheduling, etc.) of linen room is managed and controlled from the single point.

✓ The linen may be issued on *par* basis for the number of rooms assigned to each attendant or may be issued on the basic rule of "one clean for one soiled" basis.

#### Decentralized linen room channel

✓ It refers to the *satellite system* of linen management whereby the clean linen is stored in advance in floor pantries (or satellite points) in order to perform daily operation or to handle unusual occupancy situation like laundry breakdowns. After the completion of guest room cleaning, all collected soiled linen is sent to floor pantries and further transported to the laundry section via *linen chute*.

✓ Sufficient amount of linen along with *par stock* is kept in the floor pantries to serve all available rooms on a particular floor or floor section.

✓ The term "decentralized" means that all authority and responsibility is delegated with other satellite linen rooms. But staff scheduling and their reporting, attendance, work allotment, etc., is executed by centralized linen room.

# 19.8  UNIFORM

Staff uniform may be made-to-measure or directly purchased in standard sizes. But the number of sets of uniforms provided mainly depends upon the nature of work being performed by the employee as well as laundry availability (or whether laundry is on- or off-premises). Some important functions of linen/uniform room staff in relation to selection of uniform or uniform material is explained below.

## Uniform Arrangement

- ✓ Generally, in most hotels standard sized readymade uniforms are issued to operational level staff of different departments of hotel. On the other side, supervisory and managerial level staff mostly get made-to-measure option for their uniform arrangement.
- ✓ Made-to-measure staff uniforms look smart and are essential for supervisory and managerial staff. Standard size uniforms lower the total requirement of uniforms but may be ill-fitting, thus do not look as smart as made-to-measure uniform.

### Uniform Room Equipment

- Portable uniform stand
- Hangers
- Coat brushes
- Hampers
- Trolleys
- Ladders
- Cupboards
- Folding tables
- Racks
- Plastic/paper bags

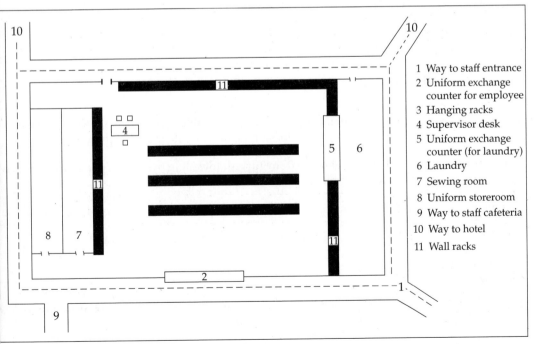

1 Way to staff entrance
2 Uniform exchange counter for employee
3 Hanging racks
4 Supervisor desk
5 Uniform exchange counter (for laundry)
6 Laundry
7 Sewing room
8 Uniform storeroom
9 Way to staff cafeteria
10 Way to hotel
11 Wall racks

**Layout of Uniform Room**

✓ The layout of linen room should be planned in such a way that work flow is n disturbed.

✓ Uniform exchange counter, sewing/mending room, working counters and hangii racks/cupboards, storeroom, etc., should be in sequential order of work flow. Ti passageway should have enough area for the movement of staff. An ideal layo of uniform room is given above.

## CONCLUSION

The term *linen room* refers to a section responsible for the in-flow and out-flow of hot linen and guest clothes whereas uniform room is related with in-flow and out-flow hotel's internal staff uniforms. But nowadays, most hotels have constructed commo room for both tasks. *Sewing room* is also an important subdivision of linen/unifor room because it is responsible for mending work and stitching of new uniforms ar linens. In hotels, clothes/linen discarded are mostly used as dusters and wiping clot The operation of linen room can be centralized or decentralized. *Centralized linen roo* is a place where all soiled linen collection and displacement of clean linen takes plac whereas decentralized linen room indicates to the satellite style division of linen roon In this system, soiled linen is placed in floor pantries which already contain enoug stock of clean linen. Whether these sections are combined or independent, both shoul contain enough hanging spaces and required small and large equipment.

| Terms | |
|---|---|
| (with Chapter Exercise) | |
| Upholstery | It refers to the fabric and stuffing used to add padding to furniture or the covering of furniture with fabric. |
| Jupone | ? |
| Software items | All in-room fixtures that are depreciable in nature like mattress, curtains, pillows, etc., but it doesn't include bed linen and bathroom linen. |
| Terry weave | A bath towel made from Turkish towelling by using a pile weave, known as terry weave. |
| Candlewick | A soft heavy cotton thread similar to that used to make wicks for candles. |
| Bidet | It is a low oval basin used for washing one's genital and anal area. |
| Soft furnishing | ? |

| Duvet | It is a quilt, usually with a washable cover, that may be used in place of a bedspread and top sheet. |
|---|---|
| Linen chute | ? |
| Quilted comforter | It is a bed cover (in informal setting) that commonly works on behalf of both bedspread and blanket. |
| ? | It is a motif of two or more interwoven letters, mainly used as an identification symbol. |
| Bolster | It is an elongated pillow which stretches the width of the bed. These may be filled with less resilient filling than pillows. Nowadays these have gone out-of-fashion and replaced by two pillows. |
| Satellite linen room | ? |
| Bathmats | It is a mat used in front of a bathtub or shower, as to absorb water and prevent slipping. |
| ? | It is a rich, heavy silk or linen fabric with a pattern woven into it and mainly used for table linen and upholstery. |
| Flannelettes | It is brushed cotton material which is cheaper and warmer than cotton sheeting. Sometimes also used as an under-blanket. |
| Pillow-top mattress | Pillow-top mattresses have extra layers built into the mattress itself for extra comfort. |
| Storage hampers | A shelving unit in a satellite linen room, with shelves adjusted to receive soiled linen, thus act as storage hamper for dirty linen. |
| Stain | A discoloured or soiled spot or smudge. |
| Dust raffle | A decorative bed linen that covers the box springs and the frame. It is a pleated cloth skirting that extends around the sides and foot of the bed. |
| Multans | ? |
| Apron | Garment usually fastened in the back, worn overall or part of the front of the body to protect clothing. |
| Hopper | A conveyor used to carry or remove soiled linen several times each day from housekeeper's cart to the satellite linen rooms. It may also be used to carry rubbish sacks from chambermaid's cart for emptying. |

| Draperies | Material that hangs in a window or other opening as a decoration, shade, or screen. |
|---|---|
| Sheer curtains | ? |
| Coverlet | Bed linen that covers only the mattress. If coverlet is used then dust raffle is also required. |
| Shams | Pillow covers that usually match the fabric used in bedspread. |

# 20 CHAPTER

# Cleaning and Polishing

OBJECTIVES

*After reading this chapter, students will be able to...*

- describe what is cleaning and polishing;
- draw the cleaning schedule and cleaning system;
- find out the areas of cleaning;
- identify cleaning agents and suitable methods for their application; and
- explain different types of polish and standard steps of polishing.

## 20.1 CLEANING AND POLISHING

The term *cleaning* is concerned with the process of removing dirt, dust, marks and stains from the surface whereas the term *clean* relates with hygiene and sanitation (or free from dirt, dust, marks and stain). In hotel industry, the task of cleaning is performed under the direct control and supervision of housekeeping department, although when some areas are outsourced, especially public areas/out-house areas. Remember, cleaning requires a systematic and planned approach, both towards the individual tasks involved and towards the hotel's cleaning operations as a whole. Collectively, it comprises two components, *cleaning process* and *cleaning schedule*.

The term *polishing* refers to the process of making the surface smooth and shiny by rubbing it. In polishing, polish is used to wipe the surface in order to make it shiny. Some polishes also contain protective coatings that inhibit tarnishing. Polish is usually applied with the help of *polish cloth* or *rags* after placing in *polish applicator tray*. Often mechanical equipment are also used for the polishing like *polishing/scrubbing machine* in order to polish the floor. Remember floor polishing is known as *buffing*.

## 20.2 CLEANING PROCESS

The term *cleaning process* means the procedure of cleaning or step-by-step course-of-action to make the surface neat and clean. The basic objective behind this step-by-step process is to remove as much dirt/dust/stains/marks as possible without damaging the surface. For effective cleaning, it is necessary that the attendant must have scrupulous knowledge of different types of stains and their remedial cleaning agents, cleaning equipment, cleaning systems, cleaning methods (or process of cleaning) and cleaning frequency.

| Activity |
|---|
| *Cleaning system* is a set of principles/procedures according to which cleaning task is performed. Three major systems of cleaning are given below. Fill in the empty spaces in the table. |

| Cleaning system | Description of cleaning system |
|---|---|
| Orthodox | • A traditional method of cleaning in which a room attendant completes all task of cleaning in one guest room before going to the next room of same section. On an average one room attendant is responsible to clean 12-20 guest rooms in one shift.<br>• ? |
| Block | • In it, room attendant moves from room to room and completes all tasks in every room before returning to begin the cycle again for the next task. Block cleaning involves blocking of several rooms/sections/floors for the purpose of cleaning. Block of several rooms is known as room section.<br>• In general, more than one room attendant is assigned work in the room section. For instance, one room attendant might perform room-make-up on a particular room section while second attendant cleans the bathroom & toilet and the third replenishes the supplies. |
| Team | • ?<br>• It is opposite of orthodox method. In team method, two or more room attendants work together and are responsible to clean one room before moving to the second room of same section. |

## 20.3 CLEANING SCHEDULE/FREQUENCY

The term *cleaning schedule/frequency* refers to the timetable of cleaning or occurrence of cleaning. It is always necessary that housekeeping department must work out an accurate *action plan* for effective and efficient cleaning operation. The action plan supports in proper execution of cleaning task, at the same time cleaning schedule assists in selecting particular staff member for a particular work during a particular period of the day. In simple terms, cleaning schedule suggests when deep cleaning or special cleaning will take place and who will perform it. Cleaning schedule (or cleaning tasks) can be classified into three broad groups, i.e., daily cleaning, weekly cleaning and periodic cleaning.

## Daily Cleaning (Routine or Day-to-day Cleaning)

- It is *routine cleaning* (or day-to-day cleaning) performed by room attendant during room make up process. Generally, daily cleaning is performed in the absence of guest but occasionally it may also be carried out during guest presence (on the request of the guest).
- As cleaning can done be daily, weekly or special but some cleaning is necessary on daily basis, for example, dusting, moping, wiping, sweeping, vacuuming (including cleaning of bathrooms, toilets, suction cleaning of floors, dusting furniture and so forth).
- Daily cleaning is the most important part of every establishment as it provides eye appealing and hygienic accommodation to the guest. The workload of housekeeping department varies from property to property due to daily cleaning. The workload depends on daily *room count, house count*, vacated rooms and going to vacant rooms.

### Activity

Daily cleaning task involves dusting, moping, wiping, sweeping and vacuum cleaning. In the below given blank spaces, you need to fill in proper description and required equipment to perform the daily cleaning.

| Daily cleaning | Description | Required equipment |
|---|---|---|
| Dusting | ? | Duster |
| Moping | ? | ? |
| Wiping | Cleaning surface by using a slightly damp cloth and then drying with a clean soft cloth. | ? |
| Sweeping | ? | |
| Vacuuming | Cleaning surface with the help of vacuum cleaner | Vacuum cleaner |

## Weekly Cleaning (Deep or Special Cleaning)

- The term *weekly cleaning* refers to the task of cleaning on a weekly basis, for example, dusting under the bed and furniture, dusting of high areas, cleaning of AC vents and filters, vacuum cleaning of upholstery, wiping down walls and base boards, and so forth. Often this weekly cleaning is also known as *deep* or *special cleaning*.

- The preparation of weekly cleaning schedule (or *weekly cleaning chart*) depends on the quality of daily cleaning, quality of furniture/fixture/metal and general wear and tear of hotel rooms. The necessity of special cleaning is usually identified during room inspection.

- The ways of scheduling special cleaning may be: assign as a part of daily cleaning, assign as an extra cleaning task for everyday or room/section/floor may be locked during the low occupancy period for the purpose of deep cleaning.

| Activity | |
|---|---|
| **Ways to organize special cleaning** | **How to organize given special cleaning programme** |
| Part of daily cleaning | ? |
| Extra cleaning task for everyday | Enhance workload by assigning extra cleaning task for everyday to each room attendant in addition to their scheduled routine cleaning. |
| Room/section/floor locked during low occupancy | ? |

## Periodic Cleaning (Spring or Seasonal Cleaning)

- *Periodic cleaning* is done every 15 days or once every month, for example, changing curtains, cleaning water tanks, removing carpet stains, polishing surface, dry cleaning of lampshades, cleaning (or washing down) walls, ceilings and windows and so forth. Often this periodic cleaning is also known as *spring, shopping* or *seasonal cleaning*.

- Business hotels arrange their spring cleaning during the slack season. Spring cleaning can be organized in two ways, i.e., by individual room and by couple of rooms. Generally, spring cleaning programme is organized in three phases, i.e., first remove furniture/fixture/fittings; second start cleaning process; and lastly refit furniture/fixture/fittings to its original place.

- Spring cleaning schedule is mostly performed during low occupancy period (by blocking entire floor or section) but may also be carried out during high occupancy (by putting room *out of order*), if required.

| Activity | | |
|---|---|---|
| **Areas** | **Spring cleaning tasks** | |
| Bedroom | Task 1 | ? |
| | Task 2 | ? |
| | Task 3 | ? |
| Bathroom | Task 1 | Remove cobwebs and dust from all surfaces |
| | Task 2 | De-scale and polish bathroom tiles, sinks, taps and shower |
| | Task 3 | Clean mirror (with Colins) and wash bathroom furniture |
| | Task 4 | Clean and disinfect toilet pan from inside as well as from outside |
| | Task 5 | Vacuum clean or mop bathroom floor |
| Banquet hall | Task 1 | ? |
| | Task 2 | ? |
| | Task 3 | ? |
| Kitchen | Task 1 | Remove cobwebs and dust from all surfaces |
| | Task 2 | Clean doors, doorframes and kitchen appliances |
| | Task 3 | Clean cabinets and cupboards (inside and outside) then polish |
| | Task 4 | Defrost and clean (behind & underneath) the refrigerator/freezer |
| | Task 5 | Scrub, polish and deodorize all kitchen sinks and waste bins |
| | Task 6 | Clean and polish kitchen tiles and clean microwave from inside and outside |
| | Task 7 | Vacuum clean or mop bathroom floor |
| Foyer/lobby | Task 1 | ? |
| | Task 2 | ? |
| | Task 3 | ? |
| Elevator/escalator | Task 1 | ? |
| | Task 2 | ? |
| | Task 3 | ? |
| F & B outlets | Task 1 | ? |
| | Task 2 | ? |
| | Task 3 | ? |

## Weekly Cleaning Schedule/Chart

This chart is entered in the floor register which is available with the floor supervisor. A tick mark is made against each room number for work done. If any work is pending in the room on that day, make it blank or cross tick it against that room number, so it can be cleaned or carried forward for the next week.

| Days | Housekeeping Attendants | Tasks | 101 | 102 | 103 | 104 | 105 | 106 | 107 | Remarks |
|------|------------------------|-------|-----|-----|-----|-----|-----|-----|-----|---------|
| Mon | Housekeeping Attendant A | 1 | Off | | | | | | | |
| Tues | Housekeeping Attendant B | 2 | | Off | | | | | | |
| Wed | Housekeeping Attendant C | 3 | | | Off | | | | | |
| Thurs | Housekeeping Attendant D | 4 | | | | Off | | | | |
| Fri | Housekeeping Attendant E | 5 | | | | | Off | | | |
| Sat | Housekeeping Attendant F | 6 | | | | | | Off | | |
| Sun | Housekeeping Attendant G | 7 | | | | | | | Off | |

The following list of tasks can be allotted on daily basis for a particular week.

Task 1   Clean the cobwebs from fans, AC vent and lighting fixtures.

Task 2   Dusting in the room & bathroom and defrosting the refrigerators.

Task 3   Clean the windows and vacuum clean the carpets.

Task 4   Clean dados (base boards/skirting) and vacuum clean under beds & furniture.

Task 5   Scrub the balcony or terrace attached to the room.

Task 6   Scrub the bathroom tiles and shower curtains.

## 20.4 CLEANING AGENTS

*Cleaning agent* refers to all those materials/substances that are used to remove dirt, dust, mark, stains, bad odours and untidiness from the soiled surface. These cleaning agents are mostly available in liquid form and may also be found in powder and paste form. But the purpose of all cleaning agents is same, i.e., to remove dirt, dust, marks, stain, bad odour and untidiness and to avoid the spreading of dirt and contaminants from one place to another.

**Name of Companies making Cleaning Agents**

- Reckitt Benckiser
- Prime source
- Carroll
- Claire
- Johnson Diversey

Cleaning agents are available in different grades, i.e., from very strong to mild. It is always desirable to use mild cleaning agents (although they may need to be used several times) in place of strong because strong cleaning agents may damage the surface; it is especially applicable with laundry cleaning agents and soft floor cleaning agents. Some cleaning agents can kill bacteria and other microbes and clean at the same time, whereas other cleaning agents may perform this task in several stages. A brief description of major cleaning agents used in hotels is given below.

## Water

It is a primary and simplest cleaning agent but water alone will be ineffective as a cleaning agent due to *surface tension*. Water alone can remove only soluble dirt but addition of cleaning agent/ detergent intensifies its cleansing power. If only water is used for cleaning (without adding any cleaning agent) then it is not an effective cleanser.

| Water Hardness | |
|---|---|
| - Soft Water | Below 01 GPG |
| - Slightly Hard | 1 - 3.5 GPG |
| - Moderately Hard | 3.5 - 07 GPG |
| - Very Hard | 7 - 10 GPG |
| - Extremely Hard | Above 10 GPG |

Water can be classified into two types, i.e., hard water and soft water. Soft water has relatively low concentration of calcium carbonate and other ions. The water that lathers with soap easily is called *soft water*. Water that has passed through ground containing limestone is known as *hard water*.

- There are two types of hardness in water, i.e., temporary hardness and permanent hardness. *Temporary hardness* is due to the naturally present salts of calcium and magnesium (mainly bicarbonates and sulphates) in water. The hardness can be removed by heating or boiling the water.

- *Permanent hardness* is usually caused by the presence of sulphates of calcium and magnesium, which do not precipitate as the temperature increases. Therefore, cannot be removed by boiling. The ions causing permanent hardness of water can be removed by using a water softener like zero light water softener.

- The Water Quality Association of the United States defines hard water as having dissolved mineral hardness of 1 GPG (grain per gallon) or more.

| Activity |
|---|
| How can a person remove the hardness of water or make water soft? Describe the below suggested ways. |

| Ways (or methods) | Name of chemical agents/ mechanical methods | Description of each chemical agent / mechanical method |
|---|---|---|
| Chemical method | - Zero light (water softener)<br>- Soap<br>- Soda<br>- Ammonia solution<br>- Base exchange method | -<br>-<br>-<br>-<br>- |
| Mechanical method | - Boiling<br>- Heating | -<br>- |

## Detergents

It is not possible to clean with detergents alone; it is always used in conjunction with water. Detergents are cleaning agents which remove the dirt from surfaces. It also holds the suspension so that dirt cannot be re-deposited on the clean surface. All detergents should have certain basic properties. Detergents may be soapy or soap-less (synthetic).

A detergent is a material intended to assist in cleaning. Detergents have molecules with one side that prefers water (hydrophilic), and another side that prefers oils and fats (hydrophobic). The hydrophilic side attaches to water molecules, and the hydrophobic side attaches to oil molecules. This action allows the oil droplets to break up into smaller droplets, surrounded by water. These smaller droplets are no longer stuck to the material to be cleaned, and are washed away. The detergents are the most important type of cleaning agents, and are usually mixed with water before use. They can be divided into three broad groups, i.e., acid, neutral and alkaline detergents. Synthetic (soap-less) detergent also plays a crucial role in cleaning.

**Basic Properties of Detergents**

- Wetting power
- Emulsifying power
- Suspending power

- *Acid detergents* are generally based on phosphoric or sulphuric acid, and their use is rather limited. They are extremely effective in removing salts precipitated from water in hard water areas, and in cleaning aluminium where the acids readily remove the white scale that forms on the surface of the metal.
- *Neutral detergents* comprise a wide range of materials that are mainly suitable for light cleaning. They are generally similar to household detergents, and their wetting ability makes them ideal for the dispersal of grease and oil. *Alkaline*

*detergents* can vary in strength from those that are only a little stronger than the neutral types to ones that are strongly alkaline, consisting almost entirely of caustic soda, thus require extreme care in use.

## Activity

Describe what synthetic detergent is and write down a brief description, properties and suitability of liquid and powder synthetic detergents.

| S. No. | Synthetic detergents | Brief description | Properties of detergents | Where is it suitable? |
|--------|----------------------|-------------------|--------------------------|-----------------------|
| 1 | Liquid | | -<br>-<br>- | -<br>-<br>- |
| 2 | Powder | | -<br>-<br>- | -<br>-<br>- |
| 3 | Gel | | -<br>-<br>- | -<br>-<br>- |
| 4 | Crystal | | -<br>-<br>- | -<br>-<br>- |

## Abrasives

- It is a substance or material capable of polishing or cleaning a hard surface by rubbing or grinding. But overtime, abrasives will dull and scratch the surface because they contain quartz or silica composed of grit which can easily damage/ scratch the soft surface. It can also damage the hard surface thus be careful while applying abrasives.

- It works on the principle of rubbing or scratching movements. The extent of cleaning depends on the nature of abrasives, types of stain, shape and size of utensils. The use of abrasives also depends on the type of surface to be cleaned.

- There are various types of abrasives floating in the markets such as glass, sand, emery paper, steel wool, nylon web pads, powdered pumice, etc.

**Hard & Soft Surfaces**

- *Hard surface* like porcelain, stainless steel, etc.
- *Soft surface* like fibreglass, laminate, etc.

These abrasives can be classified as fine, medium and hard abrasives. The fines abrasive includes precipitated whiting (filtered chalk) and *jewels rough* (pink oxide of iron). To get a good result, the abrasives should be used in the form of finely ground minerals.

| Activity | | |
| --- | --- | --- |

Fill in the below given spaces:

| Abrasive Types | Abrasive Constituents | Examples/Features/Functions |
| --- | --- | --- |
| Fine | Jewels rough and filtered chalk | ? |
| Medium | Fine minerals (like limestone and calcite) mixed with soap or detergent and alkali | ? |
| Hard | ? | Sand paper, fine ash, pumice stone, steel wool |

## Toilet and Window Cleanser

- There are various types of cleansers used for the cleaning of toilet pans. Acids are usually used to remove various stains from the toilet pan. The cleaning of toilet pan can be done either by any of the following cleansers or cleaning agents- crystalline, powdered, liquid and liquid chlorine bleach.

> **Toilet Cleansers**
> - Crystalline
> - Powdered
> - Liquid
> - Liquid chlorine bleach

- All the above-mentioned toilet pan cleansers are designed for the purpose of cleaning and disinfecting of lavatories & urinals only. It is important that these cleansers should never be mixed with any other cleaning agent because it may produce harmful gases.

- The term *window cleanser* is used for any cleaning agent used to clean windows. In general, window cleanser consists of a water-miscible solvent (most often *isopropyl* alcohol) to which a small amount of surfactants and alkali are added to increase its effect on the surface.

- Some window cleansers also contain a fine abrasive material. It is applied with cleaning rag and rubbed-off with a clean & soft cloth. Window cleansers can also be applied by spraying and the surface wiped clean. Its substitute can be made with water based solution, i.e., water with some methylated spirit or vinegar.

| Activity |
|---|
| Match the following below-mentioned cleansers with their correct description. |

- Crystalline        - Powdered        - Liquid        - Liquid chlorine bleach

| Cleansers | Description |
|---|---|
| ? | Generally, these are based on sodium acid, sulphate, a mild acid which is mixed with an anti-caking agent (often pine oil is used). Its effect can be improved by using a small amount of acid resistant anionic surfactants. |
| ? | It consists of a soluble acidic powder, chlorinated bleach and finely ground abrasives. Toilet is cleaned with the help of brush and an effervescing substance. It helps in scattering the active ingredients throughout water. |
| ? | It is a dilute solution of hydraulic acid. Thus, extreme care should be taken while using it because it can damage the surface of the pot, its surrounding area and the cleaner or attendant, if spilt. |
| ? | It is an alkaline stabilized solution of sodium hypochlorite. It should not be mixed with any other lavatory cleanser. It is used to clean and disinfect the lavatories. Sometimes, the solution may contain detergents which assist in cleansing process and increase the viscosity. |

## is/Alkaline

t refe    laundry based cleaning agents for example soda (*caustic* and *washing*) and am    nia used as a grease emulsifier and stain removal agent. Strong alkali materials such as caustic soda are extremely corrosive and poisonous in nature.

- Strong alkaline materials such as caustic soda and ammonia (either in liquid or flake form) are used to clean blocked drains and other large commercial equipment, for example, oven. These cleaning agents are high in pH value. Therefore, they must be used with extreme caution like after wearing rubber gloves and under the supervision of experts.

| Activity |
|---|
| Match the following cleaning acids with their correct description. |

- Sodium carbonate        - Sodium hydroxide        - Sodium hypochlorite

| Alkalis | Description | Hint |
|---|---|---|
| ? | Used to remove grease marks from the grills and also clean blocked drains | Washing soda |
| ? | Used to remove stains from hard and soft surface and also works as a whitening agent | Bleach |
| ? | Used to remove light grease marks and it also works as water softener | Caustic soda |

# Acids

- It refers to a substance with particular chemical properties including turning litmus red, neutralizing alkalis and dissolving some metals. It dissolves the metals and removes the stains from metal surfaces like stains that occur around the water taps, tarnish on copper, silver, brass articles.
- Acids may vary from very mild acid such as *citric* and *acetic acid* to strong concentrated *hydrochloric acid*. Strong acids are poisonous and corrosive in nature. Therefore, it should be used with extreme caution while wearing rubber gloves. Acids should be used in solutions followed by thorough rinsing.

| Activity | |
|---|---|

Match the following cleaning acids with their correct description.

- Oxalic acid    - Paraffin oil    - Citric and acetic acid    - Sulphuric acid

| Cleaning acids | Description |
|---|---|
| ? | It is also called as "diluted hydraulic acid, muriatic acid or spirits of salt". It is a strong cleaning agent used to remove tough stains like lime scale from the surfaces like sanitary ware, swimming pool, etc. |
| ? | It is an effective cleaning agent. It is used to clean bathrooms but nowadays it is hardly preferred due to its strong smell. |
| ? | It is a domestic formula (like vinegar and lemon) to remove tarnish from copper, brass and calcium deposit build-up (or mild water stains on baths). It may damage the glaze; therefore metal utensils must be quickly washed. |
| ? | It is added into domestic acidic drain cleaners then used to clean/unblock clogged pipes by dissolving greases, proteins and even carbohydrate-containing substances. |

# Organic Solvents

- The term *organic solvent* refers to a substance that dissolves a solute, resulting in solution, thereafter used for removing stains from different types of surfaces. In simple terms, they are usually spirits like methylated spirits, white spirits and carbon tetrachloride used to remove grease and *wax* from the surface. Aerosol dry cleaners are suitable for wallpaper and furnishing.
- In housekeeping department, the common uses of organic solvents are in dry cleaning (such as tetrachloroethylene), spot cleaning (such as hexane, petrol ether), in detergents (citrus terpenes) and so forth. Above given methylated and white spirit (i.e., carbon tetrachloride) are inflammable in nature and second one is also

harmful, if inhaled. One can also use turpentine as a substitute product for white spirit.

# Bleaches

- It indicates to a chemical or alkaline based solution (typically a solution of sodium hypochlorite or hydrogen peroxide) used to make substances whiter or for sterilizing drains, sinks, WC pans, etc. Remember, it is never mixed with any other cleaning agent or cleanser because the solution may damage the surface.
- Bleach is white in colour and contains germicidal properties. Therefore, it is important to take care of bleach (for spotting) while doing the cleaning. Bleach works on the principle of *oxidation*.
- Chlorine is a base for the most commonly used bleaches, for instance, the solution of sodium hypochlorite. Strong solutions of chlorine bleaches discolour the utensils like copper, aluminium, silver and stainless steel.
- Many types of bleaches have strong bacterial properties and are used for disinfecting and sterilizing. Thus, they are used in swimming pool sanitation to control bacteria, viruses and algae and in any institution where sterile conditions are needed. Sometimes a blend of surfactants is added to increase its effect or cleaning power.

| Important | | | |
|---|---|---|---|
| For manufacturing of cleaning agents, the products (R-series) of Taski or Diversey company are considered the benchmarks in hospitality industry. The letter "R" stands for *room care*. A brief description of R series is given below: | | | |
| Product code | Used for | Product code | Used for |
| R1 | Bathroom, toilet & urinal cleaning | R9 | Hard stains from bathroom walls and fittings |
| R2 | Hard-surface like walls & floors (all purpose) | TR 101 | Carpet shampooing |
| R3 | Cleaning of glasses, windows & mirror | TR 102 | Carpet detergent |
| R4 | Furniture polish | D6 | Glass cleaner |
| R5 | Room freshness/air freshener | D7 | Polishing stainless steel |
| R6 | Toilet pan/bowl cleaner | D8 | It is used as a silver dip |
| R7 | Cleanser or detergent for oil and grease | Z18 | It is a cleaner or degreaser |
| R8 | Metal polish or kettle de-scaling | | |

# Disinfectants

- It refers to chemical cleaning agents that destroy pathogenic germs on inanimate surfaces. Remember, disinfectants should only be used in places where harmful germs are likely to exist. It directly kills the harmful bacteria within the *dwell time*. Often disinfectants are also known as *sanitizers*.

- Most disinfectants have strong smell and therefore should be used in recommended amounts in areas where germ control is required. There are various terms used to describe specific disinfectants, i.e., *bactericides* (a chemical that kills bacteria), *fungicides* (a chemical that destroys fungus), *germicides* (a substance that destroy harmful microorganisms), and *virucides* (a virucidal; tending to destroy viruses).

- Remember the use of a disinfectant alone on a soiled surface is ineffective because the deposited soil protects the bacteria from the germicidal action of the disinfectant. But when disinfectant is combined with detergent, the formed solution is relatively more effective if used according to instructions/directions. Idophors, quaternary ammonium compounds, hydrogen peroxide, hypochlorite and phenolic compounds are some common disinfectants.

| Activity |
|---|
| Match the following below-mentioned cleaning disinfectants with their correct description. |

- Halogens          - Phenol          - Cationic surfactants

| Disinfectants | Description |
|---|---|
| ? | It is used in dilute or concentrated form in order to disinfect surfaces. |
| ? | It is a useful bactericide as well as deodorant. But remember it should never be used with any other cleaning agent, especially with anionic soap or soapless detergent. |
| ? | Its components include chlorine and iodine and may be used as disinfectants |

# Soap and Emulsifier

- It refers to a cleaning agent or substance used with water for washing and cleaning. It is made up of a compound of natural oils or fats (or fatty acids) with sodium hydroxide or another strong alkali and typically having perfume and colouring added. The process of making soap is known as *saponification*.

- Soap cleans by acting as an emulsifier. Actually, soap allows oil and water to mix so that oily filth can be removed during rinsing. Detergents are similar to soaps but they are less likely to form foam (soap scum) plus are not affected by the presence of minerals in water (hard water).

- The term emulsifier refers to a surface-active agent (like soap) which promotes formation and stabilization of an emulsion. An emulsifier keeps oil droplets and water droplets together, so a thick mixture of oil and water will not separate. Some detergents and surfactants are used as *emulsifying agents.*

## Activity

Technically, *surfactants* are wetting agents that lower the surface tension of a liquid, allowing easier spread, and lowering of interfacial tension between two liquids. Detergents and soaps are *surfactants.* These surfactants can be divided into three types, i.e., cationic, anionic and non-ionic. Fill in the below given blanks with their correct description/ properties/examples/uses.

| Surfactants | Description | Properties | Examples | Uses for cleaning |
|---|---|---|---|---|
| Cationic | ? | Positively charged ion or cation | Ensure chloride or bromide are included in cationic surfactants | Cellulose and its blends |
| Anionic | ? | Negative charge on their hydrophilic end | ? | ? |
| Non-ionic | It does not have an electrical charge, which makes them resistant to water hardness deactivation | ? | Ethoxylates, Alkoxylates, Cocamide | Synthetic fiber and their blends |

## Others (Deodorants and Antiseptic)

- A *deodorant* is a substance which removes or conceals unpleasant smells, especially bad odours. Deodorants counteract stale odours and sometimes also introduce fragrance in the area. They are used in guest rooms, bathrooms, and public areas. These are available in liquids, powders and crystalline blocks. Deodorants are also known as *air fresheners. Naphthalene balls* serve as effective deodorizers.

- On the other side, the term *antiseptic* is used for chemical agents that prevent the growth of disease-causing microorganisms (also known as *microbials*). Some antiseptics are true *germicides,* as these are capable of destroying microbes. Antibacterials are antiseptics that have the proven ability to act against bacteria. Microbicides which destroy virus particles are called viricides or antivirals.

| | | |
|---|---|---|
| **Activity** | | |
| Match the following. | | |

| S. No. | Cleaning agents | Used to |
|--------|-----------------|---------|
| 1 | Solvents | Kill harmful organisms |
| 2 | Acids | Soften fats, grease & oils |
| 3 | Detergents | Remove hard water scale |
| 4 | Disinfectants | Breakdown & remove fats, grease and carbon |
| 5 | Alkalis | Clean contaminated equipment |

## 20.5  CLEANING METHODS

It refers to different ways of cleaning involved in cleaning of different surfaces. For effective cleaning, it is most important to select correct cleaning equipment, cleaning agent thereafter apply correct cleaning method. Simultaneously, cleaning tasks are carried out methodically and also consider the level of activity and amount of traffic present in the area to be cleaned. A few cleaning methods along with brief description are given below:

**Dusting Supplies**

- Vacuum cleaner
- Furniture polish
- Window cleaner
- Disinfectants
- Cleaning rags
- Canister vacuum
- Duster

Dusting        Removing of dry dust with dry *dusters*, thus often it is also known as *dry dusting*.

Sweeping       Removing of dirt and other unwanted particles with *soft* or *hard* brooms.

Mopping        Removing of dust, debris and other unwanted particles from the floo walls and ceiling without raising and dispelling dust. Mopping can b dry as well as wet.

Vacuuming      Removing of dirt and other small unwanted particles from the surfac by the process of suction. The vacuum cleaner/suction cleaner is use for this purpose.

Brushing       Removing of stubborn dry or wet/impregnated dirt and debris fror hard as well as soft surfaces. In it, stiffness of bristles plays a crucial rol It must be in accordance with surface type.

| Wiping | Cleaning of surface with the help of cloth or other similar materials. For example, *glass cloth* is used to wipe mirrors and drinking glasses. Wiping can be dry as well as wet. |
|---|---|
| Buffing | Cleaning then polishing of a floor with low speed floor machine (175-350 rpm). In this machine, the *beige pads* are used for buffing. |
| Scrubbing | Cutting the soil from the surface to remove it, usually by scouring process. One can also use appropriate cleaning agents and equipment (like hard brush) to make the result more effective. |
| Polishing | Application of polish on any surface to give glossy and shiny appearance and to form a protective layer. It guards against finger marks and scratches. |
| Scarifying | Cleaning then polishing of a floor with the help of bristle tips or edge of a cutting tool. It cuts into impacted soiling and removes it by means of a chisel-like action. |
| Burnishing | Cleaning then polishing of a floor with high speed floor machine (350 + rpm). In it, the attached pad creates a high glossy (or wet look) finish on the floor surface. |
| Shampooing | Removing of dirt, dust and other similar particles from the fibre based materials, particularly from carpet. In order to make it more effective, carpet shampooing machines are used. |
| Laundering | Removing of dirt and stains from clothes and linen fabrics. In laundering process, water is the base material which works effectively when mixed with any detergent or cleaning agent. |
| Dry cleaning | Cleaning of delicate fabrics like wool or silk, by using solvents other than water. The solvents commonly used are *per-chloro-ethylene* and *tri-chloro-tri-fluoro-ethane*. |
| Swabbing | Swab is a soft and absorbent cloth, whereas swabbing is the process of wet cleaning (with swabs) of surface above floor level. |

## 20.6   CLEANING AREAS

*Cleaning areas* refers to various places that need to be cleaned. These areas can be classified on the basis of their position such as in-house areas and out-house areas as well as on the basis of traffic of people such as public areas and guest rooms. A brief description of all these areas along with standard cleaning process is given below.

## In-House Areas

✓ It refers to the various *hotel interior* areas usually available inside the main hotel building. It includes both internal public areas (for example, all outlets or sections of major departments, i.e., food and beverage, front office, housekeeping and food production) and guest rooms.

✓ Often it also includes human resource, purchase, store, account & administrative and sometime sales and marketing too. In case of *business hotels*, almost all areas are available inside the main building, except parking space and time office/security office.

✓ Remember, cleaning of guest rooms and various internal public areas (such as restaurant, corridor, lobby, indoor swimming pool and so forth) is the sole responsibility of housekeeping, except kitchen department cleaning.

## Out-House Areas

✓ It refers to the various *hotel exterior* areas usually available outside the main hotel building. It mostly includes external public areas only.

✓ Examples of out-house area include parking space, garden, pavement, lounge, outdoor swimming pool, hotel entrance (or main entry gate), golf course and other similar outdoor gaming spaces.

Conventionally, the cleaning responsibility of all these areas is of housekeeping department but nowadays, these are contracted-out (or out-sourced) but still hotel management appoints a public area supervisor who ensures that cleaning task is performed well. This kind of cleaning can be easily seen in *resort* properties.

## Public Areas

✓ Collectively it refers to the various in-house and out-house communal areas where traffic or movement of general public is high, for example, restaurant, lobby, corridor, dining spaces, gaming zones, bars, meeting halls and almost all out-house areas.

✓ In business hotels or city hotels, almost all public areas are available inside the main hotel building/premises with very limited exterior public areas like entrance and parking space.

## Guest Rooms

✓ Guest rooms are perishable product of hotel industry with limited shelf-life (or *revenue day*) of 24 hours. Most hotels configure themselves with different levels of room configuration, predominantly standard, suite and promoted set of rooms.

✓ Remember, guest rooms are always considered guests' private areas even though these rooms are almost occupied by different guests at different times.

✓ In case of resort properties, these guest rooms may split wing-wise in acres of space whereas in business hotels all guest rooms are mostly located floor-wise; this is due to limited available space. Housekeeping department is responsible to clean all guest rooms.

## 20.7 POLISHING

The term *polish* refers to cleaning agent applied to a surface to form a hard protective layer and thus guard against fingerprints and scratches. Polish is applied to give a glossy finish, to protect the surface and always used in small quantity. Excess polish should be wiped away with newspaper or *rags* before it dries. Remember, polish applicator should label with type of polish so that each applicator can be used with just one kind of polish; it should not get mixed with others. Polish falls into three broad categories: spirit based, oil based and water based plus it comes in three forms, i.e., liquid, paste and cream. Broadly, polish can be categorized as metal polish, furniture polish and floor polish.

| Activity | | |
|---|---|---|
| Match the below given polish types with probable use surfaces then fill in the given blank spaces with appropriate descriptions. | | |
| **Polish** | **Mostly used to polish** | **Description of polish type** |
| Spirit based | Metal based utensils | ? |
| Oil based | Wooden material | ? |
| Water based | Different flooring | ? |

## Metal Polishing

- The term *metal* refers to utensils/equipment made from solid materials such as silver, copper, steel and brass. Due to regular use, the utensils made from different metals get dull. Metal polishes are used to clean and brighten the metal items.
- Solvents, acids, and various fine abrasive materials (generally jeweller's rouge and precipitated whitening) are used to degrease, clean, buff, and polish metals. Desired level of shining can be obtained by rubbing the surface of metal. In order to achieve a smooth and shiny finish, you require correct polish (such as *Silvo* for silver and *Brasso* for brass metal), rubbing cloth (like chamois leather), polishers and buffers.
- Metal polishes may be foam and/or liquid based plus long-term or impregnated wadding and are normally formulated for hard and soft metals. Metal polishing can create a variety of hazards like chemical exposure, entrapment/entanglement, noise exposure, and ergonomics.

## Activity

Fill in the below given spaces with appropriate description.

| Metal types | Name of polish | How to apply/clean metal |
|---|---|---|
| Silver | Silver dip, burnishing, pink powder and polivit. | Step 1 ..................................................<br>Step 2 .................................................. |
| Copper | ? | Step 1<br>Step 2 |
| Steel | ? | Step 1<br>Step 2 |
| Brass | Brasso | Step 1<br>Step 2 |

## Furniture Polish

- Polish protects furniture from dust and grime plus creates a thin layer of resin or wax. Wax has its principal use in waterproofing. A common liquid furniture cleaner is Murphy Oil Soap.

- *Furniture polish* is available in three forms, i.e., aerosol, liquid, semisolid (cream and paste form). The polish layer gives protection against abrasion and absorption of spillage, simultaneously also provides a smooth, shiny and reflective surface. Remember furniture with synthetic resin finish does not require polishing.

- It is necessary to polish wooden furniture once a year. Some polishes may be dangerous for the furniture as they contain silicone oils and other contaminants. Remember aerosol type polishes damage the varnishes and lacquers.

- *Beeswax* is readily available as furniture polish and is also used as a wood finish. Apart from beeswax, paraffin and carnauba are also used to polish the furniture. *Silicone* containing furniture polish gives a harder and more lasting finish to furniture. It also improves the heat and moisture resistance power of furniture.

## Activity

Match the following below mentioned cleaning disinfectants with their correct description.

| Furniture polish types | Wax content (in %) | Composition | Description | Suitable to polish |
|---|---|---|---|---|
| Aerosol | ? | ? | - Expensive, quick and pressure based<br>- On glass surface, the wax build-up may make the surface smeary. | ? |
| Liquid | ? | Contains more amount of solvent (thus it induces cleaning action) | ? | French polished furniture |
| Semisolid<br>- Paste form | 25 to 30% | May contain silicon | ? | Wood paneling and other antiques |
| - Cream wax | 18 to 20 % | ? | Emulsions of blends of waxes in water and oily solvents | ? |

## Floor Polish

- *Floor polish* is used for the cleaning/polishing various types of floors. Floor polish is also known as *floor wax* which provides a temporary film over a floor surface to protect the floor from damage, simultaneously improving its appearance.

- Regular maintenance of floor and use of quality floor polish or wax enhance the durability of floor (or hardwood floor) as well as protect the floor. Floor polish is available in two forms, i.e., *water based* and *spirit based*. Water based polish is known as *emulsion polish* whereas spirit based polish is known as *solvent based polish*.

**Wax content in water based polishes**

| B.P. | Wax % | Polymer % |
|---|---|---|
| Fully | 45-60 | 20-40 |
| Semi | 25-40 | 45-60 |

**Wax content in spirit based polishes**

| Paste polishes | 25-30% |
|---|---|
| Liquid polishes | 08-12% |

- Floor polish or wax forms a thin protective layer (not thicker than 0.2 mm) on the floor surface which reduces the impact of scratches, scuffs and other damages, simultaneously repels dirt and moisture and also gives the floor a shine.

- Floor polishes comprise water, polymer, surfactants and waxes. The wax content of floor polish determines the amount of *buffing* in order to give the shine. Some floor polishes also contain plasticizers that help the film formation process and provide additional protection to the floor.

## Activity

Fill in the below spaces with their correct type/composition/description/examples/ suitable to polish furniture.

| Floor polishes | Types | Composition | Description | Examples (waxes) | Suitable to polish |
|---|---|---|---|---|---|
| Spirit based | Natural Artificial | Blend of natural waxes in spirit solvent | ? | - ?<br>- Polyethylene | - Magnesite floor<br>- Wooden floor<br>- Cork floor |
| Water based | ? | ? | It is always the liquid and wax content will determine the amount of buffing. | Carnauba Montan | ? |

## *Standard tasks involved in polishing any surface:*

Task 1     The chemicals used for polishing may be flammable or hazardous, therefore, the attendant should thoroughly read the *material safety data sheet* (MSDS) for each chemical prior to use.

Task 2     Polishing should be done in well-ventilated areas away from smoking zone/ sources of flame. Rags/clothes used to polish the surface should be properly disposed of.

Task 3     Polishing chemicals should never be mixed with other chemicals, as it may create explosion or other hazards.

Task 4     A place where polishing will be done should be properly cleaned because even the presence of suspended dust in the air can create an explosion hazard.

Task 5     Person who will polish the surface must wear protective clothes and *personal protective equipment* (PPE). Barrier cream, face mask, ear muffs, etc., can also be used to prevent any reaction.

Task 6     Correct polish should be used to polish the surface, simultaneously use correct tools and techniques. For instance, while polishing limited floor surface use rayon mop whereas for large spaces use floor polishing machine. Do not forget to clean *floor seals*.

## Activity

A *floor seal* can be either solvent or water based. It is applied to a floor surface to form a semi-permanent protective barrier which will prevent the entry of dirt, liquids, grease stains and bacteria. Depending on the traffic they (surfaces) receive, they may last for up to five years before replacement is necessary.

| Sealers based on | Example of sealers | Prefer to use on floor | Remarks |
|---|---|---|---|
| Clear solvent | ? | Wood, cork and magnesite | ? |
| Synthetic materials | One pot plastic | ? | ? |
| Coloured pigments | Pigmented sealers | ? | Provide colour and strengthen the sealer |

## CONCLUSION

Cleaning like dusting and moping is a part of daily/routine work whereas deep cleaning, periodic cleaning and polishing are carried out at a scheduled period/frequency. Room cleaning process can be undertaken either via orthodox, block or team method. But before starting cleaning task, it is most important to find out cleaning areas. Generally, housekeeping department is responsible for in-house cleaning whereas cleaning of out-house areas is mostly out-sourced. The selection of correct/appropriate cleaning agent is a primary concern because it affects degree of cleanliness and durability of the cleaned surface. Apart from routine cleaning, metal and furniture surfaces also require polishing. Polish is applied to form a protective layer which guards against fingerprints and scratches. It is mostly applied via rubbing process. Remember before applying polish, attendant should carefully refer to MSDS manual. Do not forget to clean floor seal applied to prevent dirt, liquid, grease stains and bacteria.

| Terms (with Chapter Exercise) | |
|---|---|
| Wetting power | Detergents should have the power to decrease the surface tension of the water and enable the surface of any article to be thoroughly wetted. |
| Cleaning schedule | It refers to the cleaning timetable that is prepared for routine/daily cleaning. It also shows the duty and responsibility of individual employees. |

| | |
|---|---|
| Emulsifying power | It means the detergents should have the power to break down the lubricant or grease and enable the soiling to be loosened. |
| Cleaning agent | It is a collective term for all cleaning aids/materials (such as floor and laundry cleaning agents) used to clean the surface. |
| Suspending power | It means the power of detergents to remove the re-deposition of the soil or dust. |
| Cleaning frequency | *Rate of cleaning recurrence,* for instance when deep/special cleaning will be done. |
| Murphy oil soap | It is a wood cleaner that can be used on cabinets, laminated surfaces, Formica and painted woods. Murphy oil soap is meant to be diluted with water and then applied to the surface. |
| Polish | It refers to a cleaning agent or equipment in which dry or chemical soaked cloth is used to wipe the surface to make it shiny like floor polishing. |
| Jewel's Rouge | It refers to a pink oxide or iron used as a fine abrasive for polishing silver and so forth. |
| ? | Fine abrasives, i.e., also known as *whiting.* |
| Action plan | ? |
| Wax | Use of wax or another cleaning agent to put a protective coating on wood like floor or furniture. |
| Dirty dozen | It refers to all those areas where cleaning work is not easily investigated like top side edges of wall mounted pictures, top horizontal surface of door. |
| ? | It is also known as floor *finish* and mainly available in two forms, i.e., permanent-type and penetrating solvent based. It is mainly used on concrete, marble, terrazzo or other stone surfaces. |
| pH | It refers to a measure of the acidity or alkalinity of a solution, numerically equal to 7 for neutral solutions, increases with increasing alkalinity and decreases with increasing acidity. The pH scale commonly in use ranges from 0 to 14. |
| Bleach | Chemical agent used for bleaching purpose. |

| ? | It is the method of cleaning whereby room attendant completes all tasks in one guest room then moves to the second room. |
|---|---|
| Abrasive | It is the grit material made from materials such as quartz or sand or pumice. Abrasives may be fine, medium or hard. |
| Solvent | It is a liquid that dissolves a solute resulting in a solution. |
| Block cleaning | ? |
| ? | Total number of guests available in any hotel at any particular point of time. The formula used to calculate is Previous House Count + Expected Arrivals – Expected Departures. |
| Room count | Total number of rooms occupied in any hotel at any particular point of time. The formula used to calculate is Previous Room Count + Expected Check-ins – Expected Check-outs. |

# Guest Room Cleaning, Inspection and In-Room Services

◁ **OBJECTIVES** ▷

*After reading this chapter, students will be able to...*

- describe guest room cleaning and its procedure;
- express bathroom cleaning and its process;
- elucidate turndown service and how it is different from the second service;
- enlist amenities for VIP rooms; and
- explain care and cleaning of different in-room equipment.

## 21.1 GUEST ROOM CLEANING

It is a room make-up process of guest rooms, i.e., parlour room (living room), bedroom (sleeping room), bathroom (including WCs, vanity units), kitchenette and other available spaces. Guest rooms are cleaned on daily basis usually after guest departure (or before guest arrival), during guest occupancy and also cleaned, as and when guest demands. Guest rooms requiring deep cleaning when kept aside as OOS/OOO for long. Therefore, during peak season *lettable rooms* are usually cleaned on daily basis (or after guest's departure) whereas rooms kept aside (or locked) due to low occupancy or repair and maintenance work (i.e. OOO/OOS), thus opened after an extensive period of time requiring deep cleaning first then are handed over to front office. Afterwards, these guest rooms are cleaned in accordance with requirement/guest occupancy.

Guest room cleaning can be performed either via *orthodox, block* or *team based* scheme. Room attendants are responsible for guest room cleaning. They need to carry all required accessories (such as cleaning equipment and supplies, *closet set*, *stationery set*, room and bathroom amenities, linen, etc.). Floor/room attendants stock all these items on *chambermaid trolley* and move towards the allotted floor/section. The trolley must be large enough to hold the supplies and amenities. Therefore, in many large hotels there is a storeroom on each floor (or *floor pantry*) for replenishing the supplies on the cart. The main amenities on a cart, in almost every hotel, are room and bathroom linen, amenities and stationery.

## Cleaning procedure of guest room

- First of all, confirm the guest room status, i.e., whether guest is available in the room or not.
- Secondly, put toilet cleaner (like R-1 and R-6) in toilet bowl and remove all rubbish.
- Next, room attendant is required to open the drapes/ turning on room lights.
- Lastly place all room amenities and supplies at their designated place.

## Cleaning of vacant rooms

- First of all, clean the inside of the balcony windows then outside windows. Afterwards, move the balcony furniture, clean underneath and rearrange their designated places.
- Attendant should clean the ceiling, walls, air-conditioning unit then sweep clean and dry the balcony floor.
- All required cleaning agents and equipment should be kept in the bathroom.

## 21.2 ROOM CLEANING PROCEDURE

It is the systematic process of cleaning of guest rooms, especially the occupied guest rooms. The cleaning of these rooms is performed on daily basis. As we all know that room is a perishable product, simultaneously the source to generate maximum revenue because once rooms are constructed then it only requires operational expenses along with intermittent repair and maintenance cost. All hotel rooms should be cleaned as per the frequency by using dry & wet cleaning methods by deploying adequate trained personnel and cleaning equipment/machines. Room attendant/chambermaid is responsible for cleaning all guest rooms. These rooms are to be cleaned during the hours when guests are not present in their rooms plus as and when guest demands. Room cleaning task can be performed either through orthodox, block or team based method. Generally, room cleaning is performed during the morning hours because most guests are not present at this time in their rooms.

**Complementary items**

- Coffee
- Tea
- Milk
- Sugar

**Room appliances to be cleaned**

- Kettle
- TV & remote
- Hair dryer
- Clothes press
- Crockery
- Glassware
- Telephone

## Preparation for Room Cleaning

Room attendants need a special set of tools to do their job. These include cleaning supplies and equipment, room accessories, amenities and linens. A list of major tasks involved in room cleaning is given below:

Task 1    Wear personal protective clothing like full apron, gloves, etc. Thereafter, recognize bedrooms to be cleaned.

Task 2    Choose and collect cleaning equipment like vacuum cleaner, duster, waste bin liner and take to the appropriate room safely. Don't forget to bring complimentary room items.

Task 3    Recognize all room equipment like single bed, double bed, bed linen, bedroom furniture and other in-room appliances.

Task 4    Start room cleaning from dusting and polishing of surfaces, thereafter vacuum clean the floor area and dispose of waste bin contents and place a fresh liner in the waste bin.

Task 5    Now, unclutter bedroom contents and replenish complimentary items as required. Check/replace soiled crockery and glassware with clean items.

Task 6    Check that all in-room electrical equipment/appliances are operating and report about faults, breakages, deficiencies if any to the supervisor.

Task 7    When room cleaning task is completed, report to the supervisor. Thereafter, clean and store your personal protective clothing. Finally, wash and dry hands thoroughly.

## Room Cleaning Process

Before starting room cleaning task, room attendant should stock all room items on the chambermaid's cart which every room attendant will use to clean and prepare between twelve to twenty rooms per day. These are all necessary for preparing rooms for guests. A systematic process for making a room involves the following steps.

Step 1    First, room attendant knocks at the guest room door, announces "housekeeping" and waits for guest's response. In general, if guest is not present still it is obligatory to check the room before entering.

Step 2    Now, room attendant opens the door and places the *door wedge*. After entering the room, attendant first draws the curtains and often window in order to allow fresh air into the room.

Step 3    Next, room attendant takes a look at the room. If the linen, waste basket, television or anything else are missing or furniture found damaged then report to the housekeeping supervisor immediately.

Step 4    Now attendant will check for any item the guest might have left behind in balcony, desk drawer, fridge, underneath the bed/pillow/mattress, wardrobe, behind the bathroom door/main door or any other place.

Step 5    Afterward, room attendant will strip the mattress gently, shake it carefully off the bed and check for valuables. If anything is found in the guest room then deposit it with housekeeping control desk.

Step 6    Check all in-room electronic devices if they are functioning properly or not (if not then report to maintenance department). Afterwards switch *off* the lights, television, air conditioner and other electronic devices.

Step 7   Remove waste and tray from the bedroom, bathroom, clean ashtray and waste containers. Replace soiled linen in room and bathroom with clean linen.

Step 8   Attendant should also replace the guest room and bathroom supplies and amenities. Place all collected soiled linen in chambermaid trolley.

Step 9   Remember, while cleaning the room and bathroom attendant should take care that no damage occurs to the flooring, carpet, and other equipment provided in the rooms as well as guest luggage.

### Cleaning of vacant rooms

- As these rooms were not occupied during the last night thus they do not require full cleaning.
- Apart from regular dusting of room furniture, bath and wash basin, room attendant needs to open curtains; put *off* the bed side lamp; remove the bed side card (and place on the bed side/ writing table drawers); convert night bed into day bed; change drinking water, flush toilet, etc.
- If room is vacant for many days then wash the bathroom and balcony floor (if required scrub as well), vacuum clean the carpet.

### Cleaning of stay-over rooms

- As these rooms were occupied during the previous night as well as guest is not going to leave on that day.
- Apart from regular cleaning, room attendant should clean under the guest belongings, neatly arrange in-room tables, clean wash basin and vanity units, etc.
- Other cleaning tasks include hang the cloth in the wardrobe or fold and keep on the chair; arrange shoes and slippers neatly under the luggage rack.
- Replenish room complementaries; guest room supplies and bathroom supplies.
- Remember not to open the drawer since the guest may have his personal belongings. In addition, attendant should never try to wear any of the guest's belongings.

### Cleaning of sleep-out rooms

- Apart from standard cleaning, room should be aired, dusted, damp wiped and checked on daily basis.
- Replace bedcover and offer turndown service (i.e. draw curtains and switch-*off* the bedside light).

### Cleaning of VIP rooms

- Apart from standard cleaning, room should be aired, dusted, damp wiped and checked on daily basis.
- Replace water of flowers and eaten fruits, peels; and cutlery, crockery and glassware.
- Remember, if these belong to guest then do not replace without his permission.

## Activity

Place below given points in order of cleaning preferences or priorities:

| Activity 1 | Occupied room | Vacant room | Vacant & dirty room | Sleep out room |
| Activity 2 | Bed making | Room cleaning | Bathroom cleaning | Room furniture |
| Activity 3 | Walls | Floors | Fixtures & fittings | |

## 21.3 BED MAKING

*Bed making* is preparing a bed for guest in order to make it more comfortable as well as eye appealing and pleasant. A complete and perfect bed making comprises mattress, mattress protector, bed sheets, blankets, quilts, duvet, pillows (with covers), bed cover, bedspread and so forth. Remember match the crease of all bed linen with each other for even distribution and if any label appears on any of the linen it should be shifted at the bottom of the bed.

**Items of Bed Linen**

- Bedsheets (single/double)
- Pillow with cases
- Pillow protectors
- Mattress with protectors
- Blankets
- Duvet with cover
- Bed cover

## Preparation for Bed Making

Task 1  Wear personal protective clothing like full apron, gloves, etc. Thereafter recognize the beds/rooms to be stripped.

Task 2  Choose and collect bed linen and move towards the rooms. Strip the bed and check for damage and deficiencies. If found, report to supervisor.

Task 3  Dispose of dirty linen correctly and select items of linen to be used for bed making. Now start bed making and when complete then report to supervisor.

Task 4  Now clean and store your personal protective clothing. Finally, wash and dry hands thoroughly.

## Bed Making Process (or *bed make-up* process)

It refers to the step-by-step course of action to make a bed ready. A room attendant can make a bed in various styles. But a systematic procedure to make a bed includes several steps.

Step 1  First, room attendant should strip the soiled bed sheets and pillow cases thereafter dust them separately. If any guest belongings are found then place it on the bed side table. Next turn the mattress side-to-side followed by end-to-end turning.

Step 2　Now attendant should shake the mattress protector and relay it. If mattress protector gets soiled, change it.

Step 3　After this, attendant should pick the first fresh bedsheet and open it evenly and tuck it securely at the head, foot and the sides. Hereafter, open second fresh sheet and distribute it evenly on the first laid sheet. The sheet hem should be evenly pulled up to the headboard. Tuck this sheet at the foot.

Step 4　Remember, match the crease of both the sheets for even distribution. Now open blanket and place it evenly on the top bedsheet. Next pull the blanket from the headboard then fold top bed sheet over the blanket. Now fold both once again.

Step 5　Afterwards, the blanket and the top bed sheet are tucked together uniformly on both sides while corners at the foot of the bed are neatly *mitered* at all four corners. Remember, mitered corner is different from *hospital corner*.

Step 6　Now room attendant should cover the pillow with clean pillow slips, afterwards place pillows at the top of the bed. Pillows and pillow slips should fluff and the excess of pillow slips should be neatly folded downwards. Remember, the folded side of pillows should be away from the guest's view.

Step 7　Lastly, cover laid down bed with *bedspread*/bedcover and ensure that it should evenly fall all around the bed. One extra bedspread can also be placed towards the headboard to crease in between the pillows in order to make bed attractive.

Step 8　Nowadays, many properties have replaced top bed sheet, blanket and quilt with *duvet*. Room attendant may also place spare blanket in the wardrobe.

## Activity

Differentiate between:

| Base/criteria | Mitered corner | Hospital corner |
|---|---|---|
| What is? | ? | ? |
| Where preferred? | ? | Military |
| How folded? | ? | ? |
| Importance | ? | ? |

## 21.4  BATHROOM CLEANING

*Bathroom* is a small room used for bathing and as a washroom. Like room cleaning bathroom cleaning also involves systematic cleaning process such as room attendant should start cleaning from ceiling then move to the floor. While cleaning floor, it should be cleaned from the wall farthest to the door then move towards the exit. All soiled bathroom linens should be replaced with fresh *bathroom linens*. All collected soiled bathroom linen should be placed in *laundry bags*.

### Preparation for Bathroom Cleaning

It is concerned with assembling of various equipment (like Johnny Mop), cleaning agents and similar articles that are required for cleaning the bathroom. Simultaneously also replace bathroom supplies like bathroom linens (such as towels and wash-cloth and items of *amenity package* such as soaps, powder, shampoo, bubble bath, toothpaste shaving kit and so forth. Major tasks involved in bathroom cleaning are listed below:

| | |
|---|---|
| Task 1 | Wear personal protective clothing like full apron, gloves, etc. Thereafter, make sure that area (bathroom) to be cleaned is not occupied. |
| Task 2 | Choose and collect bathroom components, cleaning equipment and cleaning materials/ agents and take to the appropriate guest room safely. Replenish tray with complimentary toiletries if required. |
| Task 3 | Now remove rubbish containers, remove dirty/used towels and dispose of correctly. Don't forget to remove partially used soap. |
| Task 4 | Choose and use cleaning material and start cleaning areas correctly. If any damaged or broken items are found, report to supervisor. |
| Task 5 | Now, dispose of rubbish safely and when task is completed, report to supervisor. Thereafter, clean and store your personal protective clothing. Finally, wash and dry hands thoroughly. |

**Bathroom components**
- Bath
- Handwash basin
- Toilet

**Cleaning materials**
- Mirror/glass cleaner
- Bath cleaner
- Handwash basin cleaner
- Toilet cleaner

**Cleaning equipment**
- Toilet brush
- Duster
- Cleaning cloth

### Bathroom Cleaning Process (or *bathroom make-up process*)

It refer to cleaning of bathroom floor, bathtub, toilet bowl, shower curtains, vanity uni water closet, bathroom mirror, soap dish, shower walls, chrome fixtures, shower head faucets, mirror and so forth. A step-by-step process for cleaning a bathroom is give below.

Step 1    The room attendant should open the windows and exhaust vents. Afterwards, collect soiled bathroom linens and shake each separately. If any guest belongings are found then place it on the bed side table. Transmit collected soiled linens into laundry bags or *linen hamper*.

Step 2    Next collect trashes from sanitary bins and ashtrays then place in garbage bag. Apart from regular cleaning (like dusting and cleaning of cobwebs), attendant should wipe bulbs and shades with a dry cloth.

Step 3    Afterwards, wash the bathtub, surrounding areas and wipe the shower curtains (wash from both sides) and its rods with wet sponge (use all-purpose cleaner). Thereafter, wipe dry each and ensure they do not have any water spots or marks.

> **General cleaning considerations in bathroom**
>
> - Upon checkout or weekly disinfect the bathroom.
> - Use appropriate cleaner to clean toilet seat, lid hinges, base, caps, sinks, vanity unit and so forth areas.
> - Reset the entire bathroom area with fresh supplies and amenities like clean bathmat, towels, soap, shampoo, tissue, etc.

Step 4    Next scrub the toilet bowl with an appropriate cleaner (like Harpic or R-1 and R-6) and toilet brush. Clean the seat, lid and outer side of the toilet bowl (check holes under the ring) and put a disinfectant solution inside. Remember to scrub dry the surrounding areas of wash basin and counter.

Step 5    All areas should be dry and spotlessly clean. Now attendant should clean the bathroom mirror (and polish chrome with glass cleaner) with a dry cloth. Pay special attention if it is a mirror TV (or TV based mirror).

| **Activity** | | |
|---|---|---|
| Differentiate between: | | |

| Base/Criteria | Sheer Curtains | Shower Curtains |
|---|---|---|
| Made from | Polyester | ? |
| Mainly used in | ? | Bathroom |
| Reason to use | ? | To keep separate dry and wet area |
| Curtain fabric | ? | ? |

## 21.5  HOLISTIC CLEANING

### 21.5.1   Care and Cleaning of Bed

As we all know that room is a highly perishable as well as highest revenue generating source of any hotel establishment. The positioned furniture, *whiteware*, electronic appliances, linen and other room accessories enhance the look of the room, especially sited bed because when guests stay in a hotel room, they are naturally concerned with sleeping and the comfort of the bed, thus it is of great importance. The bed must not only be comfortable but also be attractive and eye appealing. All these things depend on the design of the material from which they are made and their finish & appearance in the room. Major tasks involved in care and cleaning of bed are listed below:

Task 1   Before starting bed making, room attendant should ensure that attached headboards (like common headboard in *twin room*) should not be loose and mattress does not hang down and buttons or other tufting should not be missing.

Task 2   Room attendant should also check for soiling and tears in *ticking* and rotate the inner spring mattresses occasionally, from both sides and top to bottom. Attendant should also use *underlay* on bases of mattress and supply water-resistant sheet, as and when guest needs.

Task 3   Attendant should fit base covers or *valances* where bases are not covered with PVC. And periodically clean the open wire spring with duster and brush. If required, also wipe it with an oily rag. Afterwards attendant should vacuum the upholstered bases and mattress.

Task 4   Plastic and rubber foam made mattress should be wiped with damp cloth. At last, remember to launder and dry clean valances and base cover at scheduled frequency plus as and when required.

## 21.5.2   Care and Cleaning of Mattresses

It is always desirable to purchase stain resistant, non-allergic, good quality and durable mattresses. Initially, it is found to be expensive but in long run it will be cost-effective for the property. In order to protect as well as to augment life of mattresses, it must be covered with a moist or water resistant mattress cover. Major tasks involved in care and cleaning of mattresses are listed below.

Task 1   While attaining mattresses, first room attendant should turn the mattress from head to foot, next from side-to-side on routine basis.

Task 2   Afterwards check for torn or worn seams, if found then immediately send to sewing room (to *seamstress*) for mending work.

Task 3   Otherwise brush mattress with upholstery brush and vacuum clean on routine basis. Lastly, arrange mattress properly and cover with protector/mattress cover.

## 21.5.3   Care and Cleaning of Room Curtains

Room curtains can be dry dust/vacuum cleaned and wet cleaned when necessary. Attendant should attach appropriate attachment while using vacuum cleaner for cleaning curtains and its components. At regular scheduled frequencies, all attendants should reverse the curtains, so the colour of curtains will be similar on both sides. Otherwise, the colour of curtains on one side will fade through the continuous reflection of sunrays.

Remember, mending of curtains, drapes and other attached linen based items are sent to sewing room (or to *seamstress*) whereas for repair and maintenance work of attached curtain rod, pulleys and other attached mechanisms, maintenance department is called. Major tasks involved in care and cleaning of curtains are listed below:

Task 1    The room attendant should dust open the curtains then start vacuum cleaning. Curtain rod should be cleaned with wall broom or vacuum cleaner. He should attach appropriate attachments with vacuum cleaner.

Task 2    Afterwards check curtain and its components for damage/wear or tear, if found then immediately send for mending work. For instance, repair of lining, frayed edges, bent tracks, pulleys and so forth. Curtains are also sent for deep cleaning at scheduled frequency.

Task 3    In deep cleaning, curtains can be cleaned with spray extraction machine with appropriate cleaning agents or solvents. Remember, the *lined curtain* is usually dry cleaned whereas good quality *unlined curtain* is laundered.

## 21.5.4   Care and Cleaning of Bed Linen

It is concerned with care and cleaning of in-room linens that use to lie-down over the bed, for example, bedsheets, blankets, duvets, bed spreads, quilts, pillow and their covers and so forth. These are expensive linens and usually change on daily basis (in occupied rooms) in luxury hotels whereas in small organizations these may be changed on alternate days. After changing bed linen, the collected soiled linens should be kept in chambermaid trolley and transported to linen room (via *linen chute*) from where they further move to laundry section. All linens should be handled carefully and never use to wipe out any article, floor or for dusting any surface. Standard guidelines to properly handle different kinds of bed linen is given below.

| Care & cleaning of pillows | Care & cleaning of cushions |
|---|---|
| - Feather *pillows* should be shaken well whereas foam and rubber made pillows wiped down properly. | - First cushions should be shaken properly, next brush/ vacuum clean then place orderly. |
| - If pillows get shabby or scruffy (i.e., split or tear) in the ticking then mend immediately. | - If cushions get shabby or scruffy, then send for mending work immediately. |
| - Periodically (or when required) launder or dry clean the pillows after removing covers. | - Periodically (or when required) wash/dry clean after removing cover. |
| - Protect all pillows with under pillow slips. | - During cleaning check for stains, if found then clean with appropriate cleaner/solution. |
| - Pillows covered with pillow covers in increase their life. | - Cushions must have covers to enhance their life. |

### Care & cleaning of blankets

- First blankets should be shaken properly. If blankets get ragged from ends then immediately send for mending.
- Periodically (or when required) wash/dry clean blankets.
- During cleaning process check for stains, if found then clean with appropriate cleaner/solution.
- While storing woollen blankets, be cautious for *moth*.

### Care & cleaning of duvets and quilts

- First *duvets* and *quilts* should be shaken properly. If any of them is torn or worn then immediately send for mending.
- Periodically (or when required) wash/dry clean duvets and quilts.
- During cleaning process check for stains, if found then clean with appropriate cleaner/solution.
- While storing feather based duvets and quilts, be cautious for *moth*.

### Care & cleaning of bedspreads

- First bed spreads should be shaken properly. If any bed spreads get torn or worn then immediately send for mending.
- Periodically (or when required) wash/dry clean bedspreads.
- During cleaning process check for stains, if found then clean with appropriate cleaner/solution.

### Care & cleaning of loose covers

- First loose covers should be shaken properly. If any of them is torn or worn then immediately send for mending. Lastly, affix orderly.
- Routinely brush/vacuum clean and periodically (or when required) wash/dry clean loose covers.
- During cleaning process check for stains, if found then clean with appropriate cleaner/solution.

## 21.5.5   Care and Cleaning of Bathroom Linen

It refers to care and cleaning of guest room's bathroom linens and fittings like shower curtains (linen or plastic made), bathmats, towels and so forth. Like room linens, bathroom linens are also expensive items and change on daily basis, except shower curtains. The remaining procedure for sending soiled bathroom linens into laundry section is identical to guest room linens movement procedure. Bathroom linens should never be used to wipe floor, mirror, vanity unit and other surrounding surfaces. Guidelines on care and cleaning of different bathroom linens are given below.

## Care & cleaning of shower curtains

Shower curtains should be wiped then dried on daily basis. Its rod should be wiped down too with damp cloth and appropriate cleaner.

## Care & cleaning of bath mats

- Mostly bath mat is placed immediately outside the bathroom and sometimes inside the bathroom itself (but after the shower area).
- In luxury hotels, bath mats are changed on daily basis whereas in small hotels on alternate days.

## Care & cleaning of towels

Various towels like face towels, hand towels, bath towels and wash *clothes* should be replaced according to laid standard or on daily basis.

## Care & cleaning of bath sheets

It has already been stated earlier that bath sheets are bath towels but larger in size. So, care and cleaning of bath sheets is also identical to bath towels.

## 21.6  TURNDOWN OR EVENING/NIGHT SERVICE

*Turndown service* refers to evening service/night service, usually between 6:30 p.m. and 9:30 p.m. Luxury hotels mostly offer this service to their guests. Often this turndown service is also known as *second service*. But remember, second service is only offered at the request of the resident guest, for instance guest mainly requests second service when he/she has had the party in his/her room. Often it has been seen that second service is mostly given in *over rooms*. In turndown service, room attendant prepares the guest room, bathroom, closes the curtains, switches on the light and so forth in the evening hours usually before the guest goes to his/her bed. A brief guideline on turndown service is given below.

Task 1  Close the curtains and switch on the bed side lamp. Afterwards, remove the bedspread and store it in the wardrobe shelf then turn down the bed.

Task 2  Place the door knob menu card on the pillow next empty the ashtray, waste bins and take the soiled plates (on trays), if found in the guest room.

Task 3  Replace supplies and complementary items when instructed in the tariff plan. Many hotels also place a good night chocolate on the bed side table. Lastly, dust and vacuum the guest room, if necessary.

Task 4  In turndown service, few tasks need to be performed in the bathroom. For instance, replace any used towel and supplies, clean basin, bath, toilet and floor.

## Activity

Differentiate between the second service and turndown service.

| S. No. | Second service | Turndown service |
|:------:|----------------|------------------|
| 1 | It can be offered at any time of the day. | ? |
| 2 | ? | It includes closing curtains, switching on lights, folding one corner of the bed and replenishing emptied supplies. |
| 3 | ? | ? |
| 4 | Only light cleaning is done in it like removing empty bottles, replacing toilet supplies, etc. | ? |
| 5 | ? | In general, do not disturb rooms and late arrival rooms demand turndown/ evening services. |

## 21.7 OFFER COMPLEMENTARY ITEMS AND SUPPLIES IN VIP ROOMS

Hospitality desk is responsible for arranging complementary items for the due-in guests. These complimentary items may range from a fruit basket, chocolate tray, champagne bottle to flower bouquet. In group arrivals, usually a welcome drink (non-alcoholic) is given as complementary item. A *guest-count* of VIP guests and expected to arrive VIP guests is issued by the front office department to various relevant sections and departments along with an *amenities voucher*. Which item is to be offered or which not depends upon either management, guest demand or guest status (VIP). The term *guest status* means VIP, CIP or PIP guests; these guests can also be classified into three sub-categories, i.e. VIP "A", VIP "B" & VIP "C".

| Code | Clientele Type | Complementary Amenities and Supplies | |
|------|----------------|-------------------------------------|---|
| | | **In room** | **In bathroom** |
| VIP "A" | It includes president, GM corporate or hotel, directors, Ambassadors & any Government Officials (like governor, minister), owners, etc. | Hotel offers champagne, flower, fruit basket, chocolate tray, petite four, full in-room mini bar, valet service and butler service. | Bathrobe, bathroom slippers, bud vase, bubble bath, branded soaps, body moisturizer, after shave lotion, bathroom kit & hair drier. |

| VIP "B" | Manager, Branch Manager; General from army, Air force, Police, any official or any staff (like chief of government officials) | Hotel offers flower, fruit basket, chocolate tray and partially filled in-room mini bar. | Bath-robe, bathroom slippers, body moisturizer, after shave lotion and bathroom kit. |
|---------|---------------------------------------------------------------------------------------------------------------------------------|------------------------------------------------------------------------------------------|-------------------------------------------------------------------------------------|
| VIP "C" | Organizer of group, tour leader, return guest (3rd stay on), public figure (entertainer), etc. | Hotel offers flower and fruit basket as complementary. | Bath-robe, bathroom slippers and bathroom kit. |

**Note.** The above-mentioned complementary items should be set upon guest arrival, only fruit basket and flowers are changed on alternate days.

## CONCLUSION

Hotel room is a place for which guest actually pays hefty room tariff, therefore it demands extra care and cleaning efforts too. Standard room comprises sleeping room and bathroom whereas *suite room* contains one extra room, i.e., *parlour room* which means that different room types take different amount of time to clean. Generally, room cleaning process starts from vacant/dirty rooms then move to saleable rooms followed by occupied rooms. This cleaning sequence may also reform as per the demand of situation. In-room cleaning task starts from ceiling to wall followed by floor. Remember bed making is also a part of room cleaning process (or *room make-up* process). While cleaning guest room, all collected soiled bed and bathroom linens should be tied in laundry bags/*linen hamper* and kept over chambermaid's trolley, thereafter sent into laundry section. Housekeeping attendant must replace all used supplies and amenities, especially bathroom supplies. After the completion of room cleaning, floor supervisor may inspect any guest room for the degree of cleanliness. Apart from cleaning, housekeeping also offers *turndown and second service*; some hotels also provide *valet service* to their guests.

| **Terms** (with Chapter Exercise) | |
|------------------------------------|---|
| Bathroom kit | It is a set of bathroom amenities like comb, toothbrush, paste, disposable razor, etc. |
| Full in-room mini bar | It is an array of beverages in mini bar, usually all major five spirits (white + brown spirits) like- W, G, V, R & B with beer, soft drink, mineral water and salted peanuts. |
| Mattress | It is usually a rectangular pad of heavy cloth filled with soft material or an arrangement of coiled springs, used as or on a bed. |

| OOO | It refers to all those rooms that are removed (for an extended period of time) from total room inventory due to renovation, repair or maintenance work. |
|---|---|
| OOS | It refers to all those rooms that temporarily (for very short period of time) prevent the guest due to some minor maintenance work. |
| Sleeping rooms | It refers to guest room inventory. Remember, it never includes parlour room, cabana room, powder room, cloak room, etc. |
| Seamstress | It refers to a tailor who is mainly responsible for stitching and mending hotel linen and staff linen. |
| Duvets | It refers to a quilt that may be used in place of bedspread and top bed sheet. It usually has a washable cover. |
| Quilts | It indicates to a bed linen, i.e., duvet. |
| Bed cover | It is an ordinary term for bedspread. |
| Loose cover | ? |
| Moth | ? |
| Floor pantry | It indicates to a satellite (or decentralized) system of linen management whereby all hotel floors have a pantry room for *room make-up* (or maintaining guest rooms) on that particular floor. |
| Hand caddy | It refers to a handy container used for storing and transporting cleaning supplies into guest rooms. It can also be carried away from a cart. |
| Closet set | It refers to all personal use items in the closet, placed under a table for use by guests like slippers, a shoe mitt, and a bathrobe. |
| Stationery set | ? |
| ? | It refers to personal use items for guests in their bathrooms like soaps, shampoo, toothbrush, etc. |
| Amenity | ? |
| Cleaning supplies | It refers to various tools and cleaning products used to clean and sanitize guest rooms. |
| ? | It is a strong cloth used to cover mattresses and pillows. |

| Johnny Mop | ? |
|---|---|
| Over-room | It refers to the room that was not assigned for service in the morning but done by evening room attendant. |
| Mitered | It refers to a top bedsheet fold/tuck (along with blanket) at the head of the bed plus both sides of the bed. Simply, to join by means of a mitre joint or seam. |
| Wardrobe | It refers to a tall cabinet, closet, or small room built to hold clothes. |
| Apron | It refers to the garment, usually fastened in the back, worn overall or part of the front of the body to protect clothing. |
| Turndown service | It refers to an evening guest room service performed by evening room attendant. For example, open the corner of a bed so that guest can feel cared for and get in bed easily. |
| Second service | ? |
| VIP | It refers to the guest status/category that has been designated as special guest by hotel management, thus is offered special services. |
| Valet service | It refers to the special hotel staff that are responsible to offer valet service to guests plus to wash, clean and iron fine clothing. In front office, valet parking attendant is responsible to park guest vehicle. |
| Butler service | It refers to a personal server of a guest who is mainly responsible to offer all types of services to the resident guest. This service is mainly offered to VIP guests. |
| Amenity package | It refers to a group of personal use or convenience items offered to guests free of charge. |
| Guest-count | It refers to *house count*. Its formula is *previous house count plus due-ins minus due-outs*. |
| ? | Bed linen consisting of a cover for a pillow. |

# 22
**CHAPTER**

# Public Area Cleaning and Inspection

**OBJECTIVES**

*After reading this chapter, students will be able to...*

- describe public area cleaning;
- recognize different cleaning materials and equipment;
- find out the different in-house and out-house public areas and explain cleaning process for each; and
- identify suitable cleaning agents and suitable methods of their application on different surfaces.

## 22.1 PUBLIC AREA CLEANING

It refers to cleaning of *common areas*, for example, entrance, lobby, corridor, restaurant, elevators, escalators, public restrooms/waiting area, swimming pools, exercise rooms and so forth. As these areas are the division of in-house, therefore, the cleaning of these areas is usually not outsourced (due to security concern). All public areas should be cleaned on daily basis preferably during non-operational hours, mostly before guest arrivals. If required, the cleaning task may also be performed during guest movement but without affecting the guest's inflow and outflow like once in each shift or as and when required.

Housekeeping is not usually responsible to clean the furniture and fixtures of point-of-sale outlets, predominantly food and beverage outlets, gift shops, game zone and so forth. In these point-of-sale outlets, housekeeping only cleans the floor surface. In food and beverage outlets, staff usually cleans and prepares all cooking and serving areas and working surfaces themselves. The larger the hotel, the more different kinds of cleaning tasks housekeeping must perform. In general, there are two main kinds of cleaning they must do, light cleaning, and deep cleaning. If the hotel has large function or meeting room facilities then housekeeping staff may also be assigned with the responsibility to clean meeting rooms, conference centre, exhibit areas, ballrooms, banquet rooms and other hotel operated shopping areas.

# Preparation for Public Area Cleaning

Apart from guest room cleaning, housekeeping is also responsible for cleaning public areas and for back-of-the-house areas like executives' offices. A list of major tasks involved in any public area cleaning is given below:

| | | |
|---|---|---|
| Task 1 | Wear personal protective clothing like full apron, gloves, etc. Thereafter, recognize the public areas to be cleaned. | **Cleaning Equipment & Material** |
| Task 2 | Choose and collect cleaning equipment and materials and place *warning signs* in position before starting actual cleaning. | - Cloth<br>- Mop "squeegee"<br>- Duster |
| Task 3 | Shift furniture (as appropriate) and clean that area using appropriate method. Now carefully dispose of waste material and reinstate earlier shifted furniture. | - Bucket<br>- Broom<br>- Dustpan<br>- Vacuum cleaner |
| Task 4 | After cleaning, remove and properly store equipment and remaining materials safely. Don't forget to remove and put back warning signs. | - Warning signs<br>- Bin liner<br>- Floor cleaner<br>- Polish |
| Task 5 | Now clean and store personal protective clothing. Finally, wash and dry hands thoroughly. | - Water |

Remember, if cleaning is going on during guest inflow and outflow then it is necessary to place safety signage board to avoid any accident. The cleaning of public areas is equally important as guest room cleaning because it supports in framing first positive impression onto the guests. A brief description of some foremost public areas cleaning process is given below.

## 22.1.1 Cleaning of Hotel Entrance

Hotel entrance demands stringent attention since it has the most heavy traffic; therefore, it should be kept clean for both aesthetic appeal as well as for safety. The frequency of cleaning of entrances chiefly depends on guest movement as well as weather conditions. During wet weather, matting or runners at entrances can prevent footprints simultaneously wet weather will demand frequent mopping or changing of runners & matting. Remember in rainy season, runners or matting helps in keeping areas free from puddles plus prevents outside dirt and also protects guests from slips & falls. Its cleaning process involves.

Step 1 Housekeeping attendant should ensure that runners and matting must lay in proper position. The remaining entrance area should be mopped and tidied frequently throughout the day.

Step 2    Attendant should clean the fingerprints and smudges that appear on the door surfaces particularly glass areas. Door and door tracks should be cleaned before the guest checks in to avoid guest inconvenience.

## 22.1.2  Cleaning of Hotel Lobby

Lobby is also called *gateway of hotel* because in many hotels lobby is a hub of activities where guests check-in, check-out, some guests relax at waiting/seating area. Thus, it is an area with heavy traffic. Some hotels also offer the facility of window shopping at special interest or novelty stores in the hotel lobby. Therefore, hotel lobby demands continual cleaning. Remember, cleaning duties in the hotel lobby area can be performed every hour, every 24 hours or once a week. Its cleaning process involves:

Step 1    Generally, lobby is cleaned in the *graveyard shift* (or late night hours) or early in the morning (before check-in time) by lobby attendant.

Step 2    Thereafter, during the day, lot of work needs to be performed in order to maintain the aesthetic appeal of the lobby such as emptying ash trays and sand urns, straightening furniture and so forth.

Step 3    Apart from it, lobby attendant is also required to provide cleaning activities as and when demand arises. Remember, *deep cleaning* like scrubbing of lobby is done on a weekly basis during the graveyard shift.

| Activity | | |
|---|---|---|
| S. No. | Cleaning tasks on hourly basis | Cleaning tasks on graveyard shift basis |
| 1 | Polishing of railing and drinking fountain | ? |
| 2 | ? | Polishing of wooden furniture |
| 3 | ? | ? |
| 4 | Dusting furniture and table fixtures | Vacuuming/cleaning drapes and window coverings |
| 5 | ? | ? |
| 6 | ? | Vacuuming of upholstered furniture |
| 7 | Polishing door knobs and sweeping tables | ? |
| 8 | ? | Cleaning of high or hard to reach areas |

## 22.1.3  Cleaning of Corridor

It refers to the cleaning of *passageway*, especially to guest rooms' corridors. Remember, all guest rooms characteristically open towards the corridor. The major part of corridor cleaning is the cleaning of floors, vacuum cleaning of carpets; *shampoo* on a special project basis. A process of cleaning involves:

Step 1    While cleaning *baseboard*, housekeeping attendant should start the corridor cleaning from one end and move towards the second end of the hall. Afterwards work back down to another side of the hall towards his/her starting/end point.

Step 2    Attendant should dust the air supply vents, sprinklers, light fixtures such as emergency exit lights and so forth. Simultaneously, he should also check its proper functioning like fused bulb should be replaced and damaged things should be reported to the supervisor.

Step 3    Thereafter, clean the front and back of the exit door and wipe any dirt or dust from the tracks, next check and ensure that door opens & closes properly. Remember, during the cleaning process attendant should pay special attention to the fingerprints and smudges.

## 22.1.4  Cleaning of Public Restroom

The term *public restroom* refers to a room or small building which predominantly contains one or more lavatories, washbowl/basin with hot and cold water provision and mirror for the use of in-house general public. In every hotel, there is separate public restroom for men (also known as *half-bathroom*) and women (also known as *powder room*). These areas must be cleaned on daily basis. Its cleaning is easier than other public areas because public restrooms are relatively invariable from property to property. Simultaneously, it is not preferable to offer enormous conveniences in the public restroom because it will demand extra efforts and cleaning. Some hotels provoke special atmosphere by decorating with ornate fixtures & mirrors, allocating lounge space with upholstered furniture & indoor plants and by offering additional services such as hand blower or dryer. Its cleaning process involves:

Step 1    Before entering the restroom, housekeeping attendant should ensure that it is vacant. Thereafter, prior to entering the washroom, the attendant should knock at the door and announce "housekeeping" and wait for response. Generally, it is assumed that there is no one in the washroom if response is not heard after the announcement.

Step 2    While entering the washroom, the attendant should prop the door open and place an approved floor sign at the entrance which indicates that the restroom is being cleaned.

Step 3　　The procedure for cleaning includes: first, empty the dustbins/replace the waste basket liners; clean sink, *vanity units*, toilets, urinals, partition wall (between the toilets), baseboard, floor; restock dispensers (with toilet papers, tissues, paper towels, soaps, etc.); dust/polish dispenser to remove smudges/fingerprints.

Step 4　　At last, make a final check/inspect all the areas and smell the air for any unusual odours. Remember, never stand on the toilet to clean the high or hard to reach areas.

## 22.1.5　Cleaning of Swimming Pool

Swimming is perhaps the most popular of all recreational sports. Many properties, predominantly resort hotels, cater to this interest by providing swimming facilities. It can be either inside (indoor) or outside (outdoor) the main hotel block. Pool design is as varied as hotel operation and may range from basic to elaborate setting. Some pool areas include *whirlpool* and *saunas*. In many properties, it is the duty & responsibility of engineering & maintenance department to clean the swimming pool, sauna & whirlpool whereas in other properties housekeeping staff is responsible for the entire cleaning task. In these properties engineering and maintenance department is mainly responsible for repair and maintenance work, handling pumping machine/water suction system, etc. In collective terms, housekeeping and maintenance department must have close liaison to provide clean and hygienic atmosphere. Swimming pool cleaning process involves:

Step 1　　Housekeeping attendant or maintenance employee is required to start the pumping machine/system so that it ensures that it can constantly circulate the water in and out (out water is being purified before return to pool).

Step 2　　All particles floating on the surface of water are collected in a net. The pool sides need to be scrubbed with an "Antialgae" solution. In the evening time, chlorine has to be added to the water and its (chlorine content) test is conducted by the engineering department.

Step 3　　Ideally, chlorine content in swimming pool should range in between 1 to 3 parts per million. The method used to measure the chlorine residuals is to add a chemical compound called DPD (Diethyl Paraphenylene Diamine) then test either through a *chlorine comparator* or *photometric instrument*.

Step 4　　Housekeeping should also clean wall areas; sweeping then moping hard surfaces and surrounding areas; washing windows and glass areas (like in *cabana room*); and cleaning and straightening nearby furniture.

Step 5　　Apart from it, housekeeping staff is also responsible for collecting wet towels and dirty linen. Thereafter, restocking them; emptying and cleaning trash receptacles and ashtrays.

## 22.1.6  Cleaning of Exercise Room

The term *exercise room* refers to physical fitness centres like gym; it may also include sauna and pool. Since last two decades, the interest of people in physical fitness and health oriented activities is increasing which produces demand for health products and related services. In response to this trend, many properties have created various health oriented facilities as a part of their overall package. This trend is especially popular among resort hotels. Like any other area, health centres also demand care and cleaning. It is the duty and responsibility of housekeeping department to clean all fitness areas.

Cleaning of these areas largely depends on the size and scope of the room and equipment. For instance, equipment for an exercise room might include universal gyms, stationery bicycle, rowing machine, barbells and dumbbells. Almost all hotels appoint professional/trained staff in fitness centre. The design of these fitness areas often incorporates special flooring or hardwood surfaces, mirrors and unique light features. Sometimes, it also provides the extra facility of locker room, shower areas, etc. Along with housekeeping, maintenance department also plays a crucial role in proper functioning of fitness centres. Its cleaning process involves:

Task 1   For security reasons, it is exceedingly important for housekeeping attendant to follow the instructions written on different exercise equipment then start cleaning task.

Task 2   Attendant should dust all equipment, furniture and light fixtures; clean the mirrors and glasses; sweep and mop the floors; remove soiled linen and restock with fresh one; and straighten the furniture.

Task 3   Housekeeping attendant may also be responsible for cleaning shower room, locker room and replenish the fitness centres with appropriate guest amenities and supplies.

## 22.1.7  Cleaning of Other Functional Areas

It refers to cleaning of various serviceable areas like dining room (such as restaurant, banquet hall & coffee shop), employee areas (such as locker room, offices and cafeteria/staff dining hall), kitchen sections, sales and administrative offices, business centre and so forth. These functional areas are required to be cleaned on routine basis (usually dusting/vacuuming, mopping and sweeping) as well as needing deep cleaning at scheduled frequency. Deep cleaning is usually performed periodically like weekly, fortnightly or monthly. Most of these functional areas, especially public functional areas, should be clean during the graveyard shift or after the completion of function, so the guest is not disturbed. Its cleaning process involves:

Task 1   *Daily cleaning* of these areas mostly includes cleaning of tables, chairs, changing of linen, cleaning on-the-spot spills, light vacuuming and so forth. On the other side, *deep cleaning* includes shampooing of carpet, window washing, cleaning of high or hard to reach areas.

Task 2 Remember, daily cleaning also includes collection of soiled linens and replenishing with clean ones; cleaning of phone, upholstery and spot cleaning of walls; wiping of window sils; dusting and polishing of furniture and so forth.

Task 3 When cleaning employees' offices, housekeeping attendant shall not transfer any paper, file, folder, etc., from one place to another. In order to perform proper cleaning, it is always preferable to move all the chairs away from the table before vacuuming beneath the table.

Task 4 Housekeeping attendant should follow the manufacturer's instructions as well as also take assistance from engineering and maintenance department (especially for electronic equipment) for operating typical cleaning equipment.

| **Activity** | |
|---|---|

Below few major functional areas are given. Fill in blank spaces with an appropriate description.

| Functional areas | Description in relation to cleaning task |
|---|---|
| Employee offices | ? |
| Banquet hall | ✓ Its cleaning is analogous to dining room cleaning. Sometimes, it is not possible to clean the carpet spots immediately. Hence, after the completion of function remove it the next day.<br>✓ Clean carpet with vacuum cleaner; if spot appears then remove with correct stain removal like TR-101 & 102. Also shampoo carpet at scheduled frequency or as and when needed. |
| Restaurants | ? |
| Business centre | ✓ Its cleaning is also akin to dining room cleaning. Remember, it should be cleaned (like dusting of furniture) as soon as possible after the completion of meeting.<br>✓ Pay special attention while cleaning walls and floors like remove the food particles and stains from the upholstered surface. |
| Cafeteria | ? |
| Locker room | ✓ It is a guest restricted area. Kitchen stewarding staff is mostly responsible for the cleaning of locker room. |
| Kitchen | ? |

## 22.1.8   Cleaning a Sauna

The term *sauna* refers to *dry heat bath*. Its cleaning comprises cleaning of sauna room, its heater, rock, equipment and surrounding areas. Remember sauna bath is different from *steam bath*. Cleaning the sauna must be a part of a regular cleaning routine because it gets dirty after every use. It also demands regular maintenance. Its routine cleaning and regular maintenance is obligatory because in most of the properties, especially in resorts, it works as a value addition facility plus magnifies hotel's RevPAC by generating additional revenue. Therefore, in order to maintain a safe and hygienic sauna there are a series of tasks that must be accomplished periodically. The sauna cleaning process involves:

Step 1    Routinely vacuum or sweep clean the sauna room floor in order to remove hair, dust, dirt and other waste. If it is used by many guests on any day then it may require to be cleaned twice or thrice a day. Often, it is necessary to wash the sauna floor as well.

Step 2    Remember, if water spills on the floor then make sure it is mopped up before it causes damage to the sauna room floor or any accidents.

Step 3    Sauna room's surrounding areas like benches and walls also need to be cleaned with an appropriate mild detergent/cleaning agent and water solution. While cleaning sauna room, pay special attention to the areas that are most frequently touched like door handles, control panel, switches and benches.

Step 4    Remember harsh chemicals can damage the wooden surface thus avoid them. Lastly, make sure to wipe out all soap residues; simultaneously do not use too much water, as it can damage the wooden surface.

| Cleaning of sauna heater | Cleaning of sauna rock |
|---|---|
| - In sauna heater, water spots collect on the metal, plus dirt and grime can accumulate too. So clean it as per frequency schedule or when needed. <br> - While cleaning a sauna heater, ensure to use a non-abrasive cleaner so that the metal does not get scratched. Lastly, wipe out the grimy fingerprints as they look unpleasant. | - Clean sauna rock, if anything other than sauna fragrances and clean water has been decanted over it. But never clean hot sauna rock. <br> - Cold sauna rock should be cleaned with a mild detergent and water solution, then use a soft cloth or sponge to wipe it. Ensure rocks are rinsed properly thereafter allowing them to dry completely. |

## Activity

Differentiate between sauna bath and steam bath.

| S. No. | Base/criteria | Sauna bath (sauna bathroom) | Steam bath (steam bathroom) |
|--------|---------------|------------------------------|------------------------------|
| 1 | Temperature | ? | 40°C |
| 2 | Humidity | ? | ? |
| 3 | Benefits | Muscle stimulation, lowers blood pressure, improves cardiovascular health, etc. | ? |
| 4 | Heat generation | ? | ? |
| 5 | Built of | ? | ? |
| 6 | Nature | Dry heat bath | Moist heat bath |

## 22.2   CARE AND CLEANING OF HOUSEKEEPING EQUIPMENT

It refers to care, cleaning and maintenance of various small and large, manual and mechanical equipment of housekeeping department. Remember, if equipment are not properly cared for, maintained and cleaned then their functioning and efficiency gets affected which reduces productivity standard. A brief description on care and cleaning of these equipment is given below.

### Care & cleaning of brushes

- Before washing, brushes should be cleaned off all fluff and threads.
- Remember, brush should not be washed, else it may cause loss of stiffness. But they can be washed then rinsed in cold saline water/mild detergent based water. Shake off excess water and leave to dry. Disinfectants can be used for toilet brush.

### Care & cleaning of brooms (hard and soft)

- After every use, shake well to remove dust and fluff. Never use *soft broom* on wet surface, instead *stiff broom* can be used.
- Remember, broom should be stored either lying horizontally or hanging bristles downward but never make it stand on its own bristles, as it will get out of its shape.

### Care & cleaning of carts (or trolley)

- All carts should be kept neat and clean. These should be wiped on daily basis (and thoroughly cleaned on weekly basis). When not in use place them in a lockable, dry and well-ventilated place.
- Remember, cart wheels may be oiled or greased during the cleaning process, if required.

### Care & cleaning of vacuum cleaner

- After use, check the attachment heads, associated filters and lastly empty the dust bags (else it may cause damage due to excess pressure).
- Wipe the casing daily and check hose and flex before use. Its wheel should also be oiled or greased intermittently. Never use dry vacuum to clean wet surface.

### Care & cleaning of mops (dry and wet)

- After use, shake well to mop then remove detachable handle and wash it in detergent based hot water. Afterwards rinse, squeeze and dry.
- Remember, if mop is left damp then the chances of bacterial infestation will be more, so shake well to remove moisture and leave to dry in open air.

### Care & cleaning of sundry equipment

- Sundry equipment refer to miscellaneous pieces of equipment like carpet beaters, ladders, abrasive pads, choke removers, rubber gloves, airing racks, fit pumps and so forth.
- All these sundry equipment should be cleaned after every use plus as and when required.

### Care & cleaning of dusters & cloth mittens

- After use, scrupulously wash, rinse and dry the dusters.
- After washing, if *cloth mittens* are drenched with mineral lubricant then keep them covered otherwise they will attract dust.

### Care & cleaning of swabs and wipes

- After use, wash in detergent based hot water solution then rinse and dry the swabs and wipes.
- As swabs and wipes are used to clean WCs, thus these should be sterilized after washing.

### Care & cleaning of glass, floor, any wet cloth and scrim

- After use, wash, rinse and dry glass, floor, any wet cloth and scrim.
- Remember, floor cloth and any wet cloth should be washed in detergent based hot water solution then sterilized also.

### Care & cleaning of polishing cloths and rags

- After use, rags and polishing cloths are always preferred to be disposed of.
- Dispose of is necessary due to the absorbance of polishes into the rags & polishing cloths, as it may contain combustible chemicals and strong whiff.

### Care & cleaning of chamois leather

- After use, remove surfeit dirt with paper. Do not wash it regularly, if required then wash in cold water afterwards rinse and keep dry flat. When dry, rub to soften it.
- Remember if it is not maintained properly then leather may crack and damage soon.

### Care & cleaning of dust sheets, druggets and hearth & bucket cloth

- After use, shake well or use hard brush to clean stubborn soiling. If required wash, rinse and dry. Now, fold and store in cupboard.
- Plastic *druggets* can also be damp wiped. While canvas and linen should be washed, rinsed and dried.

## 22.3   CLEANING OF FIXTURES AND FITTINGS

It is concerned with care and cleaning of wall mounted appliances and decorative items, for example, wall paintings, wall lamps, control panel, switches and so forth. All these fixtures and fittings enhance the wall decoration, simultaneously support in proper functioning of guest room. A brief description on care and cleaning of these equipment is given below.

### Cleaning of stainless steel sinks

- Once a week, clean all stainless steel sinks. While cleaning always apply appropriate cleaner and buff with a non-abrasive cloth or sponge.
- Remember, if a cleaner contains chloride then rinse the surface immediately to prevent corrosion.

### Cleaning of faucets, fittings, shower door frames

- For cleaning, use water and mild soap thereafter wipe the entire surface completely dry.
- Do not use cleaners that contain bleach, ammonia or other chemicals because it can damage the finish.
- Use warm water to clean gold plated faucets and other similar fittings. Afterwards completely dry with soft cloth.

## 22.4   CLEANING OF METAL SURFACE

It refers to care and cleaning of various metal surfaces and utensils, for example, utensils made up of copper, brass, silver and so forth. The ideal cleaning technique is to always blot dry any water from metal surfaces before applying metal polish. Often water deposits are found on metal surfaces, it is due to the contact of evaporated water with metal surface. These water deposits are necessary to clean at regular intervals else it will form a layer. While cleaning metal surfaces it is important to use a dabbing action to dry metal, not an abrasive or rubbing action. Cleaning with a damp sponge and buff drying keeps your product beautiful. Brief guidelines to clean various metal surfaces are given below.

### Cleaning of brass surface (like door knob/handle)

- Clean/dry dust brass surfaces on daily basis, if required wash in *sudsy,* lukewarm water then rinse and dry.
- Polish brass utensils (with *Brasso*) at regular intervals, else it will tarnish when exposed to air. Tarnished surfaces should be cleaned with vinegar and lemon.
- Brass surfaces may get lacquered when tarnishing is avoided. Un-lacquered surfaces need to be polished with *Brasso* or similar polish.

### Cleaning of copper surface

- Daily dusting and occasional washing in lukewarm soapy water is required to clean copper utensils.
- Copper is mainly used to manufacture kitchen utensils and often found in bar counter like on *bar die.*
- Tarnished copper surfaces should cleaned with vinegar and lemon. Unlacquered surfaces need to be polished at regular intervals.

### Cleaning of silver surface

- Silver surfaces or silver made utensils should be hand polished with a high quality silver cream or Silvo or D-8.
- Silver made utensils are often found in restaurants (like silver cutlery, dessert bowls, etc.).
- The tarnished silver utensils (and EPNS utensils) should be polished with either polivit, silver dip, pink powder or burnishing method.

### Cleaning of glassware

- Due to regular use, white stains may develop at the bottom of the glass. Normal washing with water and soap may not get off all stains.
- When glass is machine washed or hand washed, each individual item must be polished and dried with a linen made glass cloth, as water leaves stains.
- To remove water stains, use vinegar/lemon juice based hot water solution and polish with glass cloth.

## Cleaning of wall coverings

- Use suction/vacuum cleaner or wall broom to wipe out surface dust from wall coverings.
- If there are tough spots/stains like grease on wall coverings then remove it with proprietary grease absorber. Whereas remove simple marks by rubbing with a soft rubber/cloth.
- If sponge paper is used as wall covering then wipe out stains/marks with a damp cloth or sponge.

## Cleaning of office equipment & air conditioner

- Office equipment (such as cupboards, wardrobes, shelves, racks, etc.) and air-condition (such as window, split and cassette) are cleaned as per frequency schedule by suitable method.
- Outer body of ACs like *Louver* should also be cleaned. Air-condition also require deep cleaning at specified time. Follow the manufacturer's instructions while cleaning air conditioners.

## Cleaning of stainless steel/ mild steel/PVC hand railing

- These are mainly provided as staircases or balconies. These must be cleaned along with the balusters by wet, dry or any other suitable cleaning method. It depends on the metal fitted like D-7 can be used to clean/polish stainless steel railing.
- Remember while cleaning these, no damage shall occur to the cleaning surface.

## Cleaning of doors/window frames

- Different kinds of panelled or glazed doors/windows like wooden, aluminium, galvanized steel sheet, fire rated, etc., should be cleaned by wet, dry or any other suitable method at scheduled frequency.
- Be careful as damages or scratches can occur while cleaning the frames.

## 22.5 CARE AND CLEANING OF SUNDRY THINGS

It refers to the care and cleaning of various miscellaneous and other supportive things. For example, care and cleaning of wall coverings, glasses, mirrors, pavement, tree guards, signage board, dustbins, water tanks and different electronic devices and appliances such as air conditioners and fans. As these are supportive items but still it is must to have provision because they add value plus enhance luxury, make differences (with competitors), support property standards, and simultaneously improve aesthetic appeal of the room. A brief description about care and cleaning of several major sundry/ miscellaneous things are given below.

## Cleaning of signage board and furniture

- Signage/notice board and different types of furniture placed in rooms and offices are to be cleaned as per the frequency schedule by suitable method/s mostly by dry dusting.
- Remember, *signage boards* should be kept neat and clean as they provide direction towards different destinations.

## Cleaning of glasses and mirrors

- Glasses and mirrors should be cleaned on a daily basis. Due care should be taken about the breakage and deterioration of glossiness of mirrors and glasses.
- Apply correct cleaning agent like *Collin*, D-6 or R-3 to clean the mirror and glass surface.

## Cleaning of fans and fixtures

- Generally, fans are not found in five-star hotels. But if available then clean as per schedule. Clean spotlessly the fan's body and its blades and remember they should not get loose.
- Indoor as well as outdoor fixtures should be cleaned as per schedule. Remember it is not damaged as well as its functioning should not get affected. So, appoint trained staff for its cleaning.

## Cleaning of pavement and tree guards

- *Pavement* area should be cleaned by dry sweeping or any other suitable method whereas during the cleaning of *tree guards*, painted surfaces should not deteriorate.
- Remember to place signage board while cleaning pavement and tree guards in order to avoid accidents.
- Guest traffic movement should also not get affected due to this cleaning work.

## Cleaning of water tanks

- Water tanks like boiler should be cleaned at scheduled frequency (it may be weekly/fortnightly/monthly) with suitable cleaning agents and fresh water.
- If required water should be stored before cleaning.
- Remember no residue plus cleaning agents should be left in the tank after the completion of cleaning.

## Cleaning of dustbins

- It is always preferable to place adequate number of dustbins in order to reduce wastage. These are mostly placed in executives' and managers' offices.
- Dustbins should be covered with plastic cover (from inside) and emptied/cleaned on daily basis.
- Spare dustbins should be kept, so attendant should not face any problem while emptying them.

### Cleaning of elevators

- Elevator floor, walls, electrical fans and other light fittings are cleaned on daily basis. Inside and outside switch panel, indication panel and communication equipment also require cleaning.
- Apart from it, inside and outside doors of elevators must also be cleaned. Follow all safety precautions while cleaning the elevators.

### Cleaning of escalators

- Escalator steps, railing (*balustrade*) and other fittings are required to be cleaned on daily basis.
- Apart from it, housekeeping attendant also requires to clean escalators as and when required. Follow all safety precautions (or manufacturer's instructions) while cleaning the escalators.

### Cleaning of fire fighting equipment

- Fire fighting equipment include fire extinguishers, smoke detectors, fire detectors, sprinklers, etc.
- All these equipment require *daily dusting* but special cleaning can be done at regular intervals. Check for date of expiry and keep them in working condition.

### Cleaning of telephone & computer & its accessories

- Telephone instrument demands daily *dry dusting* whether kept in guest rooms or at any working counter/desk.
- Computer and its accessories are required to be dry dusted/cleaned as per frequency schedule. Computer system should be in shutdown stage while cleaning, else data can be deleted.

## CONCLUSION

Nowadays, most hotels outsource their public area cleaning, especially the out-house areas like parking space, garden, golf course, outdoor swimming pool and so forth. Public area is the very first place which guests see and observe. Most of the guests perceive the value of a hotel establishment and room interior by seeing these public areas only, therefore public area cleaning demands same level of attention as guest rooms. In order to ensure effective cleaning of public areas, many hotels appoint their own employees, i.e., public area supervisor. In-house public areas like lobby, corridor, eating places, exercise room and restroom also require routine cleaning. Cleaning equipment required to clean different spaces also demand routine and scheduled care and cleaning. Wall mounted decorative items, electronic panels, escalators, elevators, glass/mirror, signage boards, door/window frames, water tanks, tree guards, fixtures, metal door handles, locks, railings and so forth sundry things (or miscellaneous things) also require regular cleaning.

| Terms (with Chapter Exercise) | |
|---|---|
| Elevator | Platform or an enclosure raised and lowered in a vertical shaft to transport people or freight. |
| Public areas | Common areas of a hotel where general public can move like lobby, corridors, restaurants, etc. These areas can be located in-house as well as out-house. |
| Special cleaning | Cleaning that uses more than routine methods and cleaning products. |
| Deep cleaning | Cleaning that uses stronger methods or special products like shampooing of carpet, removing stains from curtains and carpets. |
| Light cleaning | Cleaning that uses normal routine methods like dusting, wiping and mopping. |
| Shampoo | Use of soap to clean and rinse something that has fibres like hair. |
| Escalator | ? |
| Drugget | Cloth material made up of coarse linen, fine canvas or clear plastic. It is used to place on the floor in doorways to prevent excessive dirt, especially during bad weather. They may also be placed in the passage in between the kitchen and dining area to catch spills and debris. |
| Balustrade | Supportive structure used at different places of building, i.e., railing; it is mostly mount on staircases, balcony, etc. |
| ? | It indicates to dry heat bath. |
| Steam bath | ? |

# Laundry Services

*After reading this chapter, students will be able to...*

- describe laundry service and process of laundering clothes;
- differentiate and explain in-house and out-house laundry;
- draw laundry service cycle and explicit the concept of dry cleaning;
- identify different laundry supports/agents and explain their role; and
- explain how different cloth materials like silk, wool, cotton and synthetic are laundered.

## 23.1 LAUNDRY SERVICE

The term *laundry service* indicates to the provision of laundering clothes and linens, both hotels linens as well as guest clothes. This laundry service may be available either on-premises or off-premises; it depends on certain factors like availability of space, guest turnover rate, amount of linens needed to be laundered, water availability, standard of hotel, wastewater treatment/disposal, laundry exchange standard, available fund and manpower and so forth. The basic work profile of every laundry whether on-premises or off-premises is to wash hotel linen such as bedsheets, towels, napkins, tablecloths and so forth that get soiled every day plus guest clothes. Housekeeping staff and valet is responsible for all laundering work. Housekeeping staff is responsible for washing then supplying cleaned linen to all hotel departments and sections whereas valet is responsible for guest clothes and linen. In some hotels, valet is also responsible for uniforms of the staff. For guests, valets must provide quick and effective washing, cleaning and ironing services.

Remember whether laundry is on-premises or off-premises, housekeeping plays a crucial role in both styles of laundry options. In *on-premises laundry*, housekeeping staff collects linen and clothes and sends to adjacent laundry section for cleaning, thereafter collects and distributes to different outlets or sections. On the other side in *off-premises laundry*, housekeeping staff is responsible for collecting linen and clothes then despatching to out-sourced laundry along with list of materials. Thereafter, it receives laundry and distributes to different departments and sections.

## 23.2 LAUNDRY OPTIONS

The laundry department/section is responsible for providing laundry, dry cleaning and ironing/pressing services to in-house guests. It has been seen that some hotels maintain their own laundry (or in-house laundry) whereas others rely on out-house commercial laundry operators (or off-premises laundry). In both options, a close teamwork is necessary to assure a sturdy flow of linen to guest and respective areas like restaurants, banquet, floor pantry and recreational areas. A brief description of both laundry options is given below:

### In-House Laundry

✓ It is also known as *on-premises laundry*. In hotels, laundry is a section of housekeeping department usually located adjacent to the linen room. This

**Laundry Equipment**

- Washing machine
- Tunnel washer
- Tunnel dryer
- Tumble dryer
- Puffer/Suzie
- Drying chamber
- Calender machine
- Press/steam press
- pH scale
- Mobile convertible shelves/trolley
- Hydro extractor

section is mainly responsible for laundering soiled clothes. For the smooth and efficient running of all hotel sections and departments, it is essential to ensure a continuous supply of all laundered linen. In-house laundry option has several advantages as well as disadvantages. Some of them are listed below.

**Advantages and disadvantages of in-house laundry**

| S. No. | Advantages | Disadvantages |
|--------|-----------|---------------|
| 1 | In-house laundry creates a positive impact over guest experience, as it provides instant laundry service to in-house guests. | It demands heavy initial investment as well as regular direct (like labour and energy cost) and indirect operating expenses (like water, stationery, paper cost). |
| 2 | Linen is readily available during emergencies. In simple terms, linen can be instantly provided to any outlets/sections of hotel plus to guests. | Regular use of laundering equipment also demands timely repair and maintenance work which increases operating cost of laundry facility. |
| 3 | It inculcates daily revenue by laundering guest clothes which positively affects (or increases) hotel's RevPAC (Revenue Per Available Customer). | In-house laundry is only successful when adequate amount of linen/s is/are available with hotel, else it is found expensive for the hotel. |
| 4 | Laundry staff has full control over *wash cycle* as well as on laundry agents (or cleaning agents). Thus, wear and tear of clothes can be reduced, concurrently amplifying linen durability. | In-house laundry demands experienced and technically qualified manpower to operate laundry equipment. Availability of adequate amount of water is also a primary concern for laundry operations. |
| 5 | *Par stock* level can be minimized, at the same the chances of pilferage also reduce. | In-house laundry requires additional space in order to properly set up laundry. |

# Out-House Laundry

✓ Also known as *off-premises laundry*. It refers to a commercial or out-sourced laundry option in which laundering activities are performed outside the hotel establishment. The contract of out-house laundry is given on a fixed term contract basis to the commercial laundry operators. In hotel industry, in rare circumstances the laundry operation is contracted-out as it is a day-to-day activity thus found expensive when outsourced. Alike to in-house laundry, out-house laundry also has advantages and disadvantages. Some of these are given below.

### Advantages and disadvantages of out-house laundry

| S. No. | Advantages | Disadvantages |
|--------|-----------|---------------|
| 1 | Undoubtedly, out-house laundry saves huge capital investment such as construction cost, machinery and installation cost plus saves space which may be rented out. | The major disadvantage of out-house laundry service is that there is always a probability of delay in delivery timings of hotel linen and guest clothes. |
| 2 | It also saves operating cost of hotel such as electricity, water, detergent, manpower, labour cost, equipment cost and reduce other fringe benefits of related staff. | It is also possible that out-house laundry may use such cleaning/laundering agents that distress the durability of hotel linens. |
| 3 | The saved laundry space may be used for other activities from which additional revenue can be generated. Although, if saved space is rent out then it also generates revenue. | In the long run, out-house laundry operation is found more expensive. |
| 4 | Out-house laundry operation also helps in making the hotel organization eco-friendly because it does not generate any *grey water* that may pollute water. | In case of having huge linen and uniform turnover, out-house laundry is again found more expensive in comparison to in-house laundry. |
| 5 | Out-house laundry operation reduces the workload and saves time of housekeeping department, which housekeeping can utilize in improving service standard. | As on-premises laundry operation requires different licences and government permissions; off-premises laundry also reduces these issues for hotel. |

Complete the comparative study (in below given blank spaces) between in-house and out-house laundry operation.

| Base/Criteria | In-house laundry | Out-house laundry |
|---|---|---|
| Initial cost (or capital) | Heavy initial investment/cost in laundry set-up as well as in procuring laundry equipment. | ? |
| Operating cost | In-house laundry demands additional operating costs both direct as well as indirect, such as labour cost, material cost, utility costs, etc. | Out-house laundry creates more supplementary laundering cost in comparison to in-house laundry. |
| Profitability | It saves out-house laundry cost and generates revenue by washing clothes of in-house guests. | ? |
| Dependency | ? | For getting business, out-house laundry depends on a number of hotels or similar outlets available in the locality. |
| Repair and maintenance | In-house laundry demands more engineering and maintenance expenses and if customer flow is limited then it is found quite expensive. | ? |
| Legal requirement | ? | Need to fulfil enormous legal requirements like obtaining licences and permits in relation to waste management, water pollution, etc., from local authority. |
| Laundering time | ? | It is difficult for out-house laundry to provide round the clock service. Thus, linen collection and deliveries are available only at given/specified time. |
| Control over laundry agents | In-house laundry has full control over wash temperature, wash cycle and laundry agents which could affect linen durability and quality. | ? |

## 23.3   LAUNDRY SUPPORTERS

The term *laundry supporters* refers to a range of laundry agents or cleaning agents/ materials. The two basic principles of laundering, whether in-house or out-house, are *to remove dirt and stains from the linen* and *to restore linen to its original appearance* as far as possible. For gratifying these principles, laundry supporters (or laundry cleaning agents) and laundry equipment play a crucial role. In addition to the equipment requirement, there are countless materials in the market for the purpose of removing soil, grease, or stains, renewing fabric, restoring colour, and bleaching. These laundry supporters can be classified as *detergents, suspending agents* and *sequestering agents*. All these laundry supporters (or laundry cleaning agents) can be further classified as given below.

## Water

- ✓ It is alone ineffective because of the phenomenon *surface tension*. Alone it can only remove water soluble dirt but it has little effect on grease and oil based stains.
- ✓ But addition of detergents allows the water to penetrate and wetting the clothes thoroughly which makes the dirt more accessible and easy to clean.

## Alkalis

- ✓ It refers to a compound with particular chemical properties. In simple terms, it is a cationic metal combined with the anionic hydroxyl (OH) radical chemically known as a *hydroxide of the metal*. Alkali is combined with acids to form salt and water.
- ✓ Alkalis used in laundry process include washing soda, sodium phosphate, sodium hydroxide, sodium metasilicate, etc. Alkalis such as soda and borax can easily remove all vegetable stains (such as tea & coffee) from cotton fabrics but animal fibres and dyes may be adversely affected.

---

### Important alkalis that can be used

**Sodium hydroxide**

- ✓ It is *caustic soda* which is a common ingredient in detergents. It is a powerful detergent, used to suspend protein and convert fats to soap, and it is cheap. However, it corrodes aluminium and galvanized iron, strips paints, and presents a hazard to personnel using it.

**Sodium carbonate**

- ✓ It is *soda ash* which is not as efficient a cleaning agent as sodium hydroxide, but it is a cheap source of alkalinity and used as detergent filler. It is corrosive to aluminium and galvanized iron, and forms a scale of calcium carbonate and other insoluble salts in hard water.

**Sodium metasilicate**

- ✓ It is an effective detergent for many purposes. It is an excellent emulsifying and suspending agent and has reasonable wetting and rinsing properties. It has anticorrosive properties but will deposit on stainless steel. It may also deposit with soil as a grey-white coating if used in water above 70°C.

---

# Bleaches

✓ It includes a number of chemicals which remove colour, whiten or disinfect, often by oxidation. Bleaches are used on white clothes only. They remove colouring matter by their oxidation or reducing action and the process of changing a coloured substance into a colourless substance is known as *bleaching*.

✓ Powder form bleaches are first dissolved in hot water then used. Remember, bleaches weaken the fibres so should be used with great care. Bleaches are of two types, i.e., oxidizing bleaches and reducing bleaches. Most commonly used laundry bleaches include sodium perborate and sodium hypochlorite.

| Types of bleaches | |
|---|---|
| Oxidizing | Sodium hypochlorite, sodium perborate and hydrogen peroxide which release oxygen from themselves or other substances. |
| Reducing | Sodium hydrosulphite which removes oxygen or add hydrogen to the coloured substance. It is generally milder in action in comparison to oxidizing bleaches. |

# Acids

✓ These are oxalic acids, potassium acid oxalate (i.e. salt of lemon) and various rust removers. Acidic detergents are mixtures of one or more acids and surface active agents, and may be inhibitors too. These detergents have a reasonable cleaning effect. If the mineral deposits are removed by acids alone, the cleaning effect will be minimal and it may be necessary to remove fat and protein with an alkaline detergent before removing the deposits with acid.

✓ Acid removes metal stains most commonly iron mould or rust. Strong acids have adverse effects on fibres thus it is always preferred to use weak solution several times than use a strong solution at first. The inorganic acids are principally more corrosive than the organic acids.

| Types of acids | |
|---|---|
| Organic acid | Gluconic acid, tartaric acid, citric acid, acetic acid and sulphamic acid. |
| Inorganic acid | Phosphoric acid, nitric acid, sulphuric acid and hydrochloric acid |

# Organic Solvents

✓ It refers to a substance that dissolves a solute, resulting in a solution. A solvent is generally a liquid but can also be a solid or a gas. In hotel industry, the common uses of organic solvents are in dry cleaning (i.e. tetrachloroethylene) and in spot cleaning (i.e., hexane). Organic solvents can be classified into two broad groups, i.e., inflammable and non-inflammable.

✓ When using a solvent always work from outside of the stain to inwards, with an absorbent cloth underneath the fabric. Organic solvents are always applied on the most delicate fabrics as they do not harm the fibre as well as the fabric.

| Types of Organic Solvents | |
|---|---|
| Inflammable | Solvents which should never be used near a naked flame. For instance, acetone, methylated spirit, white spirit, turpentine, benzene and amyl acetate. |
| Non-flammable | Solvents which are harmful (if inhaled), thus should always be used in well ventilated area. For instance, trichloroethylene, carbon tetrachloride and perchlorethylene. |

## Antichlors

✓ It indicates to a substance that removes excess chlorine to stop a bleaching reaction which may cause damage to fabric that has been bleached. Antichlors are used to neutralize the residual chlorine in the bleach, particularly in polyesters.

✓ In commercial laundries, the antichlor is usually added right before the end of bleaching process. The use of chlorinated bleaches have a tendency to leave yellow deposits on the clothing.

## Whitener

✓ It is a very fine blue coloured powder or liquid dye, which gets bleached in the course of time. This laundry agent is also known as *optical brightener, fluorescent brightening agent* or *fluorescent whitening agent* and the process of whitening is known as *bluing*.

✓ Optical brightener has a fluorescent effect by reflecting the ultra-violet rays of the sun. Powder dye is less preferred in comparison to liquid dye because powder may tend to accumulate in the weave of the fabric and cause it to turn grey. Remember, some brighteners can cause allergic reactions when in contact with skin.

## Starch

✓ It indicates to a stiffening agent used to impart a better crease and crisper appearance to the fabric. Often it is also known as *clothing starch* or *laundry starch*. Most types of laundry starch are composed of a mixture of vegetable starch that is mixed with water.

✓ Nowadays, the use of starch has declined due to the minimum-iron finishes on garments and fabrics and the reduced use of cotton in favour of artificial fibres but still there are some articles that demand starch in order to properly drape like napkin. Remember, polyesters do not have the ability to absorb starch thus they are stiffened with sizing agents.

## Fabric Softener

✓ It indicates to a liquid used to soften clothes after they have been washed, thus it is also known as *fabric conditioner*. It is available in liquid, crystal and dryer sheet. Fabric softener works by coating the surface of the cloth fibres with a thin layer of chemicals which makes the fibre smooth and prevents build-up of static electricity. It also improves iron gliding, increases stain resistance power and reduces wrinkling and pilling.

✓ It has surfactants like a detergent but they do not perform the cleaning function. Remember, a fabric softener is never used on loads where starch or sizing will be used. Fabric softeners are based on cationic surface active agents, carry a favourable charge and create anti-static properties.

## Others

✓ Apart from the above given laundry supporters, there are yet enormous supporters like sour, absorbents, fillers, pH scale, inhibitors, silicates and so forth. *Fillers* such as sodium chloride or sodium sulphate may be used to make the detergents fluid or to turn fluidized detergents into powders.

✓ *Sour* such as acetic acid is used to neutralize the effect of alkaline soap residue and brings the pH value to the acidic ranges in between 0.5 to 6.5. Sour is added when starch is used and when sizing agent is used, then sour is generally not used.

### Note

✓ The digits/figures of pH scale express the acidity or alkalinity or a solution on a logarithmic scale. The pH is equal to $-\log_{10} c$, where $c$ is the hydrogen ion concentration in moles per litre. Pure water has a pH very close to 07.

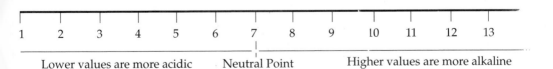

### What is detergent, suspending supporters and sequestering supporters?

**Detergents**

✓ It refers to any substance which, either alone or in a mixture, substitutes physicochemical energy for some of the mechanical energy required for removing dirt. These are designed to assist in the removal of soil from a surface and are available as powders, liquids, foams or gels.

✓ Detergents comprise *surfactants* (surface active agents); these are an important part of any detergent as they enable the detergent to increase the wetting power of water by reducing its *surface tension*. Detergents can be classified into three group i.e. synthetic, built-soap and enzyme actions detergents.

| | |
|---|---|
| Synthetic | It refers to a cleaning agent that increases ability of water to penetrate the fabric and break down greases and dirt. It acts like a soap but, unlike soap, they are derived from organic acids rather than fatty acids. Its molecules surround the dirt particle, allowing them to be carried away. |
| Built-soap | It refers to detergent cake that is made from fat and lye. As soap can easily dissolve in water, the solution becomes alkaline; and any acid present from soiled fabric reacts with soap, thus reducing its effectiveness. Nowadays, soaps are often built with alkaline products in order to counteract with acid and effect of minerals in hard water. |
| Enzyme | It refers to powdered pepsin used to remove protein stains like blood, egg, perspiration. It works effectively between the temperature ranges of 40-50°C. |

**Suspending supporters**

✓ It refers to those laundry agents which hold the dirt in suspension and prevent it from redepositing on the cloth surface. Its excellent example is CMC, i.e., *carboxyl methyl cellulose*. It is a cellulose derivative with carboxyl methyl group.

**Sequestering supporters**

✓ Sequestering supporters act along with suspending agents to hold greater amount of dirt in suspension thereby reducing the likelihood of redeposition. The amount of sequestering agents needed to be used depends on the hardness of water, the composition of detergents and soil.

✓ These supporters also have the ability to dissolve lime salts that are responsible for temporary hardness in water. Sodium polyphosphates are used as sequestering agents.

## 23.4   LAUNDRY OPERATION CYCLE

As we all know that linen is one of the most expensive items thus it is very important that staff involved in laundry operation must have elementary knowledge of the laundering processes. Because the durability of linen also depends on the care of linen in use and the kind of treatment it gets at the laundry. A step-by-step process for cloth and linen laundering is given below.

# Collection

- ✓ It is the first step in which room attendant collects linens from guest rooms and other respective areas. Certain other linens like kitchen uniforms, dusters, aprons, etc., should always be collected separately.
- ✓ If laundry is on-premises then collection of linen is directly performed in laundry section but when laundry is outsourced then usually collection is done in the linen room.

# Transport

- ✓ Now all collected linen is packaged in canvas bags lined with polyvinyl or elasticized net bags called *skips* for transporting. Room attendant can use trolley or collapsible wire cart to transport linen.
- ✓ In on-premises laundry, an in-built *linen chute* is used for transporting linen from the floor pantries to laundry section. Remember, damp linens should not be allowed to sit in carts or chutes for long, else fungus may attack.

# Sorting

- ✓ It is mainly carried out to look for heavily stained linen and to shake out any loose items/soils wrapped up in the linen. Remember, heavily stained linen should be kept separately (or *marked*), so it can be notified in the laundry section for spotting/stain removal.
- ✓ Sorting is also carried out for the purpose of counting as well as to segregate linen by load type, degree of soiling, fabric type & item, type of wash required (dry cleaning or normal wash), mending, etc.

# Weighing

- ✓ It refers to the load measurement of linen with the help of scale (weight to each load within 90 to 100% of the reco load). It is carried out to conform to the capacity of the washing machine.
- ✓ Remember, do not overfill a machine as it affects the quality of wash and repeated overloading may also cause the machine to break down. Simultaneously, under-loading will lead to wastage of effort, money, time, detergent and water.

# Loading

- ✓ It refers to the transferring of soiled linen into the machine. Loading is often done manually but nowadays automation also supports this task. Machine used to wash linen may be top-loading, front-loading or side-loading.

## Washing

✓ It refers to the actual cleaning process of linen in which soil/dirt is suspended from the linen and discharged to the drain. During washing, few points should be kept in mind, i.e., length of wash, water level & temperature, amount and type of detergent, mechanical agitation, rinsing, etc. It is also important to add detergent/ laundry agent at the appropriate time for best results.

| Activity |
| --- |
| Rearrange the below given wash cycle activities/stages in sequential order- |
| For simple soiled linen       Rinse → Suds → Flush → Extract → Sour & soft → Flush → Bleach |
| For heavily stained linen      Intermediate rinse → Break → Starch → Intermediate extract → Soak |

## Extracting

✓ It is done after draining and is concerned with the removal of excess moisture from the linen through centrifugal action like handwash wringing. This process is known as *hydro-extraction*. Linens should only be slightly damp after the wash cycle is completed.

✓ Remember, the linens which cannot be hydro-extracted must go through pumping action to draw out the extra water. Now, the collected compact mass of linen is called *cheese*.

## Drying

✓ It is concerned with transferring of *cheese* to the tumble dryer in order to completely dry linen by blowing hot air between the temperature 40°C - 60°C. Do not over-dry linen and use a cool down tumbling period to minimize wrinkles. Do not leave linens in the tumble dryer overnight.

✓ *Cheese* can be transferred manually as well as through automated system into the *tumble dryer*. For the linens that are vulnerable to damage by heat, they can simply be air-dried by circulating air at room temperature.

## Ironing

✓ It is concerned with finishing/pressing of tumble dried linen. Basically, ironing is required for those linens that demand pressed finish. But linens like towels, blankets, bedspreads, hosiery, etc., do not require a pressed finish (these are only tumble dried).

✓ In ironing, feed the linens at the proper moisture level through flatwork ironers (or *calendering machine*) and remember to keep the ironer clean. Do not iron dirty linen.

# Folding

- ✓ After drying and ironing, linens should be folded immediately in order to minimize wrinkles. Folding can be done manually as well as through machine. During folding process, attendant should look for stains, excessive wear or tear and sort out these linen for additional treatment.
- ✓ Correct folding is crucial to the appearance of linen as well as makes it convenient to store and use. Do not send an unacceptable piece of linen to guest and respective hotel areas.

# Airing

- ✓ It refers to the aeration (exposure to air) of storage space/shelves before storing linens. Airing is important as it ensures that any moisture that is likely to cause mold and mildew will be eliminated. It is especially important when the linens are to be stored in closed shelves.

# Storing

- ✓ First of all storage area must be isolated from the soiled linen and always kept clean. Secondly, pressed and folded linens should be properly stored as it will increase its life and decrease the chance of wrinkles.
- ✓ Keep the clean and folded linens off the floor. Generally, it is a standard rule that 50% of the total linen inventory should be kept in *par stock* and remaining 25% in use and another 25% in processing.

# Distribution

- ✓ Stored linen should be properly rotated and distributed, as rotation reduce the tear and wear of any particular linen. Correct rotation (particularly on FIFO principle) process assists in equal utilization of each piece of linen.
- ✓ Do not jam a stack of linens into the store or room-attendant cart (as it may tear or wrinkle them). The distribution of linen is usually done by linen trolleys.

| Activity | | | |
|---|---|---|---|
| What happens when the below given factors affect the wash cycle? | | | |
| **Factors affecting wash cycle** | **?** | **?** | **Select Result** |
| Length of wash cycle | Too short? | Too long? | Suitably clean/wear & tear |
| Temperature of water | Too high? | Too low? | Chemical may not work/damage linen |

| Water level | Too low? | Too high? | Futile to laundry agent/form protective envelope |
|---|---|---|---|
| Washing detergent | Too less? | Too much? | Incomplete cleaning process/seen as residue after rinsing |
| Mechanical agitation | Overload | Underload | Reduce speed/cause vibration |
| Rinsing temperature | Hot water | Cold water | Reduce wash load temperature/remove residue |
| Rinsing water flow | Open drain | Halt | Claim more water but effective/suspended dirt |
| Hydro-extraction time | Too short? | Too high? | Increase drying time/reduce the drying time. |

## 23.5 LAUNDRY PROCESS/LAUNDRY CYCLE

In linen cleaning process, agitation plays a crucial role as it separates dirt from the linens. In earlier times it has been seen that the linen was often twisted, rubbed or slapped against flat rocks to remove dirt particles. Even wooden bats or clubs could also be used to help with beating the dirt out. These were called *washing beetles*. But nowadays, due to the development of technology and emergence of various semi-automatic and fully automatic mechanical equipment, the laundering process is much easier and speedy. A brief description of step-by-step mechanical laundering process is given below.

Step 1   Dark coloured linen may ruin the light coloured linen, if they bleed colour. Similarly, linen like corduroy, velvets and permanent press may attract *tint* from linen like towels, sweatshirt, flannel, etc. Thus, sort linens by colour and tint. One may also sort by degree of soiling or heavily stained linen.

Step 2   Ensure that pocket is empty then close zippers in order to prevent snagging. Now pre-treat to heavily stained linens with laundry detergents or stain removers in accordance with manufacturer's instructions.

Step 3   Measure and pour appropriate amount of laundry soap into washer or its detergent dispenser. Thereafter add liquid fabric softener in accordance to manufacturer's instructions. Select water temperature (hot, cold or medium) for wash cycle, for this attendant may refer labels attached on clothes.

Step 4   Start washer and add detergent. If stain demands bleach, allow to run washer for a few minutes to mix the detergent and water then add correct amount of bleach. Next ensure correct water level then add linens. Finally close the lid of washer and allow washing for approximately 45 minutes.

Step 5   When linens are washed put them into dryer (including anti-static sheet) and select correct drying temperature such as low for delicates, medium for fabrics and high for cotton. Often delicate linens such as sweaters are preferred to hang on hangers to air dry.

Step 6    Now close and turn on the dryer for approximately one hour. Do not forget to remove lint from drier's tray. Once the linens are completely dry, remove them from the dryer, next fold and store.

**Laundry Cycle**

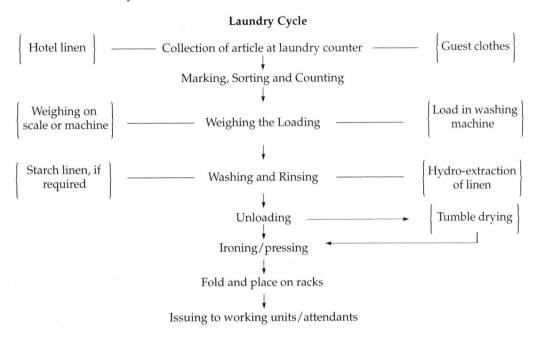

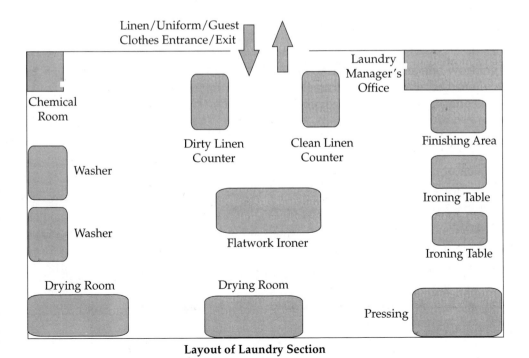

**Layout of Laundry Section**

| Tasks | Cotton | Silk | Wool | Synthetic |
|---|---|---|---|---|
| 1 | Sort out cotton clothes on the basis of their fabric texture and fastness of colour. | Wash frequently but never soak as it weakens the fabric. | First of all, shake the woollen clothes to remove any dust particles. | Soak before laundering as it loosens the dirt. |
| 2 | Determine the water temperature in the main wash. | Silk made articles must be evenly dampened but water should never sprinkle as it leaves watermarks. Generally, silk articles are dry-cleaned. | Mark the outline prior to washing and arrange it in its original size and shape after laundering and dry flat. | Do not overload machine as it forms creases which may be difficult to remove later. |
| 3 | Decide upon the need for sterilization or disinfection then select appropriate detergent to wash cotton made articles. | The detergent used should have a good surfactant (since there is no role of mechanical action in cleaning) plus they do not contain any harsh chemicals. | Avoid brush scrubbing, friction, high temperature, strong laundry agents, hydro-extraction, wringing, hanging wet fabric and prolonged soaking. | Friction and scrubbing with brush should be avoided. |
| 4 | Now, ascertain the length of wash cycle and speed of rotation of the drum. | Silk clothes should not be washed at high heat as it affects the fabric texture. Ideal wash temperature is 30°C. | In it, the wash cycle should be short and carried out at low temperature (35°C). | In it, wash cycle is short and carried out at low temperature (30°C). |
| 5 | Use appropriate bleach (if required) and optical brighteners. | Rinse silk articles with little vinegar (½ tsp for 0.5 litre) to preserve *lustre*. | Rinse cycle should be thorough and *borax* is added to prevent matting. | Rinse linens in water at room temperature. |
| 6 | Now rinse the cotton articles and decide upon whether to starch it or not. | Hydro-extraction is not recommended for silk clothes. Besides sun-drying is also not allowed as it damages the fabric and causes yellowing to clothes. | Hydro-extraction is not allowed but pumping can be done to extract extra moisture. | Hydro-extraction cycle is also short as synthetic clothes have a low absorbency. |
| 7 | If cotton made articles demand then only go for hydro-extraction process. | Iron/press on medium heat (as silk scorches easily) plus done on reverse side in order to preserve *lustre*. Lastly air the silk clothes after ironing. | Generally, ironing is not required but if needed then done on reverse side of the linen (at low temperature and when linen is completely dry) | Iron/press on low heat but quickly, as prolonged heat can scorch the fabric. |

## 23.6  DRY CLEANING

The term *dry cleaning* refers to the process by which delicate textiles are cleaned without using a solvent other than water, for example, *perchloroethylene*. Different coloured articles are dry cleaned separately but in certain instances they may be cleaned together. Simultaneously, attendant should always ensure that heavily stained articles are never cleaned with lightly soiled ones. A step-by-step process involved in dry cleaning is given below.

Step 1    Linen keeper collects the hotel linens that require to be cleaned whereas *valet* attendant collects the guest clothes to dry clean.

Step 2    Now collected articles are sent to laundry section where each article is marked with an identifying tape and checked for stain types. Afterwards, check pockets then brush to remove loose dust/dirt.

Step 3    Special stains (or heavy stains) like blood should be *pre-spotted* before cleaning, so that they must be carefully treated and easily removed.

Step 4    Now collected articles are washed in dry cleaning solvent in an enclosed machine. Nowadays, this machine alone can perform washing, extracting and drying of articles.

Step 5    The solvent left after dry cleaning may be reused; it depends on the degree of soiling, quantity of washed articles and wash cycle. Sometimes *charged system* is also followed in dry cleaning process.

Step 6    Dry cleaned articles are now spin dried in order to extract solvent and then air dried with warm air. Afterwards, cleaned articles are hung up to remove the whiff and checked for stains.

Step 7    Next cleaned articles may also require re-texturing (re-texture it before pressing), thereafter treat with a solvent containing resin to improve body and bulk.

Step 8    Finally, articles are pressed and put back in coat hangers, boxes or hampers and sent to distribution centre, i.e., linen room. Guest articles along with bill are sent to guest rooms.

Due to the development in technology, most of the hotels use computerized laundry system, *ozone washing* and *serview*. Former one is Windows-based laundry software whereas the second one is an energy efficient system that can cut laundry costs by 50%.

## CONCLUSION

Laundry is a place/section for cleaning hotel linen, staff uniforms and guest's clothes. It depends on individual hotel's guest traffic, amount of linen required to be laundered on daily basis, available space for constructing laundry, and available funds with management to determine whether to construct on-site laundry or out-source it and so forth. The decision should be taken after considering the pros and cons of each laundry option. Employees who are working in laundry section require careful selection of laundry supports/agents because certain laundry agents may create harmful effect on clothes and linens like bleach may cause white spaces in coloured clothes. It is always desirable to use mild detergents twice or thrice than using strong detergent at one shot to clean the spot. Laundering of clothes/linens made from different fibres demand different stain removal, for instance woollen and cotton based clothes require different laundry agents than synthetic and silk made materials. Remember, dry cleaning is done for cleaning expensive and delicate fabrics/materials.

| Terms (with Chapter Exercise) | |
|---|---|
| Flatwork | It is a laundry term that is used for sheets, pillowcases and table linens. |
| Washing beetles | ? |
| Cheese | It refers to the compact mass of hydro-extracted clothes. |
| Tumbler dryer | ? |
| Hydro extraction | ? |
| Chute | It indicates to an inclined trough, passage, or channel through or down from which things may pass. It is used to pass linen from floor pantries. |
| Mark | ? |
| Par stock | It refers to the standard, specific or normal level of stock. |
| Mobile shelving unit | It refers to an adjustable shelving unit which removes the need for permanent shelving in the laundry as well as in the satellite linen rooms. |
| Serview | ? |
| Drying chamber | It refers to the drying room in which low crease clothes are suspended on the hangers, thereafter steam or hot air is circulated through the chamber in order to dry the cloth. |
| Tunnel dryer | It is a fully automated system in which clothes hung on conveyor belts pass through a tunnel. Thereafter, tunnel will blow with hot air which makes the clothes dry. |

| Inhibitor | It refers to a substance that neutralizes the corrosive effect of some chemicals. The use of inhibitors depends on the composition of the detergent and the materials which will be cleaned. |
|---|---|
| Calender | Also known as flatwork ironer or roller iron in which linens pass through heated rollers for ironing. It is used for flatwork such as bedsheets, pillowcases, tablecloth, serviette, etc. |
| Solvent | It refers to a substance that dissolves a solute, resulting in a solution. A solvent is usually a liquid but can also be a solid or a gas. |
| Press | It is used for pressing small and flat linens such as napkins, staff uniform, aprons, slip clothes, etc. There can be special presses for specific functions. Press can be steam operated as well as operated on electricity. |
| Ozone washing | It is carried out by mass injecting ozone into the laundry system via cold water lines. The resulting ozonated water facilitates the breakdown of insoluble dirt. It reduces the need of hot water, chemicals and shorter wash cycle simultaneously extending the linen durability. |
| Suzie/puffer | It is used for coats and other similar clothes that could not crease heavily. In it, clothes are put onto a dummy that is inflated with steam to remove creases thereafter hot air is blown to remove moisture created by steam. |
| Dry clean | It refers to the use of solvents and special process (but never water) to clean delicate fabrics like wool and silk made clothes. |
| Mildew | It refers to a superficial coating or discoloration of organic materials, such as cloth, paper, or leather, caused by fungi, especially under damp conditions. |
| Skips | It refers to the elasticised net bags used for the collection of soiled/dirty linen for transport. |
| Wash cycle | It refers to the different stages involved in the cleaning of clothes and linens. It includes collecting, washing, drying, finishing and folding clothes and linens. |
| Silicate | It refers to a substance that may be used as anti-corrosive agent in alkaline detergents but will deposit on stainless steel. |
| Enzyme | It refers to any of the numerous proteins or conjugated proteins produced by living organisms and functioning as a biochemical catalysts. |
| Stain | It refers to a discolouration caused by some dye or contact with a different coloured substance or heat source. |

# 24
CHAPTER

# Fabrics and Fibres

⟨ **OBJECTIVES** ⟩

*After reading this chapter students will be able to...*

- describe fibre and fabric;
- elucidate principles of stain removal and methods involved in stain removal process;
- find out the different knitting patterns used in fabric making;
- identify different types of stains and their suitable removal agents; and
- explain different types of fibres, their source of origin and how they are produced.

## 24.1 FABRIC

The term *fabric* refers to the cloth or textile that is produced by either weaving or knitting technique. Fabric is used in a variety of ways throughout a hotel establishment, and may be chosen for their decorative values, for their comfort, warmth or coolness, their protective qualities, their durability and even for hygienic reasons.

Basically, fabrics are made from fibres which may be either natural or manmade. *Natural fibres* frequently preferred include cotton, linen, wool and silk whereas *manmade fibres* include rayon, nylon, terylene and acrilan. The fabric that directly comes from a loom is called *gray good*, as it has not received a *finishing treatment* and *sanforizing*. Therefore, it is not suitable for most of the purposes.

## 24.2 FIBRE

It refers to a class of materials that are continuous filaments or are in discrete elongated pieces, similar to lengths of thread. They can be spun into filaments, string, or rope and mainly used as a component of composite materials. Broadly, fibre can be classified into two major groups, *natural fibres* (vegetable, animal and mineral fibres) and *manmade fibres* (regenerated and synthetic fibres). A description of each is given below.

## 24.3 NATURAL FIBRE

It refers to the fibre derived from plants, animals and mineral sources. Botanically, natural fibre is a collection of cells of long length and negligible diameter. Natural fibres can be converted into yarn after spinning them into a fabric. An example of a most commonly used natural fibre is cotton and other examples include wool, silk, hair, fur, hemp and linen. Broadly, natural fibres are classified as given below.

**Vegetable fibre** refers to plant fibres such as cotton, hemp, jute, etc. Plant fibres contain cellulose and can be extracted from the bast (*bast fibre/skin fibre*) or stem, seed hair, leaf or husk of plant. The fibres may be chemically modified, like in viscose. A brief description of some major vegetable based fibres is given below.

### Cotton

- ✓ The term *cotton* is used for soft silky fibres that grow around the seed pod of the cotton plant (gossypium) and the colour of the soft fibre varies from white to greyish yellow. Cotton fibres are good conductors of heat and have a slight feeling of warmness. It is unsuitable for polishing as *linters* appear on the surface.
- ✓ Nowadays *mercerized cotton* is mostly used. It is a cotton which has been treated with sodium hydroxide (or caustic soda or lye) to bring out certain properties like smoothness, gloss or sheen. Mercerization is a fibre treatment which was first discovered by John Mercer in 1851.

### Linen

- ✓ This fabric is made from the cellulose fibres. It grows inside of *flax plant* (*Linum usitatissimum*). This fibre is very absorbent in nature and linen fabrics made from this fibre are valued for their exceptional coolness, freshness, dirt and abrasion resistance and finally lustre appearance.
- ✓ In hotels, many products are made of linen like aprons, tablecloth, towels, napkins, bed linen, runners, chair covers, etc. Linen fabrics are straight, smooth and almost solid. The solidity of the fibres makes linen much heavier than cotton. When linen fabric is mercerized, its strength and dye affinity also increases. The quality of linen fabric greatly depends on the *retting process*.

### Kapok

- ✓ It is light, buoyant, resilient and water resistant but inflammable fibre obtained from the seed of the kapok plant (*Ceiba Pentandra*). Kapok fibres on their own are not suitable for spinning into yarn, as they are too smooth, slippery and brittle.
- ✓ It is a popular silk cotton also known as *java cotton* and mostly used for the filling of mattresses, cushions and pillows. Kapok is similar to cotton in that both fibres are found around the plant seeds, rather than extracted from the stems or leaves.

## Other Vegetable Fibres

✓ Apart from above given natural vegetable based fibres, there is a range of vegetable fibres like jute, ramie, hemp and sisal.

✓ Jute, ramie and hemp is obtained from the stems of plants whereas sisal is obtained from the leaves of a plant. But the use of these vegetable based fibres into fabrics is limited in comparison to cotton and linen.

Animal fibre refers to the fibres derived from animals. It largely consists of particular proteins. Remember not all animal fibres have the same properties and even within a species the fibre is not consistent, for instance *Merino* is very soft, fine wool whereas *Costwold* is coarser fibre, and yet both are types of sheep. A brief description of some major animal based fibres is given below.

## Wool

✓ It is mainly obtained from the sheep. But it can also be obtained from other animals like horse, camel, goat, etc. Wool can be cultivated from the living animals (i.e., *fleece* and *virgin wools*), dead animals (i.e., *skin* or *pulled wool*) as well as remanufactured from the used wool (*shoddy* is perfect for re-use).

✓ Wool fabric is soft to touch (*Merino wool* is the softest wool in the world) and provides warmth due to which it is the preferred choice for winter apparel. Wool has elasticity and good drape and it can be easily dyed in different colours. But it cannot be dyed to the same standards of uniformity or fastness as manmade and vegetable fibres.

## Silk

✓ It is obtained from the cocoon spun by the cultivated silk worms. The most popular kind of silk is obtained from the *mulberry silk worm*. Silk can be obtained in short as well as long filament. *Spun silk* is obtained in shorter filament whereas silk obtained from *cultivated silk worm* is longer in filament.

✓ The silk that is obtained from other varieties of silk worms is called as *wild silk*. Silk is stronger than cotton because it has greater *tensile* strength. Silk fibre has a unique sheen and it is very smooth to touch. Silk is an elastic and resilient fibre; therefore it does not easily crush. But it weakens when wet.

Mineral fibre refers to a fibre made from synthetic or natural material like *asbestos fibre*. It is quite incombustible and was used for fireproof materials. The mineral fibre is generally used to refer solely to synthetic materials including glass fibre, stone wool and ceramic fibres.

## 24.4   ARTIFICIAL FIBRE

Various *manmade fibres* which can be further classified into two broad groups, i.e., *synthetic fibre* and *regenerated fibre*. The introduction of an artificial fibre increases the durability and makes laundering easier. There are enormous range of artificial fibres like nylon, olefin, acrylic and so forth but the combination of around 60% cotton with 40% artificial fibre is considered best. A brief description of synthetic and regenerated fibres and their variation is given below.

**Synthetic fibre** is the primary type of artificial fibre whose chemical composition, structure and properties are significantly modified during the manufacturing process. It is spun and woven for mammoth of products like upholstery, carpets, drapes, linens and garments. Synthetic fibre is made from *polymers*. Common synthetic fibres include nylon, modacrylic, olefin, acrylic, polyester, etc. A brief description of several synthetic fibres is given below.

> **Common Features of Synthetic Fibres**
>
> - Abrasion resistance power
> - Moth & mildew resistance
> - Elasticity & resilience power
> - Low moisture absorbency
> - Poor heat conductor
> - Electrostatic and attracts dirt
> - Great strength & thermoplastic

## Polyamide Fibres

✓ It refers to nylon and the aramid fibres. Both fibre types are formed from polymers of long-chain polyamides. Nylon is generally tough, strong and durable fibre whereas the fully aromatic aramid fibre has high temperature resistance power and it has exceptionally high strength and dimensional stability. The *sheer strength nylon* is the best example of the synthetic fibres.

✓ Polyamide was discovered in America in 1935 and almost at the same time it was also discovered in Germany and manufactured under the trade name Perlon. Basic features of polyamide (or nylon) include: produced in various forms, it is easy to wash and dry but it demands frequent wash as this fibre attracts dirt.

✓ It has great strength, elasticity and abrasion resistance power which makes it suitable for carpets and upholstery fabrics. It is also suitable for making blankets, bed linens and uniforms. Products made from polyamide fibre can be brushed or bulked, so that more air is held and fabric is warmer.

## Acrylic Fibre

✓ It refers to a synthetic fibre made from a polymer (polyacrylonitrile) with an average molecular weight of ~100, 000, about 1900 monomer units. Its typical co-monomers are vinyl acetate and methyl acrylate. Acrylic is also called *acrilan fibre*. DuPont created the first acrylic fibres in 1941 and trademarked it under the name Orlon. *Modacrylic* is a modified acrylic fibre that contains at least 35% and at most 85% acrylonitrile monomer.

✓ Acrilan fibre's features include fluffy touch, good resilience, crease recovery, good resistance power towards chemicals, sunlight, etc. It is less resistant to abrasion in comparison to nylon but it is better than wool. All these properties make it suitable for blankets, carpets and upholstery. Acrylic fibre is produced in various forms like dralon, orlon, courtelle and teklan.

---

### Forms of Acrylic Fibres

| | |
|---|---|
| Dralon | It is made from acrylic fibre and suitable for making of carpets, curtains and upholsteries. Dralon GmbH is the largest producer of dry-spun and wet-spun acrylic fibres worldwide. |
| Orlon | It is a synthetic acrylic fibre suitable for making curtains. It is a light weight, wrinkle and sunlight resistant fabric. It is available in *filament* and *staple* form. |
| Courtelle | It is based on polyacrylonitrile and resembles wool. It is suitable for twisted pile carpets and also used for blankets. |
| Teklan | It is also known as *courtaulds* and based on a copolymer or acrylonitrile and vinylidene. It is suitable for making carpets, curtains and upholsteries because it is flame retardant. It is a modacrylic fibre. |

---

## Polyester Fibres

✓ It refers to a quick-drying resilient synthetic fibre consisting primarily of polyester. Polyesters as thermoplastics may change shape after the application of heat (it melts around at 240-245°C). Thus, it tends to shrink away from flames and self-extinguish upon ignition. It is often blended with other fibres like cotton or wool to get the best of both worlds. This combination is suitable for table & bed linen and uniforms. *Dacron* is the famous American polyester.

✓ It has good resistance towards abrasion (but inferior than nylon) and sunlight (so suitable for net curtains). It is more electrostatic (in comparison to acrylic or nylon) in nature and has low moisture absorbency and has minimal shrinkage chances in comparison to other fibres. It holds creases more easily than nylon.

---

### Important

✓ Polyester sheets are usually chemically modified during the manufacturing process in order to make it smooth. This process is known as *durable press* or *no-iron effect*. Consequently, polyester fibres stay smooth while in use and even after laundering.

✓ The cotton fibre sheets can be either combed or carded before spinning. *Combed fibre sheet* is smooth and has a great tensile strength whereas *carded fibre sheet* is rough and dull in appearance. *Muslin sheet* is a perfect example of carded fibre sheet while *percale sheet* is an excellent example of combed fibre sheet.

# Polyvinyl Fibres

✓ *Saran* (plastic) is a polyvinyl fibre. Saran is the trade name for a number of polymers made from vinylidene chloride (especially polyvinylidene chloride or PVDC), along with other monomers. It does not burn but loses strength in boiling water and softens at 120-122°C.

✓ Saran is especially used as a cling film but also suitable for upholstery fabrics and deck-chair coverings. It is fully non-absorbent, therefore difficult to colour. When saran formed into a thin plastic film (or as a *cling film*), it works as an oxygen barrier and retards food spoilage and also helps to retain its flavour and aroma.

# Polyolefin Fibres

✓ It indicates to a fibre in which fibre-forming substances are available. Any long-chain synthetic polymer is composed of at least 85% by weight of ethylene, propylene or other olefin units, except amorphous (non-crystalline) polyolefins. It is suitable for making of carpet, carpet backing, upholstery, seat covers, webbing for chairs and *laundry bags*.

✓ Herculon, Marvess and Vectra are three common trade names of polyolefin fibres. It is available in two forms polyethylene (like couriene) and polypropylene (like ulstron). Basic features of polyolefin fibres include: it is lustrous white translucent fibre with good draping and it has excellent abrasion and wrinkle resistance.

| **What is Polyethylene and Polypropylene?** | |
|---|---|
| Polyethylene | It is usually made in monofilament form; although work has been done on continuous filament and staple. In it, ethylene is polymerized at high pressure; the resultant polymer is melt spun and cold drawn. Polyethylene is non-absorbent in nature and melts around 120°C and softens at below boiling temperature, i.e., 95°C. It is suitable for making deck-chair coverings, upholstery fabrics and plastic floor mating. |
| Polypropylene | It is also known as *polypropene*, it is a thermoplastic polymer used in a variety of applications. It has good abrasion resistance power and it is a light weight synthetic fibre. It is suitable for making some of the adhesively bonded felted carpets. |

| | **Activity** | |
|---|---|---|
| Match the below given synthetic fibres according to their chemical composition. | | |

| S. No. | Fibre Type | Chemical Composition |
|:---:|:---:|:---:|
| 1 | Polyamide fibre | Saran |
| 2 | Acrylic fibre | Terylene |
| 3 | Polyester fibre | Acrilan |
| 4 | Polyvinyl fibre | Nylon |
| 5 | Polyethylene fibre | Ulstron |
| 6 | Polypropylene fibre | Couriene |

**Regenerated fibre** refers to manmade fibres that are produced by dissolving a natural material like cellulose, then regenerating it by extrusion and precipitation, as with viscose. Regenerated fibre is suitable for making of carpets, bedding and canvas. A brief description of several regenerated fibres is given below.

# Rayon

✓ It refers to a textile fibre or fabric made from regenerated cellulose (viscose). It is a semi-synthetic textile filament made from cellulose, cotton linters or wood chips by treating these with caustic soda and carbon disulfide and passing the resultant solution, viscose, through spinnerets.

✓ Rayon's basic features include: it is highly absorbent, soft and comfortable, easy to dye and drapes well. The production method of rayon leads to different types of rayon fibres—viscose, cuprammonium and saponified cellulose acetate.

## Viscose Rayon

✓ It refers to a manufactured regenerated cellulose fibre produced from naturally occurring polymers. It is neither a truly synthetic fibre nor a natural fibre, it is a semi-synthetic fibre. *Viscose Rayon* is the most volatile construction because it becomes weak when wet and is the least predictable fabric.

## Acetate Rayon

✓ It indicates to a manufactured fibre in which the fibre forming substance is cellulose acetate. Acetate is derived from cellulose by reacting purified cellulose from wood pulp with acetic acid and acetic anhydride in the presence of sulphuric acid.

# Triacetate Rayon

✓ It refers to a regenerated fibre in which higher percentage of the cellulose has been acetylated, not less than 92%. The difference between acetate and triacetate fibres is in the number of the cellulose hydroxyl groups that are acetylated. For acetate fibres the number lies between 75% and 92%, for triacetate fibres it is more than 92%.

✓ Triacetate is particularly used for knitted fabrics which have elasticity with good return. These can be easily washed, dried quickly and need little or no ironing and do not wrinkle easily.

---

**What is textile?**

The term *textile* is related with cloth/fabric/linen materials like rags, dusters, tablecloth and so forth. These are graded by the *thread count* and *tensile strength*.

---

## 24.5 FABRIC

It refers to a cloth produced by especially knitting, weaving, spreading, crocheting or binding textile fibre, thereafter used in production of further goods (like garment). In reality, the term cloth may be used as synonym for fabric but it often refers to a finished piece of fabric that will be used for some specific purpose (like table cloth). Most fabrics are knitted or woven but some are produced by non-woven processes like braiding, felting and twisting. A brief description on different fabric making techniques is given below.

## Weaving (or woven fabric)

✓ It indicates to a cloth production method which involves interlacing a set of longer threads (called *warp*) with a set of crossing threads (called *weft*). Weaving can be done by hand as well as mechanically.

✓ In mechanical process, weaving is done on a frame or machine known as a *loom*, of which there are a number of types. There are several standard weaves, i.e., plain, twill, satin, damask, pile, cellular, figured, etc.

---

**Activity**

Fill in the blank spaces with the appropriate term as per the description provided.

| | | | |
|---|---|---|---|
| - Plain weaves | - Twill weaves | - Satin weaves | - Figure weaves |
| - Huckaback | - Damask | - Pile weaves | - Felt |
| - Net | - Percale | - Gingham | - Tapestry |
| - Towelling | - Velvet | - Sateen | - Cellular |

| S. No. | Description of textile with their standard weaves | Answers |
|--------|---------------------------------------------------|---------|
| 1 | In it no extra warp or filling yarns are used to create the design. *Dobby* and *jacquard* weaves are perfect examples of it. | ? |
| 2 | It is woven with one warp yarn and one weft yarn, usually with the pattern in warp-faced satin weave and the ground in weft-faced or sateen weave. | ? |
| 3 | It is tightly interlaced (over & under) to show check-board pattern. | ? |
| 4 | It requires an extra warp or filling yarn to form a mound. Mound is formed during the weaving process. It can be grouped into cut and uncut pattern. | ? |
| 5 | It is identified by the quite visible diagonal lines and is often a simple "over two, under one" weave. *Denim* is best known example. | ? |
| 6 | It is constructed by alternately combining a floating with plain weave. It is generally famous for its honeycomb effect and *Grecians* is a famous variety of it. | ? |
| 7 | It gives a loosely woven fabric which holds air in the cells between the threads, for instance cellular blankets. | ? |
| 8 | It is made up of weft yarns, rather than warp yarns. | ? |
| 9 | A fabric woven with floating yarns in the warp in many variations. | ? |
| 10 | It refers to any variety of weaves where the pattern is created by ground wefts and do not run from end to end. | ? |
| 11 | In it, fibres are directly transformed into fabric without being spun into yarn. | ? |
| 12 | In it, yarns are interlaced, interlooped, twisted and knotted to form openwork fabric. | ? |

## Knitting and Crocheting

✓ It refers to interlacing loops of yarn, which are formed either on a knitting needle or on a crochet hook, together in a line. The two processes are different in that knitting has several active loops at one time, on the knitting needle waiting to interlock with another loop, while crocheting never has more than one active loop on the needle.

# Spread Tow

✓ It refers to a production method whereby the yarns spread into thin tapes then tapes are woven as *warp* and *weft*. This method is mostly used for composite materials; spread tow fabrics can be made in carbon, aramide, etc. The plain weave is a most common weave in which warp and weft threads are perpendicular to each other.

# Braiding

✓ It refers to plaiting that involves twisting threads together into cloth. Knotting involves tying threads together and is used in making of *macramé*.

# Lace

✓ It is made by interlocking threads together using a backing and any of the methods described above, to create a fine fabric with open holes in the work. Lace can be made either by hand or machine.

✓ Carpets, rugs, velvet, velour and velveteen are made by interlacing a secondary yarn through woven cloth, creating a tufted layer known as a *nap* or *pile*.

# Felting

✓ It refers to a cloth involved in pressing a mat of fibres together, and working them together until they become tangled. A liquid, such as soap water, is usually added to lubricate the fibres, and to open up the microscopic scales on strands of wool.

# Non-woven

✓ It refers to textiles that are manufactured by the bonding of fibres to make fabric. Bonding may be thermal or mechanical, or adhesives can be used.

# Bark Cloth

✓ It refers to a cloth that is made by pounding bark until it is soft and flat. Many producers that use the term *paper clothing* are actually referring to bark-cloth.

| | Activity | |
|---|---|---|

Match the following.

| S. No. | If fabric is finished with | It makes fabric |
|---|---|---|
| 1 | Scotchgard | Shrink resistant |
| 2 | Durafresh | Mothproof |
| 3 | Proban | Flame retardant |
| 4 | Velan | Bacteriostatic protected |
| 5 | Permalose | Oil repellent |
| 6 | Dielmoth | Water repellent |
| 7 | Rigmel | Soil release |
| 8 | Calpreta | Crease resistant |

## 24.6  STAIN REMOVAL

The term *stain* is concerned with spot or block that discolours the surface. The cause of stain may be the contact of surface with any foreign material or unwanted particles that are difficult to remove. The process used to remove the spot or block is known as *stain removing process* whereas the cleaning agents used to remove the stain is known as *stains removals*. Cleaning agent is used at the time of actual washing, but as soon as stain is seen on the surface, the primary treatment should be given as early as possible. The primary treatment may be rinsing with cold water. A below given set of general principles can be adopted as a guideline for primary treatment for stain removal.

Principle 1    Attempt to remove/scuff the stain as quickly as possible; else it will be difficult to remove from the surface. Remember to wipe out the glut staining material.

Principle 2    Select appropriate stain removal agent at the time of washing. Remember, before using any cleaning agent test the agent on any unremarkable place for seeing it's after wash effect.

Principle 3    First, use milder stain removal agent twice or thrice and if stain still persists then move towards stronger stain removal agent.

Principle 4    At the end of every washing period ensure that the stain removal agent should be totally neutralized or washed away.

Principle 5    Before attempting to remove stain, read manufacturer's guidelines and given precautions. For example, often it is seen in the form of *do not wash in hot water*, *do not bleach* and so forth.

Principle 6    At the time of removing stain, first of all use a simple method to remove the stain before resorting to the use of chemicals.

Like the importance of stain removal agents in removing the stain, similarly the approach selected to remove the stain also plays a crucial role. The use of correct cleaning approach will not only remove the stain but it also enhances the durability of the fabric. Here the term *cleaning approach* is concerned with cleaning method. There are two basic methods to remove the stain.

**Stain removal by physical methods:** It refers to all those stain removing approaches in which chemicals are not used. Stains are removed with the help of corporal methods like as given below.

# Absorbent

✓ It is concerned with soaking of stained material in water. One may also use non-chemical agents like starch, French chalk powder to remove the stain. Absorbent materials absorb the stain or grease.

# Friction

✓ It is concerned with rubbing of stain in order to remove it. The rubbing process brings friction over the stain which forces stain to move out from the surface.

# Heat

✓ It is concerned with application of heat over the stained surface to remove the stain. For example, place *baby powder* over the stained (or yellow) surface of cloth then position hot iron over it to remove the stain.

# Freeze

✓ It is concerned with removing of stain by freezing technique. Freezing solidifies the stain then it can be gently lifted or scraped from the surface like candle wax & gum. Any residue may require further stain removal treatment.

**Stain removal by chemical methods:** It is related with cleaning of stains with various chemical based cleaning agents, for example, detergents, solvents, reagents and bleach. A brief description of all these is given below.

## Detergent

✓ It is a kind of compound or mixture of compounds which remove dirt from the surface. It can also be defined as a synthetic organic soap, either oil or water soluble, derived from hydrocarbons, alcohols, petroleum or other organic compounds used to remove the dirt from surface.

## Solvent

✓ It indicates to a liquid that dissolves a solid or liquid solute, the resultant product will be known as *solution*. This solution is used to remove the dirt from surface. Remember water is the most common solvent but unless it is used in conjunction with some cleaning agent like detergent, water is ineffective.

## Reagents

✓ It indicates to a group of testing solutions that help in determining which chemical should be used to kill certain bacteria/s. By reagents, one can identify the bacteria/s and their properties thereafter select appropriate cleaning agent to kill these bacteria/s.

## Bleach

✓ It indicates to an alkaline based solution of sodium hypochlorite used as a cleaning agent. Bleach is white in colour and contains germicidal properties. Strong solutions of chlorine bleaches discolour the utensils like copper, silver aluminium, etc.

| Activity |
|---|
| Match stains with their appropriate removal agents. |

| S. No. | Stains | Removal agents |
|---|---|---|
| 1 | Alcohol | Soak in cool or warm detergent solution |
| 2 | Blood (fresh) | Powdered pepsin |
| 3 | Cosmetic | Glycerine or eucalyptus oil then wash |
| 4 | Egg | Rustasol or rust remover |
| 5 | Ink (ball point pen) | Alkali or bleach |

| 6 | Fruit/tea/coffee | Bleach |
|---|---|---|
| 7 | Grass | Potassium permanganate solution followed by hydrogen peroxide |
| 8 | Grease | Scrape then use ice cube to rub |
| 9 | Ink (ink pen) | Same as of iron mould |
| 10 | Perspiration | Chlorine bleach |
| 11 | Iron mould | Scrape, soak and then wash (ammonia may also be used) |
| 12 | Chewing Gum | Sodium hydrosulphite |
| 13 | Vomit | Degreaser |
| 14 | Dyes | Carbon tetrachloride or Methylated spirit |

## CONCLUSION

This chapter is designed on two components, i.e., fibres and fabrics. There are different sources for the production of fibres which can be broadly classified as natural and artificial fibres. *Natural fibres* can be obtained from animals, vegetables and minerals whereas *manmade fibres* can be regenerated and synthetic. Thereafter, obtained fibres are knitted into different ways in order to create fabric. The knitting pattern can be woven as well as non-woven. In reality, the term cloth may be used as synonym for fabric but it often refers to a finished piece of fabric that is used for specific purpose like table cloth. Once fibre is converted into fabric and fabric is converted into finished product (or cloth), the stitching process/work will begin. Some clothes are used to stitch/make tablecloth, dusters, wiping cloths, buffet cloths, runners, etc., whereas other clothes are used to stitch bed linen, staff uniform and so forth. And due to regular use, these clothes get soiled and worn out. So during laundering process, undertake necessary mending work (if required) then use appropriate stain removal to remove stain.

| Terms (with Chapter Exercise) | |
|---|---|
| Saponified rayon | It refers to a type of rayon that is made from cellulose acetate filaments, similar to the kind used in making of acetate. These fibres are treated in a special way to produce rayon that is very strong. Fortisan is an example of famous saponified rayon. |
| Combed fibre sheet | ? |

| | |
|---|---|
| Cuprammonium | Rayon made from cellulose dissolved in cuprammonium solution. It is produced by making cellulose a soluble compound by combining it with copper and ammonia. |
| Tensile strength | It refers to the quality of fabric produced. The degree of tensile strength is determined by the amount of weight it takes to a 1 inch × 3 inches part of fabric. |
| Thread count | It refers to the total number of threads used in one-inch-square part of cloth. This term is used as a quality indicator for cloth judgement. |
| Yarn | It refers to a textile term used for fibres twisted together strongly enough for weaving purpose. |
| Mercerization (or mercerizing) | It is a treatment for cotton fabric and thread that gives the fabric or yarns a lustrous appearance and strengthens them. The process is applied to cellulosic materials like cotton or hemp. |
| Monomers | ? |
| Weft | It refers to all those threads that run crosswise (horizontally). It is also known as *filling*. |
| Retting process | It is a process of employing the action of microorganism and moisture on plants to dissolve or rot away much of the cellular tissues and pectin/s surrounding fibre bundles, hence facilitating separation of the fibre from the stems. |
| No-iron effect | ? |
| Filament | ? |
| Warp | It refers to those threads that are running lengthwise through the sheet. |
| Percale | It refers to a closely woven cotton fabric used for sheets and clothing. |
| Saran | ? |
| ? | It refers to a rough and dull looking cotton sheet made with carded cotton fibres. |
| Flax | It refers to widely cultivated plant, *Linum usitatissimum*, having pale blue flowers, seeds that yield linseed oil, and slender stems from which a textile fibre is obtained. |

| | |
|---|---|
| Carded fibre sheet | ? |
| Polymers | It refers to a class of compounds characterized by long, chainlike molecules of great size and molecular weight. |
| Durable press | ? |
| Absorbent | It refers to the cloth/linen or any material/s that absorb stain or grease. |
| | ? |
| Gray goods | It refers to the fabric that directly comes from a loom. |
| Sanforizing | It means preshrinking the cloth in order to prevent it from shrinking during regular laundering. |
| Finishing treatment | It refers to all those treatments which are required in the final touching of fabric. Simply, it is a process of washing and bleaching (with caustic soda) freshly woven fabric, also known as *mercerizing*. |
| Laundry bag | It refers to the bag provided by the hotel in which the guests put their laundry and find the laundry list. Guests tick against the given items for laundering. |

# 25 CHAPTER

# Interior Decoration

⟨ **OBJECTIVES** ⟩

*After reading this chapter, students will be able to...*

- describe interior decoration and its components;
- make the conventional style flower arrangement and rangoli;
- draw the colour wheel and find out primary, secondary and territory colours;
- identify different styles of lighting, their advantages and disadvantages; and
- explain different types of paints and where they are suitable—internal or external.

## 25.1 INTERIOR DECORATION

The term *interior decoration* is concerned with beautification of internal atmosphe of the property. Internal atmosphere focusses on guest rooms as well as public area Housekeeping department plays a vital role in creating such a pleasant atmospher There are innumerable activities that come under interior decoration, predominant flower arrangement, floor decoration, lighting, colour and painting. Some of these ca change in accordance with the requirement of occasion like flower arrangement ar floor decoration whereas remaining are usually permanent for a particular period time.

The valuable aspects of interior decoration include indoor plants, wall mounte paintings and other decoratives; chosen colour and lighting scheme, decorativ specifications; ventilation terminals; fixtures and fittings; furniture and furnishing working stations/counters; art and crafts and so forth. Often type of the hotel also pla a crucial role in determining what kind of interior decoration needs to be adopted. F example, in *boutique hotels* every room is made around some kind of theme. In certa cases, it has been notified that boutique hotels create every room different from oth rooms (or every room built around some specific theme like The Park, New Delhi). S here the interior decoration of all guest rooms is in adherence with the theme where. lobby and corridor can be decorated in any form.

## 25.2 FLOWER ARRANGEMENT

The term *flower arrangement* is concerned with collecting flowers in a decorative form for placing in different areas of property. Flowers are widely used for interior decoration; it gives aesthetic appeal to the ambience. Flower arrangement is an art and it is widely used in hotel guest rooms, in-house public areas (lobby, corridors, banquets, meeting rooms, etc.), and offices and usually replaced/changed on everyday basis. In hotels, a *horticulture* section (it is a sub-division of housekeeping department) is responsible for flower arrangement and maintaining *green house*. This horticulture section comprises four sub-sections, i.e., *floriculture, olericulture, pomology* and *ornamental horticulture*. Nowadays, many properties outsource this horticulture work because it demands extra space, additional manpower, water and other operational costs like cost of equipment and electricity.

**Horticulture Equipment**

- In-house/out-house plants
- Pesticides/insecticides
- Small & large scissors
- Grass cutting machine
- Lawn mowers/leaf mower
- Shovel, trowel & hoe
- Wheel barrow & pitch fork
- Leaf blower, pruner, tiller
- Spades, rakes and pots

| | Activity | |
|---|---|---|
| Give a brief description of below given sub-sections of horticulture section. | | |

| S. No. | Sub-sections | Brief description |
|---|---|---|
| 1 | Floriculture | ? |
| 2 | Olericulture | It deals with vegetable plants |
| 3 | Pomology | ? |
| 4 | Ornamental horticulture | It deals with ornamental trees in landscapes |

## Types of Flower Arrangement

In hotels flowers are used extensively, some are used for making outsized flower bouquets whereas others are used for undersized flower bouquets. For instance, large spectacular arrangements are kept in the lobby & restaurant whereas small arrangements are kept in rooms and suites. Remember, the extent to which flowers are used in hotel interior depends on the degree of luxury provided. In order to suit particular occasion there are different styles of flower arrangement. A brief description of some major style of flower arrangement is given below.

## Circular Shape

- ✓ Arranging flowers in circular design adds a pleasing element. It is satisfying to the viewer's eye. The circular or round shape is loved by nature since majority of flowers lie in this shape.
- ✓ In fact, it is also easier to arrange flowers in circular fashion. This type of arrangement is laid on conference tables or on buffet tables. They can be presented to high class executives or politicians on different occasions.

## Triangular Shape

- ✓ It is a most common style of flower arrangement used in personal and professional functions. Firstly, height and width is fixed with flowers and then focal point is established.
- ✓ They are placed on the buffet table or in the side station. In ceremonies also you can find such kind of flower arrangements.

**Tick for indoor or outdoor plants**

- Annals
- Massageana cane
- Ferns
- Perennials
- Dendrobium
- Peace lilies
- Schefflera
- Rose bushes
- Ornamental grasses
- Philodendrons
- Palms
- Phalaenopsis orchid
- Anthuriums
- Shrubs
- Dracaena reflexa
- Aloe plant
- Pothos

## Crescent Shape

- ✓ The crescent is asymmetrical and formal. It requires lot of skill and experience. This type of arrangement is eye catching. It is kept in the lobby of the hotel. It is used as a focal point to catch the attention of the guest.

## Fan Shape

- ✓ It is a low arrangement and does not interfere in conversation across the table. It is horizontal and generally placed in the restaurant either on buffet or dining table. Hotel rooms also have this kind of flower arrangement.

## Hogarth/'S' Shape

- ✓ This is a very graceful style of flower arrangement. It is easier to make when curved branches are used. Once 'S' shape is formed, flowers are filled at the centre.

## Famous Flower Arrangements

It has been previously discussed that flower arrangement is the responsibility of horticulture section which is a sub-division of housekeeping department. In horticulture section, flower room is responsible for *floral arrangement* in private areas (such as in guest rooms), in-house public areas (such as lobby, reception, banquet hall, restaurant, etc.) and other required areas. Due to its *floriculture* oriented work profile, the flower room should be air conditioned so that flowers remain fresh. The *flower room* must have a working table, a sink (to wash flowers, leaves), water supply and other required equipment.

Every country has its own paradigm of flower decoration which shows some meaning or reason. For instance, in Japan flower arrangement is in practice from the last 100 years. The Japanese style of flower arrangement is slightly different from others because people in Japan use flowers to symbolize season. They represent ideal harmony between earthy and eternal life. In each Japanese style of flower arrangement there is an imaginary triangle, in which the tallest line represents heaven, towards the heaven is man and the lowest line looking up to both is earth. Four famous Japanese styles of flower arrangement include the following:

**Flower Room Equipment**

- Assorted flowers
- Assorted leaves
- Cello tape
- Scissors
- Paper cutter
- Flower pots
- Foam (for placing flower)
- Tray (for *tray gardening*)

| | |
|---|---|
| Ikebana | It is a Japanese term which means *making flowers live*. |
| Ukibana | It indicates floating style of flower arrangement. |
| Moribana | It indicates basket style of flower arrangement. |
| Seika style | It indicates formal style of flower arrangement. |

## 25.3  FLOOR DECORATION (RANGOLI)

It is concerned with beautification of floor with the use of range of materials for instance different colours, flower petals, ornamental stones, coloured sand, dry flour, coloured rice, broken pieces of glass and so forth. In Indian culture, floor decoration particularly with an array of colours is often termed *rangoli*. *Rangoli* is often made with contrast colours which match with colour of surrounding walls, ceilings, furniture and wall mounted decorative. Nowadays, floor decoration is becoming a part of public area activity, often in-house public area such as hotel lobby. Major factors which affect floor decoration include space, time to prepare, decoration theme, experience of the *rangoli* maker, etc. A brief description of different approaches to decorate a floor is given below.

**9 Famous Indian Style Floor Decorations**

- Chowk rangoli
- Dotted rangoli
- Free hand rangoli
- Flower petals
- Alpana
- Wooden
- Floating
- Glass
- Traditional

## Material Based

- ✓ It predominantly reflects floor decoration with materials like flower petals, broken pieces of glass, ornamental stones, coloured sands, different colours and so forth.
- ✓ Floor decorator can also use a combination of different sets of materials. Few materials also support in creating 3D effects.

## Prototype Based

- ✓ In this kind of floor decoration, the selection of colour/s, pattern of decoration and way of presenting, all should be in accordance with particular hallmark/ trademark; nothing can be made on the basis of imagination. For example, floor beautification is based on the logo or monogram of the company.

## Colour Based

- ✓ The term *rangoli* is an ideal term for colour based floor decoration. It indicates use of different colours, often colours that contrast to each other.
- ✓ In India, such type of floor decoration is particularly found during the festive occasions. The panorama of embossing depends upon the imagination of the decorator.

## Subject Based

- ✓ It is floor decoration based on some kind of subject matter/theme. Therefore, it is also known as theme based floor decoration. Nowadays, global warming, green earth, save water, save life, stop deforestation, etc., are a few famous internationally recognized and floating subjects for floor decoration.
- ✓ This kind of floor decoration approach is often found in institutional competitions where event organizer provides topic of the theme and all participants decorate the allotted space around the given theme only.

## 25.4   HOTEL LIGHTING

At the time of planning for lighting, a planner needs to consider both- artificial as well as natural lighting. Natural light helps in reducing the energy costs while artificial light can transform a room into a seamless combination of functionality and style plus support to make an eye appealing interior decoration. One can soften the effect of artificial lights by using them invisibly or *in built* in coves like- light shelves and indirect pendants. Coloured light sources or filters can be used to create elements of excitement and surprise, form an unusual mood, or simulate the evening sky.

Remember different lighting levels can set different moods. For example, orange
r yellow tones create a warm environment while dim lighting or candles create a
mantic atmosphere. When lighting scheme is implemented properly, it can even create
markable effects that enhance both architectural and interior elements. Use special
tificial lights such as LEDs or fibre optics to create special lighting effects on the floors
r imaginary shape patterns on the ceiling. A balanced combination of natural light with
tificial light is a prerequisite of any organization as it affects energy conservation and
ntilation.

*atural lightening*: Straightforwardly, it refers to sunlight. The intensity of natural light
 controlled in different ways. Sunlight is controlled by shades, curtains, draperies,
d blinds. Some skylights direct sunlight into specific areas to provide light to dark
llways or rooms. This natural light is not a matter of choice for any organization; else
is a crucial part of energy conservation as well as ventilation. Natural light is free of
st and everybody can use it in accordance to his/her requirement.

*rtificial lighting*: It is manmade/unnatural lighting. Broadly, there are mainly two
pes of artificial lighting, incandescent and fluorescent. *Incandescent light* is used for
ament light such as light bulbs, halogen and tungsten. The light from tungsten bulbs
slightly yellow, while halogen offers a whiter light. *Fluorescent light* has a bluish tinge.
otel management must develop proper synchronization between lighting and décor
heme.

## ncandescent Lighting

✓ It is a warmer light but less efficient to operate than
  fluorescent bulbs of the same wattage. It can be
  easily directed to specific spots such as a particular
  table or any centre of attraction piece. It may make
  surroundings cheerful and inviting. It is also called
  as "filament light".

| Balance Light | |
| --- | --- |
| Incandescent | 30% |
| Fluorescent | 70% |

## uorescent Lighting

✓ Its main virtue is its low operating cost, but it is often criticized for giving a dull
  and lifeless illumination. It gives diffused lighting with flat appearances, textur
  interest and highlights can be provided by supplementing with filament ligl

## Activity

Write down the advantages and disadvantages of the given lights.

| Light types | Advantages | Disadvantages |
|---|---|---|
| LED lights | | |
| CFL lights | | |
| Mercury | | |
| Halogen | | |
| Tube light | | |
| Optic fibre light | | |

## Uses of Lighting

Irrespective of public or private areas and in-house or out-house areas, illuminatio is required at every place. It is also an important constituent of safety and security. balanced combination of light plays a great role in decoration as well as in safety an security. For example, use *incandescent* lights for clarity and sparkle whereas *fluorescer* light for soft or cool light. Light must be chosen in accordance with area specificatio A brief description of suitable lights for interior and exterior purpose is given below.

*Indoor lighting:* It indicates to lighting fixtures built inside the main hotel buildin; in private rooms (such as guest rooms) and in-house public areas (such as restaurant corridors, lobby, etc.). A combination of different sets of lights most preferably used f interior illumination is given below.

## Chandelier

✓ Chandelier is primarily found in hotel lobbies. It gives general illumination as we as also provides sparkling addition to the room décor. It adds value to interi decoration. Chandeliers demand deep cleaning at scheduled frequency.

## Dimmers and Down Lights

✓ The term *dimmer light* is concerned with adjustable style of light where illumination can be adjusted in accordance with the requirement. It is often se as a part of chandelier to balance its illumination and accent lighting levels. Down lights are used to highlight the centrepieces or to light artwork plus serving tables along the walls.

## Emergency Lights

✓ These lights supply battery back-up illumination when power goes off in the hotel or works during the black-out and brown-out situations. When light of entire hotel building goes off it indicate to *black out* situation whereas when few phases of light go off it reflects to *brown out* situation.

## Wall Scones and Exit Signs

✓ *Wall scones* are wall mounted lights used to decorate the walls as well as provide some illumination to the given décor. Exit signs (or lights) are equipped or placed on the upper side of exit doors, so guest can locate the nearest exit point during the emergency situations. It consumes almost no electricity.

*Outdoor lighting:* It refers to lighting fixtures outside the main hotel block, thus also known as external lighting. Apart from normal illumination, external light is used to make the premises safe and secure particualrly during night, to invite guests and to create a change in emphasis and atmosphere for leisure visitng in the evening. A brief description of major exterior lights is given below.

## Security Lights

✓ It refers to the boundary lights built along the boundary walls or fences. These lights support the security personnel in surveillance of the surrounding area, particularly during night shift. Nowadays, these lights are motion sensor and light sensor based. The senstivity is controlled/determined by an individual.

## Focus and Decorative Lights

✓ *Focus light* refers to the spot light fitted to give illumination to the centre of attraction. The main function of this kind of lighting is to attract attention of general public towards the fitted piece of attraction.

✓ *Decorative light* refers to the various high levels or above the ground level lamps/ lights that are built along the in-house pavement. The lighting panels are generally of glass or plastic. It looks attractive as well as offer illumination.

## Concealed and Tiered Light

✓ *Concealed light* refers to the covered strip light mostly found above the entrance gate as well as in curtained windows. Nowadays, concealed lights are almost found in all internal areas of hotel operations.

✓ *Tiered light* is also known as *pagoda light* that refers to a set of lights in a black or green casing, to keep the focus on the plants. They may feature as either plastic or metal fixtures.

## For Illumination

✓ It covers a range of external lightings that principally fit to give illumination o. for example half brick light, lantern type light or wall mounted lights. Nowadays, in hotels these lights are replaced by decorative lamp style lights.

| Activity | | |
|---|---|---|
| Comparative study between incandescent and fluorescent lights. | | |
| **Criteria/Base** | **Incandescent** | **Fluorescent** |
| Investment cost | | |
| Operating cost | | |
| Filled with mechanism | | |
| How is light produced? | | |
| Shelf-life | | |
| Voltage effect | | |
| Preference of fitting location | | |

## 25.5   COLOUR

Colour can be defined as "the property of an object of producing different sensations to the eye as a result of the way it reflects or emits light". Every colour has its own unique value, importance and role. In psychology, every colour has different meaning and they can also create either positive or negative impression. Colour in any room affects the reflection and use of light in a room. It also reveals the visual impression of space and other emotive responses. The range of colours which forms the spectrum of white light may be represented as a circle to show the effects and relationship. This colour circle is known as *colour wheel*. The colour wheel shows the relationship among the primary, secondary, and complementary colours.

......................................................................................
### Activity

ll the below given spaces with appropriate colour name to match with their psychological eaning

Red        - Blue        - Green        - Yellow        - Orange        - Purple

| olour | Psychological meaning | Colour | Psychological meaning |
|---|---|---|---|
| ? | Colour of social communication and optimism | ? | Colour of loyalty and integrity |
| ? | Colour of optimistic and cheerful nature | ? | Colour of balance and growth |
| ? | Colour of energy, passion, action, ambition and determination | ? | Colour of imagination and creativity |

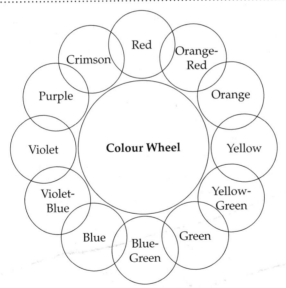

From the colour wheel, one can develop many colours and create combinations of ours that can be used in decoration but they should have some form of association ween colours to produce a comprehensive colour scheme. The following colour eme can be developed with the help of colour wheel.

## onochromatic

✓ It refers to the colour scheme derived from a single base hue and extended by using its tints, shades and tones. *Tints* are achieved by adding white whereas *shades* and *tones* are achieved by adding a darker colour (grey or black).

✓ In this colour scheme, the colour must change in tone and intensity. There mu be enough contrast within the values of the chosen colour if one does not wa the scheme to be look dull or boring.

## Complementary

✓ It refers to *contrasting colour scheme*. In it, any two colours are considered whi are directly opposite to each other, for instance red and green. This colour scher creates maximum contrast and maximum stability.

✓ In accordance with complementary colour schemes, bright tone colours can used for small areas whereas grey colour tones can be used for larger areas.

## Spilt Complementary

✓ It refers to a split-contrasting colour scheme because it is a variation of t complementary colour scheme. In addition to the base colour, it uses the tv colours adjacent to its complement.

✓ Any hue or neutral colour may be dominant, the others serving to accentuate. Th colour scheme has the same strong visual contrast as the complementary colo scheme, but with less tension.

## Analogues (colour scheme moves like wall clock)

✓ It refers to *harmonious colour scheme*. In it, there can be any three colours which a side by side on a 12 section colour wheel, for instance, yellow-orange, yellow a yellow-green.

✓ Generally, it shows the adjacent colours being used together. The contrast colo of any one of the groups can be used as an accent colour.

### Triad

✓ It refers to a colour scheme which uses colours that are evenly spaced around t colour wheel. In simple terms, it shows three primary, three secondary or thr tertiary colours used together.

✓ In triadic colour scheme, the colour should be carefully balanced. For accentuatic one colour is used intensely whereas the others are subdued by tainting or shadir

| Colour Grouping on Colour Wheel | |
|---|---|
| Primary colours | Colours at their basic essence; these are the colours (red, yellow and blue that cannot be created by mixing other colours. |
| Secondary colours | Colours created by a mixture of two primary colours (i.e., green, orang and purple colour). |
| Tertiary colours | Colours created by a mixture of primary and secondary colours. |

## .6  PAINT AND PAINTING

ie term *paint* refers to any liquid, liquifiable or mastic composition that, after application
a substrate in a thin layer, converts to a solid film. On the other side, the term *painting*
the process/practice of applying paint, pigment, colour or other medium (like brush)
a surface. Paint is mainly used to protect, colour or provide texture. Therefore, paint
i kind of *wall covering*. As a wall covering paint offers a wide choice of types, colours,
grees of gloss and even design if murals are painted on the walls. It is important to
nember that before applying paint, prepping of the surface is necessary including
me/primer. The selected *primer* should match with the top coat and to surface it is
id on: oil based primer should be used for oil-based paint and latex based primer
iuld be used for water/latex based paints. There are multiple variety of paints floating
the market, among these some major varieties are briefly explained below:

## l Bound Paints

✓ It refers to a solvent thinned paint that contains alkyd and other resin oil binders
  with a drying solvent and pigment. *Polyurethane* and silicones are often included
  to give a more scratch-resistance surface.

✓ It is mainly used with undercoat. Oil-based paints are durable, semi-flexible
  wearing surface with glossy, semi-glossy or egg shell finish.

## ulsion Paints

✓ It refers to water based paints that are made from a mix of synthetic resins (like
  polyvinyl acetate), a colour pigment and water. Emulsion paints are washable and
  alkali resistant. Being alkali resistant, emulsion paints are suitable for use on new,
  possibly damp plaster surface.

✓ Generally, it is suitable for interior decoration with an emphasis on painting walls
  and ceilings. Unlike gloss paints that give a very glossy finish these paints tend
  to give a matt effect and are often used to cover large areas. These paints have a
  longer life in terms of their colour and durability.

## amel Paints

✓ It refers to oil-based as well as water-based paints. Oil-based enamel paints are
  *alkyd paints* based on synthetic resins combined with a vegetable oil, like linseed
  oil. Oil-based paints have a strong solvent odour and are easily cleaned by using
  paint thinner or mineral spirits.

✓ Water based enamel paints are *latex* or acrylic paints that are easy to use, dry
  faster and have a fairly low odour. Enamel paints are hard, washable and usually
  glossy in appearance plus available with heat-resistant properties. *Polyurethane*
  and silicones are often included to give a more scratch-resistance power.

## Multi-colour paints

✓ It refers to suspension of paint globules in spray medium (usually emulsion bas paints). In simple terms, these are usually dispersions of cellulosic colours in wa and each colour creates separate blobs. Therefore, the resulting effect depends the number of different colours, the degree of contrast between them and the s and distribution of blobs.

✓ It is suitable for speckled effects and for applying in entrance halls, cloakroor WC, etc. It is extremely hardwearing and the multi-colour effects help to ma surface irregularities and imperfections.

## Texture Paints

✓ It refers to plastic paints that will hold textured designs like stippling, even afte is dry. Texture paint comes in several varieties which are smooth, sand or coar This paint is preferred due to various reasons because it covers imperfectio cracks, cement walls, old panelling, uneven dry walls or holes, etc.

✓ It creates dramatic and unique wall designs. There are variety of tools that can used to create various kinds of design, for example, sponges, stamps, styluses specialized brushes, palette knives, strippers, etc. The correct texture is obtair by working over the material while paint is wet.

## Micro-Porous Paints

✓ It refers to paint with rubberized base that is good for wood, plastic as well as environment too. Micro-porous materials are classified as materials which h a pore size of less than 2 nanometres which will allow the free passage of wa vapour but do not allow water molecules. Thus, also known as *breathable pain*

✓ The rate at which water moisture passes through a coating is controlled by permeability of the coating, the moisture content gradient, the film thickness, a the temperature. It was apparently invented by Dulux. This paint gives little gl but offers elasticity, thus allowing movement when the surface extends.

### Activity

Paint can also be classified as interior and exterior paint and both may be eith *water-based latex paint* or *oil-based paint*. Carefully read below given few things abc the paint then fill the blank spaces with appropriate answers.

✓ Paints come in four or five levels of glossiness, called *sheen*. From dullest to me shiny, they are matte or flat, eggshell, satin, semi-gloss, and gloss. Some pa lines do not have an eggshell sheen available.

✓ The higher the gloss, the easier the paint is to keep clean. Usually, higher she paint, such as semi-gloss or gloss, is used on trim, while a more matte finish used on the walls and ceilings.

| Paints | Description | Suitable for area |
|---|---|---|
| Varnishes | | ? |
| Chlorinated rubber | | ? |
| Lacquers | | ? |
| Flame retardant | | ? |
| Plastic emulsion | | ? |
| Cement | | ? |
| Multi-colour | | Children gaming zone |

## CONCLUSION

The term *interior decoration* is concerned with beautification and embellishment of hotel's interior. The major components of internal beautification include placement of diverse style of flower arrangement in guest rooms or any in-house public location; floor decoration with rangoli; use of artificial and natural lighting; selection of primary, secondary and territory colours; and painting of walls and ceilings. Collectively all these components create a profound atmosphere which enhance aesthetic appeal and please guests. Flowers can be arranged in various traditional ways from Ikebana and Moribana to modern Fan, Hogarth and crescent style. Each style of flower arrangement depends on the choice of the guest and demand of situation. Remember at the time of choosing artificial light keep balance between fluorescent and incandescent lights. In general, 30% incandescent with remaining 70% fluorescent is considered a perfect combination. Similarly, at the time of selecting paint hotel needs to choose either oil based or water based paint. The paint selection should be made after comparing the pros and cons of each paint type.

| Terms (with Chapter Exercise) | |
|---|---|
| Colour wheel | Circle that is divided equally in 12 sections, with each section displaying a different colour according to its pigment value. |
| Latex | Colloidal suspension of very small polymer particles in water and is used to make paint and rubber. |
| Greenhouse | Specific area under horticulture section which makes and maintains greenhouse to foster specialized plants in garden. It should have wooden racks to store pots, etc. |

| | |
|---|---|
| In-door plants | Various in-house plants that usually do not grow with photosynthesis process. |
| Out-door plants | ? |
| Floral arrangement | Flower arrangement for the purpose of beautification of atmosphere. |
| Incandescent | Filament type lights. |
| ? | Soft and cool light. Its main virtue is its low operating cost. |
| Sheen | This term is used for different levels of glossiness of paints. |
| Breathable paint | ? |
| ? | Range of colours that show in such a way that their effect and relationship can be easily compared. Simply, it shows relation among primary, secondary and territory colours. |
| Black out | ? |
| Brown out | Reduction or restriction of electrical power in a particular area. It may be done intentionally or unintentionally, for instance brown outs may be used to reduce load in an emergency situation. |
| Alkyd paint | Alkyd based paints. Alkyds are the dominant resin or binder in most oil-based paints. |
| Draperies | Fabric material that hangs in a window or other openings for the purpose of decoration, shade or screen. They are attached to traversing rods with special pin hooks. |
| Blinds | These are a type of curtain (or used in place of curtains) which are mainly available in two styles, i.e., vertical blinds and horizontal blinds. |
| ? | Tiered or lantern style light often set in a black or green casing, to keep the focus on the plants. They may feature either plastic or metal fixtures. |
| Tray gardening | ? |
| Moribana | ? |
| Ikebana | Japanese style of flower arrangement/decoration. Ikebana means *making flowers live*. |

# Flooring and Wall Covering

ƏS

Ɉᴉ ᴇᴉ

ɘɿɐ

.ɿoʇ

Ɉɛoɔ

ᴎo ɘɔᴎɐᴎɘɈᴎiɐm ɘʜɈ

.ɘ.ᴉ ɛɘʜɔɐoɿqqɐ

A .ǫᴎᴉɿooʅʇ

---

⟨ **OBJECTIVES** ⟩

*After reading this chapter students will be able to...*

- describe flooring and wall covering;
- draw the cleaning schedule for wall covering and flooring;
- find out the materials used to make wall covering;
- identify wall paper printing pattern; and
- explain different types of wall covering and flooring and their importance.

## 26.1 FLOORING

The term *floor* is often called the fifth wall of a room whereas *flooring* is a general term for a permanent covering of the floor surface with a suitable material/s. Thus, flooring is also known as *floor covering* and excellent flooring must meet the perceived aesthetic requirements and must match with wall and window coverings as well as room's furnishings. Nowadays, the concept of variant flooring is most popular. *Variant flooring*, which combines more than one flooring material, can be used for a dramatic identification of space; to add colour, texture and style to a room; and comes in a huge variety of price ranges. It is an excellent way to delineate space in open floor plan. Flooring can also be resilient and non-resilient.

## 26.2 WALL COVERING

The term *wall covering* is concerned with casing of wall with different materials order to make it more appealing and durable. Simply, wall covering is any material such as wallpaper or textured fabric used as a decorative covering, predominantly for interior walls. Consequently, the material selected for wall covering should be cost effective, fireproof, soundproof, attractive and acoustic. Thereafter, the maintenance of wall covering resembles floor maintenance and comprises three distinct approaches i.e. interim maintenance, restorative cleaning and spot removal.

| | Activity | |
|---|---|---|
| From the below given options, fill in blank spaces under the cleaning schedule. | | |
| - Daily/weekly cleaning          - Periodic cleaning         - Need based cleaning | | |

| Cleaning methods | Description | Cleaning schedule |
|---|---|---|
| Interim | Dusting and vacuuming of wall coverings and ceilings. | ? |
| Restorative | Use of detergents and solvents for cleaning wall coverings and ceilings. | ? |
| Spot removal | Removing of stains with the help of appropriate cleaning agent/s. | ? |

| | Activity | |
|---|---|---|
| From the given options, fill the blanks: | | |
| - Resilient flooring          - Non-resilient flooring | | |

| Flooring type | Description | Examples of flooring |
|---|---|---|
| ? | Flooring materials that give underfoot and when dented, it will eventually rebound wholly or partially to its original form. | Asphalt tile, carpet, linoleum, rubber, vinyl tile and wood. |
| ? | Flooring materials that do not give any degree of underfoot. | Concrete, ceramic tile, *epoxy*, marble, terrazzo and all other stone floors. |

## 26.3   TYPES OF FLOORING

It is concerned with selection of flooring material/s to cover the floor. Nowadays, there are different designs and types of flooring available in the market but the selection of particular flooring type is determined after *need analysis*. The major parameters used for selecting correct flooring type include the amount of traffic the floor area will receive, its durability, its matching with surrounding areas, budget of organization, perceived aesthetic appeal of the place and so forth. Flooring comes in various types. It can be hard, wood, fibre or sheet. Flooring can also be classified as resilient and non-resilient flooring. A brief description of each is given below.

| Activity | | |
|---|---|---|
| **Pros** | **Natural Flooring** | **Synthetic Flooring** |
| Durability | ? | ? |
| Versatility | ? | More choice for interior decoration |
| Sustainability | Ecofriendly | ? |
| Choice of variety | ? | ? |
| **Cons** | | ? |
| Expenses | ? | ? |
| Resistance | ? | Less resistance than natural marble flooring |
| Vulnerability | Susceptible to damage | ? |
| Wear and tear | ? | ? |

**Hard flooring:** Hard floors are durable, elegant and do not require much maintenance and are resistant to heat. Common types of hard floor are natural stones and synthetic/ceramics. Ceramic is made from clay that is fired in a kiln. But these differ from brick, as a coating is applied on one side of the ceramic tile then fired which creates soil and liquid impervious surface. More details of ceramic tiles are given later in this chapter. A brief description on natural and synthetic flooring is given below.

| Base/Criteria | Natural | Synthetic |
|---|---|---|
| Definition | Natural stones like granite, marble, limestone, travertine, slate, etc. | Artificial floorings like flooring with-brick, tile, concrete, terra-cotta, glass, acrylic and metal. |
| Preference | Commonly used in living room as it adds *motif* and style to the room. | Ceramic tiles for walls and countertops, bricks for interiors and terracotta for kitchen floors. |

| Features | It has natural beauty, elegance, durability and creates a dramatic appearance. | Although it is a type of hard floor but can be molded in accordance with demand and requirements. |
|---|---|---|
| Expenses | It is not affordable like any other flooring and it also requires sub-flooring. | Synthetic flooring is less expensive in comparison to natural stone flooring. |
| Cleaning procedure | *Daily dusting* with a neutral pH detergent and mopping. Remember, mop till it should not let water/chemicals remain on the floor. | Bricks are dried/wet vacuum cleaned with bristle brush whereas ceramic tiles and concrete must be cleaned via dust or damp mopping and light scrubbing. |
| Sealing | Seal with moisture permeable sealers. | Seal mortar between the bricks and grout between the tiles. Concrete floors seal with *permeable sealers*. |
| Finishing | Finish with one or two thin layers then buff. | No need to finish bricks & ceramic tiles. Finishing may be applied to concrete floors, but they should be compatible with porous floors and permeable sealers. |
| Stripping | Use either neutral or mildly alkaline solution. | No need for stripping on bricks & ceramic tiles. For concrete floors, use an alkaline stripping agent that has been properly diluted with water. |

**Wooden flooring:** Wooden floors are long lasting and require simple cleaning as they don't retain dust particles. But they require frequent vacuuming and sweeping. Although they don't retain dust particles, but wooden floors are prone to scratches and dents. Common types of wooden flooring include natural and synthetic wooden flooring. A brief description of each type of wooden flooring is given below.

| Base/ Criteria | Natural | Synthetic |
|---|---|---|
| Definition | Parquet square, plank, strip floor and so forth. | These are laminated floorings and are the most common type of synthetic woods. |
| Preference | It is preferred in foyer, boardroom and lounges as plank gives a classic look to the flooring. | This type of flooring works well in areas exposed to moisture. |

| Features | Parquet square is replaceable whereas strip floor is prone to buckling. | This type of flooring is fire resistant and also does not fade nor stain. |
|---|---|---|
| Expenses | Expensive in nature as many natural floorings demand sub-flooring. | As these are laminated floorings so they do not require finishing which reduce its cost. |
| Cleaning procedure | Daily dusting/vacuuming/buffing; intermittently damp mopping and preventive maintenance. Remember, do not use an oily dust mop. | Same as natural wooden flooring. |
| Sealing | It can be sealed with oil-modified urethane, moisture-cured urethane, the *Swedish-type* or water-based sealers. Remember, use same seal every time and do not seal persistently. | Same as natural wooden flooring. But remember, most modern wooden floors can tolerate only three to five times sealing. |
| Finishing | Most of natural wooden floorings require an occasional waxing and buffing. | Often modern wooden flooring does not require refinishing. Surface finishes such as *urethane*, *varnish* and *shellac* are not recommended. |
| Stripping | Usually stripping is not required. | Usually, stripping is not required. |

**Fibre flooring:** Fibre flooring is more comfortable than any other type of flooring as it absorbs noise and looks cosy. It is available in different colours and styles. It does not scratch however it is not as durable as hard flooring. Fibre flooring requires regular maintenance as it gathers dust particles and when wet with water or any liquid, a stain is formed. Common types of fibre floor are natural and synthetic (or carpet flooring). A brief description of each flooring type is given below.

| Base/Criteria | Natural | Synthetic |
|---|---|---|
| Definition | Coir, sea grass, sisal or jute based fibre flooring. It may also include plant fibre carpet. | *Carpet* such as cut-pile-carpet, twist, loop pile, shag and carpet tile. |
| Preference | It is mainly preferred in gardens, lounges and other open space areas. | It is mainly preferred in banquets, corridors, and rooms. But nowadays hotels do not prefer it. |
| Features | Natural fibre flooring materials give a natural look and are eye appealing. Best for home. | Carpet flooring gives cosy feeling and comprises three elements, i.e., *pile, primary*, and *secondary backing*. Often *padding* may also be accompanied. |

| Expenses | It is not as durable as other natural flooring thus recurring cost is more. | Carpet flooring is often found expensive (it depends on the quality of carpet material). |
|---|---|---|
| Cleaning procedure | Plant fibre carpets are cleaned with ICS. Simply, carpets are cleaned without using water. For instance, combination of the duo brushing machine and duo-P carpet cleaning powder. | Most carpets require daily vacuuming and sporadically shampooing with carpet cleaning machine (also known as *hot water extraction machine*). |
| Sealing | Sisal or jute based flooring should be sealed with *sisal guard* or *jute sealers* only. It reduces the absorbency thus protects from stains. | Carpet is available in three standard sizes, i.e., broadloom, runners and tiles. It should be laid from wall-to-wall and sealed with appropriate adhesive. |
| Finishing | Perfect sealing leads to perfect finishing. It acts as a barrier and reduces the chances of spills. | During carpet installation, carpet should not engorge and distend. |
| Stripping | Natural grass ground strip-up with grass cutting machine or lawn mowers. | Usually, carpets are not stripped. But after carpet installation, installer should check for any buckle or bulges. |

**Ceramic tiles:** Earlier, it was seen that ceramic tiles were only used in kitchens and bathrooms but nowadays with the emergence of new designs of ceramic styles, it is possible to use these in living rooms also. Ceramic comes in three basic types– glazed, unglazed and porcelain. Ceramic tiles are easy to maintain as well as they are durable too. These tiles have limited chance to dents, dings and scratches but remember ceramic tiles should never be used over a floor with structural movements. These tiles can be slippery when it gets wet.

Match the below given description of ceramic tiles with their correct type.

| S. No. | Description of ceramic tiles | Type of ceramic tiles |
|---|---|---|
| 1 | These ceramic tiles are the hardest as it has a higher breaking strength | Unglazed |
| 2 | These ceramic tiles have different texture and are very hard | Porcelain |
| 3 | These ceramic tiles are designed with glass coating with mineral stains and available in semi-gloss or high-gloss finishing. | Glazed |

## 26.4  TYPES OF WALL COVERING

It refers to any material or assembly used to cover the wall (or wall finish) but do not form an integral part of the wall. The wall covering material may be a flexible sheet of appropriate sized paper, fabric, plastic, etc., in decorative patterns, colours and surface textures. These are usually laminated and printed with a repeat pattern for the purpose of directly being pasted on a wall as decoration and protection. Most wall coverings consist of three layers, each of which perform different important functions. Starting from the surface of the wall covering and working to the back, the layers are given below.

### Decorative Layer (first/top layer)

✓ It refers to the actual face of wall covering. It is that thinnest layer that comprises the inks applied to the top of the intermediate layer. This decorative layer supplies the look of the product and is normally the major reason behind wall coverings selection.

✓ The decorative layer may also have a protective polymer coating to provide added performance characteristics.

### Intermediate Layer (second or middle layer)

✓ It refers to the middle layer (i.e. a layer between decorative and wall base layer) which works as a ground layer for the formation of decorative layer. It provides the surface upon which the decorative layer (ink) is printed.

✓ It also provides the background colour, while often off-white; it can be any colour depending on the design. This layer can range in thickness from less than 01 mil to as much as 10 mils as in heavier-weight, *solid-vinyl* products. Note that a mil is 1/1000 of an inch.

### Third Layer (third or last layer)

✓ It refers to the substrate or backing (or last layer) that goes against the wall. The surface of wall-covering products usually commands majority of attention paid to wall-coverings, but the backing of these products is important from a functional perspective.

✓ This backing can be made up of a wide variety of materials ranging from woven and non-woven fabrics to lightweight paper products. There are various types of backings, for instance paper, woven fabric, non-woven fabric and latex acrylic backings.

There are many different grounds and substrates used to make wall coverings. The combination of these materials provides individual characteristics typical in each type of wall coverings (degrees of strength or durability, colourfastness, scrubability, washability, stain resistance, abrasion resistance, etc). Remember, each type of wall covering has certain advantages and disadvantages. A brief description of several major wall coverings is given below.

| **Wall paper printing pattern** |
| --- |
| - Screen |
| - Rotary |
| - Gravure |
| - Digital |

## Wall papers

✓ It refers to the paper based wall covering often coloured and printed with design and pasted to a wall as a decorative covering. It is mainly used for building interiors and usually sold in rolls and is put onto a wall by using wallpaper paste.

✓ It comes in various styles like lining paper or textured paper (such as *Anaglypta*), oatmeal paper, wood chip paper, lincrusta (a thick embossed wall paper), metallic papers, wood grain and paper-backed (like hessians, felts, woven grasses, wools, etc.). In terms of making method, wall paper types include painted, hand printed (stencil or block-wood), machine printed and flock wall paper.

## Plastic

✓ It refers to the wall covering with plastic sheets like laminated plastics, expanded polystyrene, vinyl flock papers, etc. Plastic based wall coverings are non-porous in nature thus there is an extreme propensity for the growth of moulds. These moulds can be prevented by using fungicides in the adhesive or also by applying fungicidal solution on the wall before hanging plastic sheets.

✓ Many other plastic sheets are also used for sound insulation but these are expensive. Plastic sheets are easy to clean and more hardwearing. Nowadays UV plastic sheets are also used in hotel industry mainly pasted on glass windows.

## Glass

✓ It mainly refers to the mirror-like tiles which reflect light and can also alter the apparent size of the room. Coloured opaque glass sheets/tiles are often a preferred choice in bathrooms. Glass wall coverings offer style, pattern options, strength and durability as well as these are easy to clean.

✓ Nowadays, glass wall coverings are available in various forms like fibreglass, magnetic glass, glass fabric, glass-less mirror, etc. Glass wall covering is free of odour and plasticizers. It is an ecologically sound product made from natural materials. It is also cost effective in the long run.

## Metal

✓ It refers to wall coverings made up of solid material which is typically hard, malleable, shiny, fusible and ductile, with good electrical and thermal conductivity, for instance, copper, anodized aluminium and alloys such as steel. Metal foil can be elegant if used sparingly as a wall covering and it is also available in various colours.

✓ In hotels, metal based wall coverings are found in bars, kitchens and various other areas. In bars, it may be used for decorative effect where the metal is in combination with rows of bottles whereas in kitchen, stainless steel tiles may be used as they are durable and easy to clean.

## Leather

✓ It mainly refers to the animal skin based wall coverings which are found extremely expensive. Nowadays, animal skin free variation is also available like *faux leather*. As these are expensive in nature thus are only found in bars and restaurants of luxury hotels.

✓ Leather based wall covering demands careful cleaning (like during suction cleaning) and polishing at regular intervals. During polishing, polish must be applied sparingly and rubbed up well. Heat may create adverse effect on the leather wall coverings, thus never prefer to hang in kitchen and similar places.

## Fabric

✓ It refers to wall coverings that have a woven substrate of fabric or a nonwoven synthetic substrate. In either case, the substrate is laminated to a solid vinyl decorative surface. The aesthetic appearance and durability of fabric based wall covering depends on the fibre and weave used in manufacturing.

✓ The selected fabric should not attract dust and dirt as well as not be liable to stretch, buckle or sag when hung permanently. In most hotels, hessian, linen, acetate (or *viscose*) fabrics are used for wall coverings. Silk may also be used but it is expensive. Remember, wool based wall coverings may be attacked by moths.

## Wood

✓ It refers to the wall covering with timber material. These wall coverings are mostly laminated to fabric backing. Wood wall covering is also known as *wood panelling* and it may be solid or veneered. These are usually made into sheets which are 24 inches wide and provided in any length up to 126 inches long.

✓ Due to characteristics relative to environmental and grain matching, wood veneers are used mostly in the offices or conference rooms, assembly halls, staircases, restaurants, entrance halls, etc. Remember, wood panelling demands precautions in respect of dry rot and woodworms.

### Activity

Fill in the below given blank spaces with appropriate description against each type of flooring.

| Flooring | Description | Flooring | Description |
|----------|-------------|----------|-------------|
| Acoustical | ? | Terrazzo | ? |
| Cork | ? | Brick | ? |
| Digital | ? | Bamboo | ? |
| Linoleum | ? | Kota stone | ? |
| Rubber | ? | Vinyl | ? |

## CONCLUSION

*Flooring* is related with permanent covering of the floor either with natural or synthetic material. Flooring can be done with marbles, bricks, fibre, carpet, wood, ceramic tiles and so forth materials. Each flooring type has its own set of advantages and disadvantages over another. For example, marble and ceramic tiles based flooring is always more durable as compared to fibre and wooden flooring. Similarly, natural flooring also has enormous pros and cons over synthetic flooring. For example, natural flooring is sustainable, low flammability, less health hazardous, requires less chemical treatment, has unmatched beauty and quality whereas synthetic flooring offers variety of choices, affordability, easy replacement and versatility. Apart from flooring, wall covering is also an important part of room decor. There is an array of wall covering patterns/styles. The selection of specific wall covering depends on individual's choice, room size, pocket power, type of hotel and other related aspects of room interior. Most popular wall coverings include wall paper, plastic, glass, metal, leather, fabric and wood.

| Terms (with Chapter Exercise) | |
|---|---|
| Epoxy | Synthetic, seamless flooring material which is long lasting and extremely durable. |
| Need analysis | ? |
| Resilient flooring | Floors that give underfoot. For instance- flooring with asphalt tile, carpet, rubber, linoleum, vinyl tile, wood and so forth. Remember, dents are a usual problem with these floors. |
| Non-resilient flooring | Floors that do not give underfoot. For instance, flooring with concrete, terrazzo, marble, ceramic tiles, epoxy and stone based floors. Dents are not a problem with these floors. |
| Viscose | ? |
| Acoustic | This term is concerned with auditory feature (i.e., soundproof facet) of wall covering. |
| Faux leather | ? |
| Pile | Threads of yarn found on the surface of a rug. In it, the term *nap* is an indicator of quality, i.e., pile density and weight. |
| ? | It is concerned with the carpet surface into which fibres are stitched in a tufted carpet. This carpet component is mainly made from polypropylene. |

Flooring and Wall Covering

| | |
|---|---|
| Secondary backing | ? |
| Metallic papers | Wallpaper made up of range of *resemble metals* like aluminium, brass, copper, etc. |
| Woven grass | ? |
| Padding | Carpet component, i.e., a layer of material placed under carpet to increase its resiliency. Remember, carpet padding can be made from a number of synthetic and natural substances. |
| Shellac | *Varnish* made up of alcohol and refined lac (a sticky substance made from the deposits of insects). |
| Varnish | ? |
| Urethane | Urethane or polyurethane to a strong plastic resin that resists acids, fire and putrefy. Therefore, it is preferred as a substitute for foam rubber and varnish and for insulation too. |
| Woodchip | Also known as *ingrain* wallpapers. It consists of two layers of paper with wood fibre in between. It was invented by German Pharmacist Hugo Erfurt in 1864. |
| Hot water extraction machine | ? |
| Lincrusta | Deeply embossed wall covering. It was invented by British inventor Frederick Walton in 1877. These are rolled sheets made from a paste of gelled linseed oil and wood floor. |
| Carpet flooring | Flooring with fibre based materials. Generally, there are five types of carpet, i.e., silk, woollen, cotton and synthetic carpets. |
| Anaglypta | ? |
| Es | *Impregnated compound system.* For instance, in Britain there is a combination of the Duo brushing machine and Duo-P carpet cleaning powder, this system safely removes spoilage without the risk of shrinkage or "area out of use" problems. |
| Variant flooring | Flooring with different colours of the same material or a combination of different materials in matching or contrasting colours, in accordance with the need. |

# 27 Pest Control and Waste Management

**CHAPTER**

⟨ **OBJECTIVES** ⟩

*After reading this chapter, students will be able to...*

- describe pests and their types and families;
- understand control of infestation by different pest control methods;
- identify insecticides, pesticides and germicides to prevent the attack of different pes and
- explain what waste is; its type and how it should be controlled.

## 27.1 PEST AND PEST CONTROL

The term *pest* refers to a destructive organism or any insect that has a harmful effect humans, food, clothes and so forth. It spreads disease, causes destruction or is otherw a nuisance. Some examples of pests are mosquitoes, rodents, cockroaches, lizards, ra weeds, etc. Regular spray of *insecticides* and *pesticides* is the best way to control pests a insects. Both are the substances meant for attracting, reducing, destroying, or mitigati any pest. They are a class of biocide.

Sanitation is an important part of control. Successful pest control should begin with go housekeeping. Conventionally, pest control is the job of housekeeping department b nowadays it has been outsourced in most of the hotel establishments and housekeepi department assists the pest control team in spraying pesticides and insecticides. T foundation of successful pest control lies in a sound knowledge of habits, need and l history of the pest. Food is the basic attraction for pests, predominantly for rats, m and similar insects. For effective pest control, remember different kinds of pests requ different kinds of treatment.

## 27.2 CONTROL OF RODENTS

The term *rodent* refers to the gnawing mammal of the order Rodentia that includ rats, mice, squirrels, hamsters, porcupines, and their relatives, distinguished by stro constantly growing incisors and no canine teeth. Rats and mice perform incalculal damage to all kinds of property, destroying and fouling food material, plus they are al

active agent in spreading many diseases. An effective control guideline for rodents
1 their family insects is given below.

## dent Infestation

✓ Rats and mice (usually three species, i.e., common rat, black rat and house mouse)
not only damage food commodities but they also spread diseases like HPS
(Hantavirus Pulmonary Syndrome). It is transmitted by infected rodents through
urine, droppings and saliva. This disease can spread either via inhaling dust that
is contaminated with rat urine/droppings, direct contact with rat faeces/urine, or
due to the bite of rat.

✓ There are enormous ways to control rodent infestation like use of traps,
electromagnetic (or ultrasound devices) and feeding with poisonous substances
like ANTU, arsenic, barium carbonate, phosphorous paste, fluoroacetamide,
strychnine, zinc phosphide. Fumigation is another way to deal with rodent
infestation. Be careful while selecting and using any of these materials as they
may also affect food commodities.

| Ways to deal with rodent infestation | |
| --- | --- |
| Trapping | Device that is used to catch rodents. Traps are placed in areas where rodents have been seen, but they are only effective against rats when these rodents are present in small numbers. |
| Fumigation | Introduction of toxin gas such as sulphur dioxide, carbon monoxide or hydrocyanic acid gas into rat burros. |
| Poisoning | Feeding of poisonous substances to rodents. It should be applied with precautions and care must be exercised in selecting the sites for leaving the poisons. |

## CONTROL OF INSECTS

term *insect* refers to a special group of arthropods. All insects share several, easily
gnizable external characteristics like three-body segments (head, thorax, and
omen); three pairs of jointed legs, all of which are attached to the thorax; antennae
ough there are a few species of insects with no antennae) and wings (though there
some flightless insects). A standard guideline to control various insects is given
w.

### s

It refers to a eusocial insect of the family Formicidae which further belong to the
order Hymenoptera. Ants can be black, red, brown or yellow in colour and have
a pinched waist and elbowed antennae and are either wingless or winged.

Good hygiene, sanitation and cleaning up spills is the best way for preventing ant
infestations. Use of baits and insecticides is another effective way to deal with ant
infestations.

# Cockroaches

✓ It refers to insect belonging to Kingdom Animalia, Phylum Arthropoda, Cla Insecta and order Blattodea, sometimes also called Blattaria. Mainly four variet of cockroaches are found: American, German, Oriental and Brown Band Cockroaches are nocturnal in habit and attack food, clothing, shoes, and de insects.

✓ Select and investigate/inspect areas where the possibility of cockroaches is mo like dark corners (such as under sinks and around drain pipes), behind calend and wall pictures, under electrical appliances such as ovens, blenders, refrigerat and so forth. Use of Baygon Spray and regular use of insecticides is the best w to control cockroaches.

# Lizards

✓ It refers to a reptile (small reptile) belonging to the family of Reptilia and inha all the continents except Antarctica. They are found in the warmer climates arou the world. In hotels, lizards are mainly found behind wall pictures, shelv window frames, wall mounted fittings and fixtures.

✓ As lizards thrive on insects, thus the control of insects is most effective way controlling lizards. Use of pesticides (such as spray of dichlorophos or malathi directly on the body of lizards will kill them instantly. Fumigation of alumini phosphide is another effective way to control lizards.

# Flies

✓ It refers to almost any small flying insect that belongs to Diptera family a characterized by the use of only one pair of wings (for flight) on the *mesotho* and a pair of halters, derived from the hind wings (to knob) on the *metathor* Regular use of flies spray is the best way to control flies.

✓ The spray is available in three forms, i.e., space, surface and combined. *Space sp* is applied as a mist into the air whereas *surface spray* is applied on the surf as a wet spray. In space sprays, the insecticidal ingredients (like Pyrethrum Allethrin) are used whereas surface spray forms a toxic layer (as it conta Malathion or Diazinon) of either fine crystals or film on evaporation of the carr Baygon can also be used.

# Bees

✓ It refers to flying insects closely related to wasps and ants, and are known for tl role in pollination and for producing honey and beeswax. Honeybee belongs t and it also includes many solitary as well as social kinds.

✓ Use of insecticides (such as bendiocard, carbaryl, diazinon, porpoxur malathion) and spray is the best way to control bees. The dust or spray formulat of these products is preferable when bee and wasp nests are in enclosed places. Remember, nest should be treated at night to avoid getting stung. The nesting space should be opened and the comb removed, else untended honey may run down which can attract other insects.

## Spiders

✓ It refers to air breathing anthropods (spiders belong to Araneae family) that have eight legs and chelicerae with fangs that inject venom. Spiders' webs are mostly found around doorframes, windows, corners, roof tops, storage areas, etc.

✓ Periodically sweep or vacuum under furniture, behind wall mounted pictures, around windows, doorframes, etc., to prevent spiders and their web formations. Apart from it, dust formulation with space spray and surface (or residual) spray are effective insecticides for spiders. These sprays should be based on bendiocarb, bromine, chloropyrifos, DDVP diazinon, malathion, propetamphos, propoxur, pyrethrum, resmethrin, and runnel.

## Silverfish

✓ It refers to fishmoth which is also known as Lepisma saccharins. It is a small, wingless insect that belongs to Thysanura family. Silverfish are a cosmopolitan species. They inhabit moist areas, requiring a relative humidity between 75% and 95%.

✓ Conventionally, insecticides such as DDT, chlordane, dieldrin and lindane were often used for controlling silverfish. But nowadays liquids (in visible areas), dust or baits (in areas like attic, basements, bathtubs, sinks and at less potentially hazardous areas) are preferred to be used instead of insecticides. Also inject small amount of liquid or dust into crevices and cracks formed by shelves, loose drawer glides, loose floor tiles and so forth.

## Activity

Fill in the below given blank spaces with appropriate description/answers.

| S. No. | Insects | Belongs to family | Their cause and effect | Chemical to prevent infestation |
|--------|---------|-------------------|------------------------|---------------------------------|
| 1 | Termites | | | |
| 2 | Bed bugs | | | |
| 3 | Mosquitoes | | | |

Insects are responsible for enormous amount of food spoilage and because of their breeding habits, they frequently transmit pathogenic bacteria. It is better to prevent insect infestation than have to control it. Prevention is assisted by correct building design, efficient maintenance, speedy removal of kitchen waste and adequate ventilation. The main method of control is by using chemicals known as *insecticides*. These insect poisons are classified into three groups: contact, systematic and respiratory poison.

| Groups of Insect Poisons | |
| --- | --- |
| Contact insecticides | These are absorbed via external surface or cuticle of insect. |
| Systematic insecticides | These enter the insect body by way of its alimentary canal. |
| Respiratory insecticides | These enter the insect body by way of its respiratory system. |

## 27.4   CONTROL OF WOOD ROT

The term *wood rot* refers to a wood-decay fungus. There is a variety of fungus that digests moist wood, causing it to rot. Some species of wood-decay fungi attack dead wood such as brown rot, and some, such as Armillaria, are parasitic and colonize living trees. Wood rot can be divided into two parts: dry and wet.

### Dry Rot

✓ It refers to the wood decay caused by fungi/fungus (Serpula Lacrymans) that digest parts of the wood which give the wood its strength and stiffness. Dry rot breaks down the wood fibres and renders the wood weak and brittle and reduces it to a dry crumbling state.

✓ Moisture (more than 20%) is the root cause of dry rot. Thus, *boric acid* or borate wood preservatives can be used to treat and prevent fungal growth in some situations. Borate is available as a dry powder (like TimBor) or a glycol based liquid (like Bora Cave) concentrate.

| Control of Dry Rot |
| --- |
| Step 1   Locate and stop the source of the moisture into the wooden material/substance. |
| Step 2   Replace any damaged wood that has become structurally weak. |
| Step 3   Treat new or existing wood with a borate wood preservative to prevent growth of the dry rot fungus or to kill any fungus already in the wood. |

### Wet Rot

✓ It refers to a brown fungal rot that affects timber with a high moisture content (between 40-50%). It is a common cause of timber damage and can often be seen in bathrooms and kitchens. It is similar to dry rot, but not as hard to repair as dry rot. Affected timber exhibits significant loss of weight.

✓ Wet rot can be sub-grouped into brown rots and white rots. Generally, *brown rots* cause cuboidal cracking and shrinkage of the timber whilst *white rots* tend to reduce the timber to a stringy and fibrous texture. The primary solution of wet rot is to control the environment (i.e., source of moisture) that allows the growth of wet rot.

---

**Control of Wet Rot**

Step 1    Locate and stop the source of moisture into the wooden material/substance.

Step 2    The affected timber piece should be cut away and a new piece of fungicide treated timber must be jointed to that remaining piece.

Step 3    If there is any possibility of the dampness persisting, the remaining timber should also be treated with appropriate wood preservative.

---

## 27.5   WASTE MANAGEMENT

The term *waste* refers to a material, substance or byproduct that is eliminated or discarded which is no longer useful or required after the completion of a process. On the other side *waste management* is concerned with the generation, prevention, characterization, monitoring, treatment, handling, reuse and residual disposal of waste materials. Remember, waste reduction/waste control is also a kind of profit maximizing technique, because it controls/prevents wastage of available resources (as resources are always limited), prevent unnecessary manpower cost required for waste handling, etc., which will lead towards more profit margin, simultaneously waste reduction also supports our eco-system. There are so many bases for classifying waste materials, but here waste is classified into two broad groups, i.e., solid waste and liquid waste.

## Solid Waste

✓ The term *solid waste* is concerned with hard ravage material which may be recyclable or non-recyclable. *Recyclable* solid waste indicates to biodegradable waste materials, for instance paper, wood, cloth, food waste, glass and so forth. Whereas *non-recyclable* solid waste material is related with non-biodegradable wastes, for instance, plastic materials.

✓ Government has laid so many rules, laws and guidelines after seeking the harmful effects of non-recyclable materials. Nowadays in hospitality industry, *eco-friendly hotels* are excellent examples of properties known for their recyclable amenities and supplies plus effective waste treatment.

✓ Solid waste can also be classified as dry and wet solid waste. In hotels both are collected separately in garbage storeroom and then disposed of properly.

## Liquid Waste

✓ The term *liquid waste* is concerned with water waste or water stage-based waste materials. Similar to solid waste, water waste can also be classified into two broad groups, i.e., grey water waste and white water waste. Here the term water waste is purely concerned with *water pollution*.

✓ The term *grey water* is concerned with sewage, drainage, sinks, washing machines and other similar kinds of water waste which can be recycled but demands huge cost in the form of installation of water re-harvesting plant. And nowadays due to lack of water availability, most hotel organizations have this facility, especially built to re-harvest collected rainwater.

✓ On the other side, the term *white water* is related with wastage of regular tap water such as water wasted during refilling of water bottles, drinking, leakage/pilferage and other similar kinds of waste. In hotels liquid waste is mainly produced from sinks, WCs, baths, lavatory basins and so forth.

### Activity

In below gives spaces, write down any 5 sources from housekeeping department which produce waste.

| S. No. | Solid waste | | Liquid waste | |
|:---:|:---:|:---:|:---:|:---:|
| | Recyclable | Non-recyclable | Grey water | White water |
| 1 | | | | |
| 2 | | | | |
| 3 | | | | |
| 4 | | | | |
| 5 | | | | |

## CONCLUSION

*Pest* is a basic cause of diseases, commodity wastage and often also found as a reason for food related diseases like food poisoning. Pests can be seen in the form of rodents, ants, lizards, cockroaches, silverfish, flies, spiders, etc. Their reason of growth is particularly poor hygiene and sanitation of the place. Routine cleaning and scheduled/periodic cleaning of hard/difficult to reach areas stop their infestation. Apart from it, scheduled use of insecticides, pesticides and germicides also helps to reduce/prevent the growth of pests and insects. Alike to pests, wood rot (dry as well as wet) is another serious problem because it damages the wood furniture and floor. Use of *boric acid* and fungicide treatment are best ways to deal with wood rot. Often, generated wastage also becomes a reason for pest and rodent growth. So, apposite treatment of solid and liquid waste can reduce the chance of pest and rodent infestation. Improper waste treatment may lead to water and air pollution.

| **Terms** (with Chapter Exercise) | |
|---|---|
| Baits | Method of pest control for certain kinds of species, especially ant species. For instance, plastic or metal bait stations to trap ants and Baygon bait for cockroaches. |
| Insecticides | ? |
| Pesticides | Chemical compound used to prevent infestation of pests. |
| White water | ? |
| Grey water | Relatively clean waste water (but not fit for human consumption) from baths, sinks, laundering and from the cleaning of kitchen appliances and dishwashing/pot washing. |
| Recyclable waste | Waste materials, whether in solid or liquid form, that is biodegradable in nature. |
| Non-recyclable waste | ? |
| Solid waste | Waste materials found in solid form, for instance, trash, paper waste, food waste and so forth. |
| Waste reduction | Activities/processes which start in lieu of reducing wastage (or leftovers) of resources/materials. Remember, it is also one of the best ways to increase yield. |
| Eco-friendly hotels | Green hotels (or Ecotel hotels) that use only eco-friendly products for every manoeuvre. For example, Uppal Orchid, Mumbai. |
| Water pollution | ? |
| Allethrin | Synthetic clear or amber-coloured viscous insecticide, $C_{19} H_{26} O_3$, similar to pyrethrin. |

| Diazinon | Amber liquid, $C_{12}$ $H_{21}$ $N_2$ $O_3$ PS, used as an insecticide. |
|----------|------------------------------------------------------------------------|
| Fumigant | Chemical compound used in its gaseous state as a pesticide or disinfectant. |
| Malathion | Trademark used for the organic compound. |
| Maintenance | Repair work in order to keep something in good condition/shape or to prevent problems of later stages. |

# 28
## CHAPTER

# Safety and Security

OBJECTIVES

*After reading this chapter, students will be able to…*

- describe safety, security and their differences;
- expound different threats & their viable causes and effects;
- understand first-aid treatment in case of minor accidents;
- explicit different types of hotel room keys and their uses; and
- clarify the importance of key control.

## 28.1  SAFETY AND SECURITY

The term *safety* refers to the protection against accidental events whereas the term *security* is related with protection against intentional damages. In simple words, safety often refers to predictable behaviour in a situation while security tends to be protection and conservation from premeditated damages. For instance, hotel should take care of guest's safety by offering lifeguards around swimming pool and security of assets (such as hotel, guest and their belongings) from external and internal customers. In every hotel, security department is predominantly responsible for safety and security of guests, staff and the hotel.

Apart from day-to-day activities like cleaning of guest rooms and in-house plus out-house public areas, housekeeping also plays a crucial role in the safety and security of guests, their belongings; staff & their belongings; and hotel belongings.

Therefore, it is not wrong to say that housekeeping and front office act as eyes and ears of the hotel because they see most of the hotel activities on everyday basis. To prevent possible problems, it is better to stop them before they start. Housekeeping works on many items of routine maintenance, or at least finds the problems and passes them on to other departments, especially safety and security related issues to front office, maintenance and security department. Bell desk staff notifies lobby manager about the *scanty baggage*, thereafter room attendant support in confirming about scanty baggage to lobby manager because room attendant is involved in the cleaning of guest room.

Safety and security has always been an elementary concern for hotels worldwide. The recent increase in terrorist acts have had its toll on travel and tourism worldwide. Whilst there is no indication that hotels are a primary target for the perpetration of terrorist acts, hoteliers must ensure that their properties are safe and secure. Therefore, they execute anything to give a sense of safety and security to guests and staff whilst at the same time protecting their investment. There are two types of security threats hotels should be concerned about:

Guest Related Threats    Threats that might affect a guest's health, comfort or wellbeing. These threats may either emerge from hotel side or among the guests or by outsiders like terrorists.

Hotel Related Threats    Threats that affect the hotel directly, in particular its fixtures and fittings, its revenue and its reputation. It may emerge from hotel employees as well as from guests.

Threats can also be broadly classified as internal threats and external threats. A brief detail of various internal and external threats are given below. Hotels are generally occupied 24 hours a day but they are still vulnerable to *walk-in* theft and also theft of valuable goods at night. The guest always comes to a hotel with an understanding that he and his belongings, both, will be safe and secure during his stay at the hotel. Hotels have a legal responsibility for employees, guests and other persons using the facilities who may be injured by their actions. They also have a responsibility for the safety & security of hotel assets and guest's belongings. Thus, security of the site/premises should be appraised as a whole, the term hotel site comprises *hotel exterior* and *hotel interior*.

---

### Activity

The scope of safety and security covers the entire hotel premises which can be broadly divided into two parts, i.e., hotel exterior and hotel interior. In the below given table, you need to describe these areas/sites and suggest appropriate safety tools for each respective area/site.

| Hotel Site | Description of Site | Safety Tools |
|---|---|---|
| Hotel Exterior | | |
| Hotel Interior | | |

## 28.2 INTERNAL THREATS

The term *internal threats* associate with all those intent harms that originate inside the hotel organization. The intention behind threat is to inflict pain, injury, damage or other hostile action either on guest, employee or property. The probability of its occurrence is rare and hotel can execute effective measures to control them because they are driven by internal environment. Here, security prevention for guest, their belongings (or assets) & rooms plus hotel and its assets is given below.

**Security of Guests & Guest Rooms:** People expect their hotel accommodation to be safe and secure; therefore hotels must adhere to a long list of rules and regulations. They must have safe and secure door locks, fire-prevention measures, exit strategies and plans, pool safety, first aid, security guards and security cameras. Apart from it, hotel must also look towards the following criteria also.

# Guest Room Door Locks

✓ Earlier, metal keys were used to prevent unauthorized access to the room. But nowadays, an electronic key has replaced this metal key system. Electronic key system is more safe and secure in comparison to metal keys.

✓ For example, if electronic key is lost or misplaced, front desk agent can reset the allotted key code, so the lost *card key* will not work in the particular door. Due to security concern, *key card* is always issued with metal keys. Guest has to present this key card every time he/she takes back room key from the reception counter.

✓ Some hotels are also equipped with facility that informs the front desk when a lost key was inserted in the room door lock. Most guest room door locks have *deadbolt mechanism* that allows the guest to provide extra security by turning a lever and extending a thick piece of metal into the door frame.

# Room's Connecting Doors

✓ Guest rooms (like interconnecting rooms) also include connecting doors and balcony or patio doors which are usually not electronically locked because it will create an extra capital and running cost for the hotel. So, extra care must be taken to ensure that they provide guest with good security.

✓ Balcony and patio doors locks may include not only the lock on the door but also an extra lock that is controlled from the inside of the room. When connecting doors are not needed, it is important to close and lock these doors (from both sides). Only authorized hotel employees should have the keys of these doors.

# Swimming Pool

✓ Pools and spas should have lifeguards; however, most do not. If lifeguard is not there then hang *do not dive* or *no lifeguard* signage board. Hotel may also place provisional barrier around the swimming pool, during the absence of lifeguard.

✓ Hotels with pools should enforce pool hours and hotel staff should be trained in CPR and first aid. Hotel guests should be the only ones with access to the pool and spa facilities. Entrance to the facilities should require activated room card keys.

Personal safety should be a key concern for hotels and there are many things to consider. *Card key* locks are essential for hotels. The card key is assigned to a room from check-in stage and automatically expires at a specific time, regardless of check-out. Smoke and carbon dioxide detectors, fire extinguishers, emergency telephones and emergency exits are all very important in ensuring the safety of guests. The hotel staff should be equipped with evacuation plans, first-aid kits and breathing assistance, such as a respirator, in case of an emergency.

**Security of Assets:** The term *assets* refers to the useful and valuable items/articles of a guest and hotel. The hotel management is responsible for the safety and security of all assets, regardless of guest, employee or hotel assets. Security measures/guidelines for the safety of guests' and employees' assets is given below.

## Security of Hotel Assets

✓ Hotel assets start from a small lighter or matchbox to big articles like TV, AC, computer & so forth. The control of all small as well as large articles is crucial. A hotel asset can be stolen by a hotel employee as well as by guest. In order to protect hotel assets from employees, a daily checking of every employee (except top level) should be done at the entry/exit point, i.e., time office or security office at the time of checking out.

✓ Most hotels do not allow their employees to bring more cash; they have a set limit for bringing cash. If any employee brings something valuable then it should be deposited at security office during arrival and must be recorded into the register and handed over to staff at the time of departure.

✓ Sometimes guests can also try to steal valuables from the hotel rooms like internal components of TV, AC, etc. In order to prevent these incidents the front desk should inform the housekeeping & maintenance department about the guests' departure. So, when front office completes the last minute formalities like collecting room keys, handing over bills, etc., meanwhile housekeeping & maintenance department can check the availability and proper functioning of room equipment and assets.

## Security of Hotel Cash

✓ Nowadays, there are enormous measures to protect the hotel cash like most of the hotels have more than one safe. For example, a hotel can have a main safe, where cash and other important records are secured. One may also use *cashier safe drop* it is a special safe that allows items to be dropped into it.

✓ The only way to remove the item is to open the safe whose control key may be with management or some authoritative person. These are used to temporarily store the cash received in the day's transactions from the front desk and other business entry point/*revenue centres* like coffee shop, restaurant, etc. During peak season collected cash is usually deposited on each day with head cashier, thereafter head cashier deposits in the bank.

✓ *Cashier bank* is also an important point to protect. It is usually controlled by bank audit, which may be done on a scheduled basis or without notice. The purpose of the bank audit is to make sure that the contents of the bank are correct and money is not missing. It prevents hotel employees from making loans to themselves. Robbery alarm system is also an effective device for security.

## Employee and Guest Assets

✓ It is the duty of the hotel to protect the assets of its employees while they work in their shifts. So, hotels provide a locker facility to their employees, consequently they can keep their valuables in these lockers. Employee assets include employee cloth, shoes, slippers, etc., and other valuables for example, watch, mobile, purse, credit card, money, etc.

✓ When an employee arrives at work, he/she receives his/her uniform and a lock for the locker. This locker is used to store their personal assets, simultaneously when they leave the hotel after their shift, the uniform must be put in the locker.

✓ The term *guest assets* refers to the goods or luggage of guests who are residing (or resident guests' assets) in the hotel rooms. So, it is the responsibility of the hotel to protect the guests' assets until they leave the property. Guests' valuables/assets are usually kept separately in either *the room safe* or *safe deposit box*.

| Activity | |
|---|---|
| Fill the below given blank spaces with appropriate description. | |

| Safe Options | Description |
|---|---|
| In-room safe box | |
| Safe deposit box | |

## 28.3 FIRE AND BOMB THREAT

**Fire Threat:** It is one of the major threats among others. It may occur due to short circuit, from kitchen or any other reason. In order to prevent it, many hotels do not lay wall-to-wall carpeting and some hotels are also bound by law to take the necessary fire safety precautions. It is the duty of front office manager to ensure that he is satisfied with a hotel's fire safety precautions. In order to deal with fire-like situations, hotel must always have fire safety policy, emergency doors, firefighting equipment like portable fire extinguishers, fire detection system/alarm (smoke detector) and automatic sprinklers. Simultaneously, hotel management should also *drill* (or train) their staff to deal with such critical situations. There are various types (classified on class type basis) of fire (along with fire extinguisher) which may occur in a hotel.

### Activity

In the table, match correct class of fire with cause of fire (or used to control particular material based fire) and correct fire extinguisher (or how to recognize correct fire extinguisher with the help of symbol).

| Class of Fire Extinguishers | Fire Extinguisher Used to Control (or Cause of Fire) | Recognition Symbol of Fire Extinguisher |
|---|---|---|
| Class A | Electricity based fire like fire caused by short circuit, fuse, etc. | *Red square* with....?....inside and a *black square* with a burning gas |
| Class B | Combustible material based fire like fire from paper, wood, cloth, plastic, etc. | *Yellow star* with ...?... inside |
| Class C | Kitchen based (or oil based) fires like fire from mustard oil, refined oil, etc | *Green triangle* with....?... inside + *black square* with burning wood and trash |
| Class D | Flammable liquid & grease based fires like fire caused by kerosene, diesel, oil, paint, etc. | Rare to find and still hotels/restaurants use Class ....?.... fire extinguisher |
| Class K | Flammable metal based fires, like fire caused by sodium, magnesium or titanium. | *Blue circle* with....?.... inside + *Square* with a burning plug and socket. |

**Bomb Threats:** After receiving information about bomb in your hotel, you should call security department/bomb squad. The important precautions and measures that may be taken during the incident of bomb threat include:

✓ Conducting security checking programme (via security nets and body search) and search especially for those guests who are not known to the hotel staff. But try not to inform (about bomb threat) all your hotel employees (except senior managers), as your staff may leak it publicly, subsequently creating panic.

✓ Security people should check all public areas especially banqueting suites and other non-public areas. These areas must be thoroughly checked and then locked. One of your hotel employees must accompany the bomb squad, as squad is not aware of all hotel areas.

✓ The goods and bags (by hotel) received during the preceding working hours should also be thoroughly checked and kept tidy. All cluttered particles should be removed as they may interrupt in security checking.

✓ After getting bomb alert, hotel should never solicit guests to evacuate the hotel building, as bomb may be at any place in the hotel else we may send guests into a danger zone.

✓ Room maids and housekeeping supervisors should be trained to conduct regular security checks in the guest rooms – as their work profile mainly moves around the guest/guest room.

✓ If bomb threat is received on telephone then telephonist/receptionist must record all the pertinent information like – what exactly was said, time the call was received, language & accent of the caller, background noise (if possible) and so forth.

✓ Hotel must have a proper procedure to ensure the nature of the threat recorded.

In order to deal effectively with any kind of threat/incident, hotels need to train their staff occasionally and must be well defined in advance, what they should do during a bomb threat or any other probable incident. At the same time, hotel should work in close coordination with the local police department and keep them updated on susceptible events of the hotel. Police department may also assist the hotel in giving effective staff training to deal with various probable occurrences.

## 28.4   EXTERNAL THREATS

The term *external threat* is concerned with all those harms that originate from outside the hotel organization. The intention behind every threat is to inflict pain, injury and damage to hotel as a whole. In broad spectrum, hotels do not have any control over such probabilities as these threats may be an act of god or driven by external forces/environment. Up to some extent, a hotel can hamper human generated external threats by strongly implementing internal threat control measures. Some crucial security measures for the prevention of external threats is given below.

> **Safety & Security Equipment**
>
> - Key card lock
> - Security guards
> - Lifeguards
> - Security cameras
> - Fire alarms
> - In-room safe
> - Safe deposit box
> - Burglar alarm
> - Escape plans
> - Defibrillation unit
> - Luggage scanner

### Terrorist Threat

✓ Terrorist threat may also be a bomb threat that has already been discussed in above lines. Unfortunately, it is a fact that during any assassination attempt (like terrorist attack) nobody would expect that front office staff will come and do anything rather than take cover. But fortunately, these are very unlikely occurrence incidents in hotel industry.

✓ In order to consecutively deal with terrorist threat/attack, hotel staff would be expected to keep their eyes open at all the time and report anything they found suspicious. For instance, a visitor who does not seem to *fit* in the hotel or to an unattended piece of luggage.

## Control Over Robbery

✓ Big robberies do not occur on daily basis but can happen like once in several months or years. Generally, a robbery alarm system is used to protect the hotel from these kind of incidents. Usually, it is a silent device that is placed out of site, such as near an employee's (especially cashier or receptionist) foot where it cannot be easily seen from the guest side of a counter.

✓ Device sends a signal to the local police department. This device may work with the hotel surveillance system, so that when it is activated, a camera records what is going on, at the alarmed location.

## Control of Unknown Persons

✓ Due to the nature and demand of hospitality industry, we all know that hotel holds themselves open for public accommodation. So, practically it is very difficult to stop any guest from entering into the hotel but at the same time, hotels are private properties, therefore, people can be excluded if they have no specific reason for being on the premises. Hotel can identify these guests by using either motion detector or video camera.

✓ Nowadays, most of the properties use surveillance system to control unknown guests. It uses video cameras, motion detectors and other methods to identify these unknown people. Cameras are usually connected to monitors in the security office, front desk, telephone department or other areas where there is someone to watch them. Generally, these systems operate from security department.

### Activity

Along with internal and external threats, there are some additional threats towards which hotel management should be very careful. Example of some additional threats is given in table below, you need to describe them and give probable solution to deal with them.

| Example of Additional Threats | Threat Description | Probable Solution to Deal with Them |
|---|---|---|
| Guest privacy | | |
| Opportunists | | |
| Conmen | | |
| Premeditators | | |

## 28.5 BASIC FIRST AID GUIDELINES

Every year many accidents occur in the world. With some care, they can be prevented. In case of an accident you can help the guest by providing primary treatment, i.e., first aid. You will be able to give vital first aid to protect your guest by following some important instructions like - do not move unnecessarily, keep him/her warm, make a diagnosis, decide the primary treatment and treat. At the same time ensure that his/her airways should be clear which means that he is breathing and his blood circulation is normal. There is also a facility of in-house doctor in every luxury hotel. Generally, in-house doctor works during a specific shift but also available on-call during emergency. In the meantime, hotel employee must follow the below mentioned specific guidelines for different types of situations in relation to first aid.

### F.A. Equipment

- Clinical thermometer
- Cotton wool
- Bandage
- Adhesive dressing
- Sterilized gauze
- Tweezers
- Bed pan
- Urine bottles
- Rubber sheeting
- Safety pins

### F.A. Medicines

- Tincture iodine
- Tincture benzoine
- Dettol
- Burnol
- Crocin tablets
- Aspirins
- Antiseptic creams
- Junction violate solution

### Respiration

✓ If it is falling with any guest in the hotel, you need to give him/her artificial respiration, mouth to mouth or mouth to nose.

### Heart Attack

✓ Immediately call the in-house doctor. Do not move the guest anywhere. If critical, try artificial respiration by mouth to mouth or mouth to nose.

### Shock

✓ Rest the guest by lying on the floor and loosen the clothes (again be careful with lady guests). Give warmth by hot beverage and hot water bottle and keep the surrounding area silent.

### Bleeding

✓ If any guest gets injured and suffers from the problem of bleeding then first of all arrest the bleeding and protect the wound. Apply direct or indirect pressure. Cover with a dressing; apply a pad and firm bandage. Elevate him/her and keep at rest.

## Fractures

✓ This kind of incident may occur when guest is using staircases, or fall in bathroom, room, kitchen (in guest room), corridor, etc. In this situation, it is impossible to carry him/her with a well padded stiff support reaching the joints on either side. Thus, apply bandages on either side of the joint.

## Burns/Scalds

✓ This kind of accident occurs in those rooms which are equipped with the facility of kitchenette like in resort cottages. A *burn* is caused by dry heat whereas *scalds* by moist heat like hot water, steam, or oil. In this case immediately cool the area with cold water for 15 minutes, till pain subsides. Do not break blisters, or apply anything on the burns. Cover with a sterile or clean cloth, pad and bandage.

## Nose Bleeding

✓ Make the guest sit up facing the breeze and the head slightly forward. Ask him to breathe through the mouth and to blow the nose. Apply a cold compress over the nose. The soft part of the nose may be pinched close with the finger for 10 minutes. Cold application on the back of the neck and forehead may help.

## Fainting

✓ Make the guest lie down and loosen the clothing around the chest and waist (but be careful with lady guests). Turn head to one side. The legs may be raised a little. Do not attempt to give any solids or liquids. On recovery a small quantity of a drink may be given and allowed to sit up and move after rest.

The above-mentioned accidents are often found in hotels, especially around guests as well as employees like respiration, bleeding, fainting, etc. But along with this, some unusual incident can also occur but these are very rare to occur in the hotel premises. Some of them are given below.

## Bee Sting

✓ In case of bee sting, a hotel employee should not press the bag (of the sting). Use forceps (tongs) and remove the sting then apply cold or weak ammonia.

## Animal Bites

✓ It hardly occurs in any hotel. In this situation, you have to wash the location of bite with soap and plenty of water, loose bandage may be applied and provide quick medical aid.

## nake Bites

✓ First, keep the guest calm. Wash with plenty of water and soap. Do not rub hard. Apply a constrictive bandage on the heart side of the bite. But do not apply continuously for more than 20 minutes. Do not incise or attempt to such a wound. Call in-house doctor quickly.

## 8.6   KEY AND KEY CONTROL

t is the procedure that authorizes certain personnel and registered guests to have n access to keys or rooms. It is the most important aspect of security in the hotel. Jowadays, the metal keys are hardly found in any of the hotels because of emergence f new technology and better system. These systems aid in making the hotel and guest ooms safer and secure than conventional metal key system. In hotels, guest room keys re issued by front desk agent during check-in stage, thereafter received back from eparting guests by either front desk or bellboy.

Apart from room keys, front desk must also ensure appropriate key control procedure or safe deposit box facility. Hotel employees, particularly housekeeping department nd top level management, also hold a separate set of guest room keys and keys of ther private and public areas, such as managers' offices and revenue outlets. This is pecifically found in conventional metal key system. Often, you have also heard the erms *key card, card key, key fob* and *key jacket* in reference to key control. Remember, all hese terms seem alike but in reality each term serves a different purpose, thus there is lifference among them. Under metal key system, there are usually five types of keys vailable with hotel staff for a guest room, these are as follows.

| | |
|---|---|
| Room keys (guest keys) | • It refers to keys of individual guest rooms controlled by front desk. It is available with front desk agent to offer to check-in guest during arrival stage. It is always issued thereafter controlled with *key card*. |
| Section key (room attendant's key) | • It refers to a key that can open a *limited number of room/s* (or known as section/floor section) on a particular floor. Thus, it is equivalent to several room keys and mainly used by room attendants. |
| Floor key (floor supervisor's key) | • It refers to a key that can open all guest rooms on a particular floor. Thus, it is equivalent to several sectional keys and mainly handled by floor supervisor. |
| Master key (HK executive's key) | • It refers to a key that can open all guest rooms of all floors (of individual wing/block) except double-lock doors. It is equivalent to several floor keys and mainly handled by assistant or executive housekeeper. |

| Grand master key (key of departmental heads) | • It refers to a key that can open all guest rooms of all hotel wings/block including double-lock doors, but except offices. It is mainly handled by managers or departmental heads. |

| Great grand master key (key with top officials of hotel) | • It refers to a key that can open all guest room doors of all wings including double lock and executives'/managers' offices. Thus, it is also known as *emergency master key* because it is mainly used during emergencies like flood, fire, etc. It is handled by high management officials. |

## Activity

The below given room keys and their required matching is of a resort property in which rooms are split section- & floor-wise and building split wing/block-wise. You require matching the descriptions given on the left side with their appropriate keys on the right side.

| S. No. | Description | Keys |
|--------|-------------|------|
| 1 | A key that can open a particular guest room only but its copy is available with hotel management. | Floor key |
| 2 | A key that can open several guest rooms (or group of rooms) on a particular floor. | Section key |
| 3 | A key that can open all rooms of all sections but of a particular floor only. | Grand master key |
| 4 | A key that can open all guest rooms of all sections and of all floors but of a particular wing/block or building only (including double locked but except offices). | Master key |
| 5 | A key that can open all guest rooms of all sections and of all floors but of a particular building only (including double locked as well as offices). | Great grand master key |
| 6 | A key that can open all guest rooms of all sections and of all floors of all wings (including double locked as well as offices). | Card key |
| 7 | It can open guest room section of the building, including all store closets in that building. Thus, also known as in-house building key. | Key card |

| 8 | It is issued at the time of room allotment (along with room key) and guest requires to show it every time he comes back to collect his room key. | Supply key |
|---|---|---|
| 9 | It is also known as electronic key punched with fresh code after every guest departure. | Guest/room key |

## Electronic Key Card

Electronic key card is a plastic card, similar to any charge card, containing data on an embedded magnetized strip that can electronically unlock a door. This *card key* (or often termed *key card*) is powered with new code every time a room is sold (during every new check-in). It reduces the cost of loss and repairs as well as more safe & secure than metal keys. During lost & found, front desk agent can modify room access code on door lock console, due to which lost/stolen card key cannot open the room door. And issue a new card key to resident guest after punching a new code. In certain electronic door locking systems, when anybody tries to get access to the room by swiping/inserting the lost/stolen card key, the door lock will place an alarm and intimate to front desk agent that an unauthorized person is trying to get access to the guest room.

| Activity | | | |
|---|---|---|---|
| Differentiate among the following- | | | |
| **Key card** | **Card key** | **Key fob** | **Key jacket** |
| | | | |
| | | | |
| | | | |

## Key Security

Individual hotel locks (metal keys based) can usually be changed a maximum of four times before the grand master key has to be changed, it is ı prohibitively an expensive job. In practice, the individual locks are infrequently changed even when copies of the keys are lost. Although, the regulations of most large hotels clearly provide for strict control of keys, the strictness varies widely in different hotels according to enforcement measures taken by the Manager and the Chief Security Officer. Loss of keys can lead to the following problems:

✓ If it is found by an outsider/another guest then the guest valuables can be stolen and similar problems can occur. The hotel needs to replace the lock which could be a considerable expensive if not minimized in the long run.

✓ Wrong issue of keys can lead to problems of guest entering wrong rooms.

✓ Hotel staff may also take benefit of improper key control by stealing guest valuables and other expensive hotel assets like electronic gadgets & appliances.

✓ During the incident of theft or any other similar activities, everyone will play blame game, just because of improper key control. Thus, staff key control is an important aspect of safety and only the executive housekeeper and the front office manager passes a master key.

## CONCLUSION

Apart from accommodation units (plus hospitality, luxury amenities, high standard of services and so forth), safety and security of hotel guests and their belongings; hotel employees and hotel assets are also of primary concern for every hotel organization. A breach of safety and security may create a drastic impact on the entire organization. For instance, a terrorist attack on 26 November 2008 on Taj Hotel in Mumbai is an excellent example of violation of safety and security. Lack of security may also create a golden chance for skippers, conmen, premeditators and so forth fraud players. Even erroneous control and management of hotel room keys, whether metal keys or electronic keys, may lead hotel towards critical situations. But remember, electronic keys are safe and secure than metal keys because in case if guest looses his electronic key card, the guest room door lock can be instantly and easily modified by the front desk agent. Simultaneously, take care that first aid is also an essential component of hotel's safety and security.

| Terms (with Chapter Exercise) | |
|---|---|
| Hotel interior | It refers to all those hotel areas (public as well as private areas) which are situated inside the main hotel building like restaurants, lobby, coffee shop, guest rooms, etc. |
| Hotel exterior | It refers to all those hotel areas which are situated outside the main hotel building like parking zone, outdoor swimming pool, garden, golf course, portico, cabana room and so forth. |
| Hotel premises | It is a collective term for both hotel exterior and interior areas. |
| Drill | It refers to training sessions that organize to make the people aware of what to do during emergencies. |
| Key fob | It refers to a plastic/metal tag attached with metal keys. This tag contains room number, hotel's name and often address too. |
| Lifeguard | ? |

| Key card | A card issued to guest at the time of key handover (during check-in). It contains room number, name of guest, date of arrival and expected date of departure and signature of front desk agent who issued. |
|---|---|
| Scanty baggage | Actually, the term is used for light luggage but in hotel industry it refers to a guest who carries light luggage. Such guests may become skippers; but not always. |
| Fire alarm | An alarm system that buzzes during fire. Nowadays, the system is interfaced with water sprinklers. |
| First aid | Primary medical (or treatment) kit which contains basic medicines, cotton, scissors and so forth. |
| Double locked | A door lock which can be locked by revolving a small button (from inside the room) without using key. |
| Escape plan | An emergency exit plan which is a legal requirement for the hotel. It helps during emergency situations like fire, flood and so forth. |
| Burglar alarm | A theft alarm system placed to prevent/deal with robbery like situations. |
| Defibrillation unit | A treatment by stopping fibrillation of heart muscles (usually by electric shock delivered by a defibrillator). |
| Conmen | Confidence trickster who is in an attempt to defraud a hotel by gaining its confidence. |
| Opportunists | It is similar to premeditators; they show that they will pay but their intention is always of not settling the account. |
| Premeditators | A person characterized by deliberate purpose and some degree of planning for a *premeditated crime.* |
| Deadbolt mechanism | A locking mechanism distinct from a spring-bolt lock because a deadbolt cannot be moved to the open position except by rotating the lock cylinder. |
| Cashier bank | *Cash bank*, initial cash bank, cash float or often simply known as bank. It means cash put into the cash box at the beginning of the day or week to allow change to be given to guests. |
| Cashier safe drop | A temporary safe drop used to keep/drop daily collected cash from/by front desk cashier and other point of sale outlets. Its control key is always kept with the management. |

| | |
|---|---|
| Threat | Something that could possibly hurt a person or property. |
| ? | It is a card on which guest's name and room number is filled in by a front desk agent. It is issued after the successful completion of guest registration, along with room key. Guest needs to show it every time he/she comes to front desk/information desk to recollect room key. |
| Card key | Computer coded electronic key card which nowadays is found in almost all luxury hotels. |
| In-room safe | ? |
| Safe deposit box | Safety provision for valuable articles whereby guests can keep their valuables (in safe deposit boxes) at front office under the direct supervision of cashier or any other front office staff. |

# 29 CHAPTER

## OBJECTIVES

*After reading this chapter, students will be able to...*

- describe sale and understand how it is different from marketing;
- elucidate seven Ps of marketing mix;
- find out the most effective tools of hotel marketing;
- identify various room selling approaches and their functioning; and
- explain how up-selling is different from suggestive selling.

## 29.1 SELLING/ROOM SELLING

The term *selling* refers to the process of delivering or exchanging something against money. For instance, in *room selling* front desk sells guest rooms to walk-in guests during arrival phase whereas reservation agent sells guest rooms during pre-arrival phase. In both cases (and in majority of situations), money (in the form of advance deposit or paid-in-advance) is a medium of exchange between hotel and prospective guests. *Room selling technique* associates with all those methods which sales agents acclimatize to generate revenue for the organization.

In hospitality industry, there are predominantly three techniques of room sale, i.e., up-selling, down-selling and suggestive selling. Always remember that selling is different from marketing. Under *selling phenomenon* exchange takes place between buyer and seller, whereas *marketing phenomenon* is related with intimating to buyers about the available products and services. Marketing is first phase while selling is second phase of product/ service development strategy. Therefore, marketing should always be well-built so it can make selling task successful in later phase. For successful marketing, Neil Borden suggests a *marketing mix* tool. Initially, this marketing mix tool had four Ps which was later on added with three extras Ps. Product oriented industries have four Ps whereas service oriented industries have to follow seven Ps.

| | Activity |
|---|---|

Fill in the below given blank spaces with appropriate Ps of marketing mix (comprising of product mix and services mix).

**Product mix**     - Product     - Price     - Place        - Promotion
**Services mix**     - Process     - People     - Physical evidence

| Four Ps (of product) | Core competencies | Three Ps (of service) | Core competencies |
|---|---|---|---|
| ? | Design, technology, value, quality, branding, etc. | ? | Employee, management, culture, guest service, etc. |
| ? | Skimming, penetration, psychological, cost plus, loss leader, etc. | ? | Especially relevant to service sector, how services are consumed. |
| ? | Direct sale (reservation, CRO, front desk), indirect sale (involvement of supply chain) | ? | Decoration, luxury, comfort, facilities, amenities, etc. |
| ? | Special offers, advertising, free gifts, joint venture, etc. | | |

## 29.2 SEVEN Ps

The term *seven Ps* refers to *four Ps* of product oriented industries plus remaining *three Ps* of services oriented industries. The concept of four Ps is predominantly designed for those industries involved in manufacturing of tangible things whereas the concept of remaining three Ps is particularly designed for service delivering industries. Collectively, the seven Ps model is known as *marketing mix*. The term marketing mix was pioneered by Neil Borden which was later on refined by E. Jerome McCarthy. A brief description of each of these seven Ps is given below.

## Product

✓ In terms of product, hotel offers two things, i.e., accommodation and food & beverage. Accommodation is concerned with hotel room that is a perishable product and available in different grades/configurations (such as promoted, standard and suite). A hotel room has a limited shelf-life of 24 hours known as *revenue day* in hotel jargon.

✓ Hotel's *room inventory* generates maximum revenue whereas food & beverage generates additional revenue. Therefore, before accepting transient or group booking, a hotel needs to consider ancillary revenue generation through food and beverage from prospective booking.

✓ Apart from room division and food division, hotels also have additional *revenue centres* which offer value added services like spa, gift shop, gym, game zone, etc.

# Price

✓ The basic characteristic of hotel's pricing policy is that, the same room (or product) can sell at different rates to different guests at different occasions. Therefore, hotel management must design a well-planned *room-rate-matrix* in accordance with *room-sales-matrix*, *guest-segment-matrix* and *time-horizon-matrix*.

✓ Ideally, the room pricing policy is determined on the basis of room grade/room type and each room grade is offered at different prices (like rack rates, discount rate, family rate, etc.).

✓ The room grading is the basic reason behind developing different types of room rates, as each grade offers different set-of-facilities, services and luxuries. The key features of any hotel's basic product/service are non-transferable, invariable, inseparable and perishable.

# Place

✓ Service is inseparable from the provider/supplier. Inseparability refers to simultaneity which is considered an important property that basically states that the services are produced and consumed at the same time and cannot be isolated from their providers.

✓ In hotel industry, the guests have to come to the hotel to consume the offered services (like accommodation, food, beverages, nightclub, spa, etc.). Therefore, hotel should use appropriate distribution channel to reach the targeted market segment.

✓ Distribution channel is a set of independent firms involved in the process of making a product/service available to the end user. This distribution channel may be direct (like reservation department and hotel's sale division office) as well as indirect (like travel agents, tour operators, internet distribution system and hotel booking agencies).

# Promotion

✓ Promotion is the way; a hotel communicates what it does and what it can offer to guests. It includes in-house as well as out-house promotional activities. In-house promotion tools include brochure, tariff card, prospecting calls, courtesy calls and promotional campaign, whereas out-house promotion includes advertising, public relation, corporate identity, cold calls, promotional calls, sales management, special offers (like discounts), contest, gift vouchers and so forth.

✓ Promotion must gain guest attention, be appealing, give a consistent message and above all give the guest a reason to choose your hotel rather than competitors' hotel. Promotion should communicate the benefits that a guest obtains from the product/service, not just the features of those products/services.

## People

✓ Hospitality industry is more service oriented than product oriented industry and the basic characteristic of this service is that it cannot be separated from the employee who provides it. Thus, skilled, experienced and qualified staff is a basic necessity of any hotel to deliver good service.

✓ All hotels rely on their employees to generate revenue and customer flow. The range of employees starts from front-line-staff to back-line-staff and from top level management to bottom level employees. Guests judge and deliver their feedback more on the basis of employee's behaviour and attitude, especially with whom they interact.

✓ Thus, hotels must select right employees because they are as much a part of their business offering as the products/services they are offering. Subsequently, all staff members (especially front-line-staff) should have appropriate interpersonal skills, aptitude and service knowledge.

## Process

✓ The term process is related with systematic service delivery channel. In simple terms, it is the course-of-action of giving service and directly relates with the behaviour of those who deliver it. Problems such as waiting times, the information given to guest and the helpfulness of hotel staff are all vital to keep guests happy

✓ Therefore, to ensure the same service standard is repeatedly delivered to the guest many chain/large size hotels offer *blueprint* of services to their employees. This blueprint provides information about the hotel's service delivery process and welcome message.

✓ This kind of guest treatment will foster guest loyalty and confidence in the hotel Many hotels have also started their *loyalty programme* to keep their guest loyal and permanent such as *Golden Circle Programme* by Shangri-La Hotels and Resorts.

## Physical Evidence

✓ A service cannot be experienced before it is delivered but guest can make perception (up to certain level) by seeing the hotel room and surroundings Physical evidence is the physical element of the *service mix* which allows the gues to make judgements on the hotel organization.

✓ Guests make their perceptions on the basis of surrounding areas, fixtures, fittings lights and other wall mounted pieces. Facilities & services like clean, tidy, well equipped, well-manned and well-decorated front desk, waiting room, travel desk, GRE, etc., enhancing the guest's perception level.

✓ Thus, every hotel should join intangible services with tangible things to boost up guests' *moment of truth*, experience cycle and satisfaction level. Remember physical evidence always adds value to offer core products and services of every organization.

---

### Activity

List four P's that you think could influence the sale of product oriented as well as service oriented industry. Now think of the remaining three P's of service oriented industry and justify how these three P's really affect the sale of hotel rooms.

---

## 29.3 ROOM SELLING STRATEGIES

There are numerous selling strategies by which hotel can encourage its room sales, especially through repeat business and direct room sale via reservation, reception and other similar channels. When it comes to room sale then there are three basic room selling strategies, top-down, i.e., strategy, bottom-up strategy and mid-range strategy. A brief description of each strategy is given below.

**Direct Room Selling Channels**

- Internet
- Sales letter
- Telephone
- E-media
- Print media
- Hoarding
- Bill boards
- Emblem

### Top-down Strategy (or down-selling)

✓ This is the most commonly used sales strategy in which reservation/front desk agent starts by quoting a rate from promoted or suite room configuration and moves down to a standard room configuration, until guest does not accept. This strategy is used in situations where the hotel wants to run *drive rate*.

✓ This strategy is very much similar to airline's yield management concept which believes that flying with vacant seats is always better than flying with passengers paying at least something towards operating cost. Because both airline seats and hotel rooms are perishable products and the loss of revenue will never be recovered in future. Therefore, selling strategies are also very much identical in airline and hospitality sector.

✓ Generally, this strategy is not successful in a highly competitive market with low guest room demand. It is successful, though, when buyer's confidence is high. Hotels with a reputation for high quality and service are usually successful in invoking the buyer's confidence.

## Bottom-up Strategy (or up-selling)

✓ The reservation/front desk agent starts by quoting a rate corresponding to the standard room configuration or lowest room rate. The agent then lets the caller know that better room configuration is also available. Agents inform the guest of the incremental rate increase corresponding to the next level of guest room.

> **Indirect Room Selling Channels**
>
> - Tour operator
> - Travel agents
> - Airline crew and commuter
> - Incentive program
> - Tourist Information Centre
> - Independent Booking Agencies

✓ This strategy gives the guest a choice of different room rate options, hence, it could also be called "room quoting" strategy. As each successive rate is quoted, the agents have another opportunity to influence the guest that this is the room type or configuration best suited for him/her.

✓ This strategy is also successful when buyer's confidence is low. Because guest always shop for lowest room rate and may not give any chance to agent to up-sell to the next level of rooms. Hotels or agents who use this strategy must be very careful that they should never convey any room configuration as inferior because it can lead towards the aggressive attempts to up-sell.

## Mid-range Strategy (or Suggestive Selling or Substitute Selling)

✓ It refers to a room selling strategy which starts from middle level and moves towards either up or down a tier as per the acceptance or opposition of the guest like– starts from promoted room configuration and moves towards either standard or suite room configuration.

> **Special Room Selling Techniques/Channels**
>
> - Contest & coupon
> - Gift and free samples
> - Push money
> - Discount/Rate cutting
> - Package plan
> - Promotional literature
> - Contract rooms
> - Overbooking

✓ Hotels or agents using this strategy must have the flexibility to mould their approach to the guest and the progress of the call. Experienced agents are best suited for this strategy, because it requires experience and the ability to implement either the top-down or bottom-up strategy as needed.

✓ This strategy is also often called *substitute selling strategy*, as it includes both approaches, i.e., up-selling and down-selling.

Room sale agents may have varying success with each strategy as it mostly depends on an individual's attitude, dealing & experience. Hence, it is best to allow the autonomy to find the selling method best suited for each agent. In certain situations, the above-stated strategies cannot be implemented like when the hotel's yield management strategy dictated a specific course-of-action (e.g., *drive rate* or increase occupancy). Hence, in order for successful implementation of any strategy, it is most important that the front office manager should make every effort to inform the reservation agents and other sales channels when changes in strategy will occur. If new *rate triggers* are to be imposed, agents should be informed to avoid surprises.

---

### Activity

There are many hotels in the market that offer different configured rooms. Some offer explicit room configuration for some selected clientele like presidential suites, press room, etc. Browse the web sites of some distinctive hotels and see how many different types of rooms they are offered.

---

## 29.4   HOTEL'S MOST EFFECTIVE SELLING TOOLS

### Room Tariff Card (or Rate Card or Tariff Plan Card)

✓ Room tariff card is the structure of charges made for the various hotel rooms in relation to different meal plans and services & facilities provided by the hotel. It also shows the applicable taxes, service charge and extra charges for cribs or for acquiring extra bed or accommodating extra person in the same room. It is one of the most effective selling tools for selling rooms.

✓ The term *hotel tariff* is most commonly seen in British English, while speakers of American English may use *rate sheet* or *hotel rates* to describe a similar document.

✓ The basic purpose behind offering this rate card is to intimate prospective guests about available rooms, meal plans, extras and their respective prices. So, during check-out there should not arise any problem between the guest and hotel while settling or zero balancing the guest's account.

### Hotel Brochure (an Information Catalogue/Booklet)

✓ A brochure is a promotional printed material which is designed to communicate existing and potential visitors. Hotel brochures are designed to inform the prospective guests about the various available hotel services & facilities in an attractive manner.

✓ These brochures are available in two forms, i.e., hard copies and soft copies. Hard copies are mostly found at front desk and in hotel rooms in room compendium whereas soft copies are sent to guest via electronic mail during reservation request.

✓ Brochures are generally designed with one of the two basic functions in mind, i.e., to provide practical information which visitors may use during their stay and/or to establish an image of the hotel as a viable alternative when planning a future destination.

✓ Brochures are usually colourful with photographs and description of the hotel rooms, standard of services & facilities which hotel offers.

### Activity

There are many selling techniques that can influence the sale of hotel rooms. Consider several service oriented industries like telecom, marketing, airlines, tourism sector, etc., as examples. Now can you think of some factors that have influenced the sale of their services/products. Look at its market segment and customer/sales mix ratio. How have these two things had an effect on your hotel room sales?

## CONCLUSION

The chapter is designed around different marketing concepts and room selling techniques by which targeted market segment can be attracted to sell rooms and earn predetermined revenue objectives. There are enormous selling techniques which must be appropriately applied, for instance, with whom front desk agent/reservation agent has to apply top-down and bottom-up room selling approaches. But before this, it is important for every marketing related employee to understand the *marketing mix* and *service mix* of the industry. The term *marketing mix* indicates to 4 Ps whereas *service mix* refers to remaining 3 Ps of service marketing. These seven Ps are must to comprehend in order to achieve *perfect room sale*. Apart from room selling strategies, room tariff card, brochures, pamphlets, banners, hoardings and other similar tools also play an important role in selling rooms. Nowadays, most hotels also allow their front office employees, sales and marketing employees and other front-line-staff to stay in hotel rooms (especially in off-season) so that they can better explain the room and its features.

| Terms (with Chapter Exercise) | |
|---|---|
| Substitute selling | Selling approach (or en-place approach) whereby front desk/reservation agent assists the guest in choosing the product/service. |
| Buyer confidence | Guests who do not feel a specific need to stay at any hotel may have low buyer confidence and vice versa. |
| Suggestive selling | Selling approach whereby front desk/reservation agent assists the guest in selecting product/service by offering suggestions. |
| Rate cutting | Sales technique regarding how and up to which level, a discount could be given. |

| | |
|---|---|
| Blueprint | ? |
| Rate trigger | Signal programmed into the reservation computer system that instructs it to change the rate based on preset criteria. |
| Up-selling | Selling approach whereby front desk/reservation agent tries to sell expensive product/service to guest by offering suggestions. It is different from suggestive or substitute selling. |
| Loyalty programme | Corporate sponsored membership club (or programme) for frequent guests and is similar to airlines' frequent flyer loyalty programmes. |
| ? | A loyalty programme offered by all Shangri-La Hotels & Resorts to their highly valued guests. |
| Perfect room sale | Situation whereby every hotel room is occupied and no guest has walked out. |
| ? | Selling tool that provides almost all major information/detail about room and supportive services. |
| Marketing mix | Business tool used in marketing. This term was coined by Neil Borden. |
| ? | It means firstly hotel tries to agree with the guest for promoted or suite room configuration but if guest is not ready for that room then bring him down towards the standard room configuration. |
| Prospecting calls | Prospecting is a process of making enquiries (via making telephone calls) in order to generate future business for the organization. It is a vital factor in developing new business for the organization. |
| Cold calls | Process whereby a salesperson makes call to gather information in order to determine if there is any possibility of business. |
| Promotional calls | ? |
| Crib | Baby and the bed provided for him/her known as baby cot plus applied extra room rate/charge is known as crib rate. |
| Service mix | Three Ps of marketing (apart from four basic Ps), i.e., people, process and physical evidence. |
| ? | Selling technique in which agent demotes/downgrades the room selling approach when guest is not ready for an expensive room. |

| Marketing mix | Four Ps of marketing, i.e., product, price, place and promotion. |
|---|---|
| ? | Room selling technique to achieve full occupancy. Simply, a hotel may sell more rooms than the available number of rooms in *inventory*. |
| Rate shop | Guest's room purchasing approach whereby guests are simply searching for the lowest room rate and may not give any chance to agent to up-sell to the next level of rooms. |

# 30

**CHAPTER**

# Night Auditing

⟨ **OBJECTIVES** ⟩

*After reading this chapter, students will be able to...*

- describe night audit and who is responsible for it;
- explain the importance of night auditing;
- describe how night audit functions in the hotel;
- understand the job profile of night auditor; and
- explain the various modes of operating night audit.

## 30.1 NIGHT AUDIT

*Night audit* means cross-verification of generated and posted charges into respective guest accounts, non-guest accounts and house accounts (or *management accounts*). For cross-verification, first of all night auditor accumulates all supportive vouchers and keeps them in chronological order, this systematic process is known as *audit trail*. Next balances out all guest accounts, non-guest accounts, house accounts and departmental accounts then generates related report for each account type. Meanwhile check accounts with *high balances* (reached to assigned *house limit*), verify room discrepancy and perform reconciliation of accounts. Lastly, perform *date roll*. Straightforwardly, night auditor checks the accuracy of all posted monetary transaction of a *revenue day*.

The night auditor must ensure that the sum of revenues due to accounts received from the various departments (i.e., food & beverage outlets, rooms division, gift shop, etc.) found on the *department control sheets* equals the sum of the charges made to the respective guest folios. The process of night audit is called *night auditing*.

Generally, night auditing starts in *graveyard shift* when most of the revenue outlets of hotel are closed. Thus, *night audit watch* starts from 11 p.m. and ends at 7 a.m. the next morning (this time gap or work shift is called *audit work time*). Remember, night auditor mainly focuses on three areas, i.e., posting of transactions, error correction and making reports. Apart from night auditing, night auditor also works as a night manager and assists front desk employee in dealing with night check-ins and check-outs. The night auditor performs many functions, but broadly includes:

- Verifying daily hotel transactions and determining *end of the day* (also known as *date roll*)

- Generating summary and statistical data for compiling into the night audit report.
- Assuming responsibility of front office staff during the night shift.
- Processing check-in and check-out during the night shift (optional).

| Activity | |
|---|---|
| Briefly explain the below given particulars. | |

| Particulars | Brief description |
|---|---|
| Posting of transactions | ? |
| Error corrections | ? |
| Report writing | ? |

## 30.2   FUNCTIONS OF NIGHT AUDITOR

The heading is related with the work-profile of night auditor. Remember, the work of night auditing must be in a sequential order. Nowadays, due to the development of property management system/software, the frenzied and time consuming work of night auditing is simple and easy to perform. Some major works of night auditor are shown below.

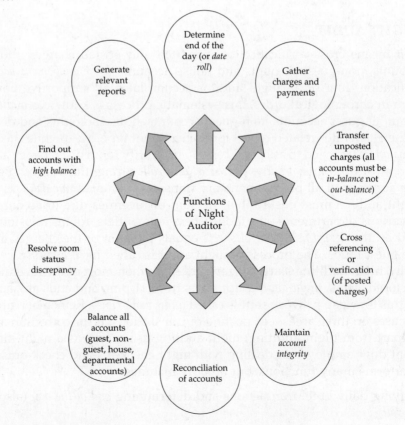

## 30.3 STEP-BY-STEP PROCESS FOR NIGHT AUDITING

Step 1  Initially, night auditor will assemble all charges and payments made by resident as well as non-resident guests. Guests can pay either by cash as well as use their plastic money/charge card, traveller's cheque, vouchers and other negotiable instruments.

Step 2  Now, night auditor will post all outstanding entries and un-posted charges of resident guests as well as non-resident guests into their respective accounts.

Step 3  Thereafter, he will verify all transactions, assigned/allotted room rates and offered discounts/special rates to various resident and non-resident guests and also verify the house and departmental accounts.

Step 4  Now he will total and balance all verified accounts (i.e., guest, non-guest, house and departmental accounts) and *ledger (front office ledger* and *city ledger)* and prepare *trial balance.* He uses *audit posting formula* to calculate *net outstanding balance.*

Step 5  Thereafter, he will crosscheck/cross-verify the room status report prepared by front desk clerk with housekeeper's room status report in order to find out room discrepancy. At the end perform *date roll* or *open new date.*

Step 6  Now, he will prepare all required reports (keep backup of all) like *early bird report, night report,* accounts with *high balance report* and *room discrepancy report,* if discrepancy persists and send all reports to management and departmental heads for review and future planning.

| Night audit process | | |
|---|---|---|

| Steps | Manual System | Semi-Manual System | Automatic System |
|---|---|---|---|

$\downarrow$

| Assemble Charges & Payments | | |
|---|---|---|

| Step 1 | Collect handmade supportive documents like vouchers. | Collect pre-printed filled supportive documents | Respective accounts will retrieve all transactions from point of source terminal. |
|---|---|---|---|

$\downarrow$

| Posting of Outstanding Entries and Un-posted Transactions | | |
|---|---|---|
| Step 2 | All entries are manually posted into respective accounts by referring to supportive documents like cheques and vouchers. | All entries are posted into respective accounts with the help of electronic/accounting posting machine. | All entries are automatically posted (or retrieved by) into all respective accounts as and when entered at point of source terminal. |

↓

| Verification of Posted Transactions and Room Rates and Applied Taxes | | |
|---|---|---|
| Step 3 | All posted transactions are verified and cross-checked for account integrity with the help of *daily and supplemental transcript* and prepared *recapitulation sheet*. | First of all, verify *pick-up errors*. Thereafter, voucher needs to imprint for those charges that are transferred into respective accounts. | Night auditor needs to verify posted day room charges & their taxes. Remaining room tariff & taxes can be easily posted (after verification) by clicking single option (in PMS). |

↓

| Total & Balance all Accounts and Prepare Trial Balance | | |
|---|---|---|
| Step 4 | Now night auditor will balance all verified accounts; total them and transfer all balance figures into trial balance to check arithmetic accuracy of all accounts. | Night auditor totals and balances all accounts (with the help of calculator) and produces *front office cash report* and *D-card*. | In fully automatic system, all account balances, thereafter their totals are automatically performed by the property management system. |

↓

| Verification or Reconciliation of Room Discrepancies | | |
|---|---|---|
| Step 5 | Night auditor also cross checks/refers the room status shown by housekeeping and front office department in order to find out any discrepancy, if occurring. | Night auditor cross-checks/refers filled room status pre-printed format of housekeeping department with front office department. | Night auditor cross-checks/refers on-line room status shown by housekeeping department with the front office. PMS may also directly detect such room discrepancies. |

| Prepare Report and Keep Backup Ready | | |
|---|---|---|
| Step 6 | Prepare summary reports with multiple carbon copies for backup. | Fill pre-printed formats of all relevant reports and make photocopies of all for backup. | Prepare reports or computer system will automatically generate all relevant reports and also save a back-up copy. |

## CONCLUSION

Night auditing is concerned with following tasks: posting of un-posted charges, cross-verification of posted transactions, balancing of all accounts, establishing account integrity and preparing high balance reports, room discrepancy report and other relevant reports like early bird report, flash reports, cashier's report for management review. The task of night auditing is performed during *graveyard shift* (evening/night shift) which mainly begins at 11:00 p.m. because most of the point of sale outlets are closed by 11 p.m. Night auditor is responsible for night auditing and mainly hired from an outside agency. He works for both, accounts division as well as for room division. Many chain hotels randomly also send their own auditors from sister hotels. At the end of the auditing, night auditor also changes the date (it is known as *date roll* or end of the day) else all transactions will be posted on the same day. There are different modes of performing the task of auditing/night auditing, i.e., manual, semi-manual or fully automatic and each of these modes have several advantages as well as disadvantages.

| Terms (with Chapter Exercise) | |
|---|---|
| High balance report | Report that intimates about those guests who have reached their pre-determined *house limit* (or maximum credit limit/charge purchase limit). It is prepared by night auditor. |
| ? | Also known as *audit log*. It is a chronological order of audit records, each containing evidence, directly pertaining to charges or resulting from the execution of a hotel process or system function. It serves as documented history of transactions for a given revenue day. |
| Trial balance | An account statement prepared on the basis of ledger balances and mainly used to check the arithmetical accuracy of all prepared ledger accounts. |

| Pick-up error | Incorrect picked-up opening balance (or previous closing balance) due to which debit and credit side of respective account is not tallied. It usually occurs in semi-automatic front office where electronic posting machine were used. |
|---|---|
| D-card | It is a card that shows the opening balance of the system which provide a running record of all charges and credit postings through the machine then shows a final balance. It is found in semi-automatic front office. |
| Cash report | ? |
| Departmental total | Aggregate (or sum of collection), including cash and other negotiable modes of money, by different hotel departments. |
| Account integrity | It refers to honesty or correctness of the account. Account balances (of guest or non-guest) should be *in balance* with supportive documents; i should never *out balance (out-of-balance)*. |
| Flash report | Snapshot report that discloses the whole day hotel's financial performance at a glance. |
| Early bird report | ? |
| Sister hotels | Hotel of same group and often associates with chain hotels only. |
| PMS | Property management system/software used to manage entire organization on one software. |
| Date roll | It is also known as *end of the day* that designates a certain point of time in the night to establish a change in date. After date roll, all charges are posted as next day charges. |
| Room discrepancy | The situation whereby actual room status shown by housekeeping department looks different from front office department. |
| Voucher | Supportive document used to record original transactions and later or support in verifying the posted transactions. |
| Electronic posting machine | ? |
| Audit work time | Time period (usually 11 p.m. to 7 a.m.) in which audit takes place. |
| Night auditing | Process of verifying, recording, correcting, posting, summarizing and balancing all accounts. |

| Ledger | Collection of accounts. |
|---|---|
| House account | Management of accounts in which transaction of complementary guests are entered. |
| Guest account | Account of resident guest whereas the non-guest account indicates to the account of non-resident guest. |
| In-balance | It means guest's or non guest's account balance should match or tally with departmental source documentation. In simple terms, it means that accounts are correct. |
| Out balance | It means guest's or non-guest's accounting balance is not equivalent to supportive documents due to some error/mistake which may be either in guest's or non-guest's account or in supportive documents. |
| Transactional documentations | ? |
| Auditing posting formula | Previous balance + Debits – Credits = Net Outstanding Balance. |
| Daily transcript | Detailed report of all guest accounts in which each charge transaction is shown that affects a guest account during the revenue day. |
| Supplemental transcript | Detailed report of all non-guest accounts in which each charge transaction is shown that affects a non-guest account during the revenue day. |
| Recapitulation sheet | Detailed report that contains both data of daily transcript and supplemental transcript. It is prepared to provide a summary of daily front office activities. |

# 31 CHAPTER

# Coordination of Front Office with Other Departments

$\overline{\text{OBJECTIVES}}$

*After reading this chapter, students will able to...*

- define the value of coordination among hotel departments;
- understand the basic work profile of various departments;
- recognize the role of supportive departments; and
- identify the various sections of front office departments.

## 31.1 COORDINATION AND COOPERATION

*Coordination* is an orderly arrangement of efforts to integrate effectively energies of different groups to provide unity of action, whereas *cooperation* reflects the collective voluntary efforts of employees working together in an organization. But the basic purpose of both is same, i.e., to achieve stated organization goal.

In hotels, front office is the hub to propagate information all guest related transactions, to relevant departments and sections. So often it acts as a centre for collection and circulation of guest information. Such information may help other departments in providing the best service to guests throughout the different stages of the guest cycle.

A hotel has many departments which may be classified as operative and administrative departments. The activities of all these departments are closely interlinked with each other and if this link is broken down at any point, the whole revenue is affected. Hence, the coordination is of utmost importance among all these operative and administrative departments for the smooth, efficient and concurrently profitable running of an organization.

The basic objective of management is to assign work, organize people & resources to achieve the stated organizational goals. It is very difficult to achieve the stated goal without proper cooperation & coordination of the staff. Remember, an employee of every level of management plays an equivalent role in managing this coordination & cooperation. Coordination is not only important among employees but also among all interrelated as well as intrarelated departments and sections.

## 31.2 FRONT OFFICE COORDINATION WITH INTERRELATED DEPARTMENTS

The front office has to maintain good relations with other major departments of hotel such as housekeeping, food & beverage and food production. This coordination is important because often the work of one department depends on other departments. If there is no proper coordination among the departments, the hotel can suffer from huge revenue and guest losses. Front office is the centre of information in a hotel; therefore front desk usually acts as a centre for collection and distribution of guest information.

| Points of Coordination | | |
|---|---|---|
| **Department** | **Front office provides information to** | **Front office receives information from** |
| Housekeeping | - Movement list (to plan staff rosters and room cleaning schedules). <br> - Receive guest's special request (urgent room cleaning request) and complaints. | - Get up-to-date room status information (like occupied, vacant & ready, OOO, etc.) <br> - Receive housekeeper's room report to cross-check room discrepancy. |
| Food & Beverage | - Occupancy forecast (to estimate provisional requirements in point of sale outlets). <br> - Movement list (arrival & departure list to control guest charge privilege). | - Restaurant booking (on the request of guest, concierge gives table reservation into restaurant). <br> - Bill transfers/deferred payments (unpaid guest bill transfers into respective guest account). |
| Food Production | - Room count (idea about how many rooms are going to check-in and check-out). <br> - House count (it aids in planning mise-en-place for meal cooking and preparations). | - Provide information about menu dishes and their ingredients (so front office can provide to guest). <br> - Provide confirmation if special requests can be fulfilled or not (on guest's request via front office). |

## 31.3 COORDINATION WITH SUPPORTIVE DEPARTMENTS

It is concerned with all those departments that do not directly generate revenue but support in generating revenue. Therefore, often these departments are also known as indirect revenue generating departments such as human resource, engineering & maintenance, sales & marketing, account & administration, purchase & store and security. Apart from these departments, there are also several small departments which are usually not popular as others (yet play a key role) like information technology, premises management, control office and so forth. Points of coordination is given below.

| Department | Front Office Provides Information | Front Office Receives Information |
|---|---|---|
| HRM | - Manpower requirements and desired qualification, experience and speciality.<br>- Front office staffs' training needs, problems, concerns and performances. | - Information on all human resource policies towards employees.<br>- Information about career opportunities and training (on-job or off-job). |
| E&M | - Repair, maintenance and replacement information.<br>- Time frame for breakdown, preventive and routine maintenance work. | - Require confirmation that guest's requested repair has been solved.<br>- Require up-to-date information of OOO rooms to update room status board. |
| S&M | - Room position status, so that S&M take steps to sell vacant rooms.<br>- Guest data (for marketing strategy towards key market segments). | - Information about special promotional efforts (special rates and inclusion).<br>- Information about marketing campaign (to anticipate increased demand). |
| A&A | - Paid out information with relevant records in order to clarify cash bank.<br>- Guest billing (like city ledger) information for credit control and revenue reporting. | - Requires clear policy, procedure and authorization for handling & recording transactions.<br>- Requires list of credit approved guests and company. |
| P&S | - Purchase requirement (in purchase order) to purchase department (via store).<br>- Give requisition/indent sheet to store in order to get required things. | - Purchased requisitioned items from purchase department (via store).<br>- Requisitioned items from store. |
| Security | - Reports of security breach and alert about suspicious person and activities.<br>- Reports about lost property and takes assistance when security deals with first aid. | - Warnings (to evacuate premises) and incidents reports (for forecasting).<br>- Assistance in opening of interconnected rooms, guest safes and guest room door locks. |

## 31.4  COORDINATION WITH INTRARELATED SECTIONS

It is concerned with coordination and cooperation among various subdivisions of hotel front office department. There are a number of routine reports and notifications that need to flow from various subdivisions of front office departments. For example, list of current house-count (or guest index), movement list (list of expected arrivals and expected departures along with stay amendments), list of reservation amendments & cancellation, list of overstaying guests, list of guest room amendments (like room change), list of departed guests (in order to timely up-date room status board), and so forth. Points of coordination among different front office subdivisions are given below.

| FO Sub-Divisions | Provision (By Front Office Sub-Divisions) | Requirements (From Front Office Sub-Divisions) |
|---|---|---|
| Front desk | - Prepare ANS & DNS; thereafter send to relevant sections and departments, simultaneously up-date room status board. Provide room change information. | - Movement list (to verify the booking of arriving guest and to keep up-to-date room status statistics). Receive VPOs (from concierge or bell desk) to enter in respective guest account. |
| Bell desk | - Provide information about scanty baggage, guest room change and collected room keys from checked out guest. | - Movement list (to facilitate up-bell and down-bell activity to guests) and room change request of guest via front desk. |
| Cashier | - Open guest account, thereafter provide information to various POS outlets regarding charge privilege or assigned house-limit to various guests. | - Guest reservation, registration and room change information (to open account, charge correct room rate, allow discount, handle advance deposit & PIA, VPOs, deferred accounts, etc.) |
| Concierge | - Arrange movie tickets or fulfil guest's similar requirements, thereafter raise VPO and submit to cashier. | - Movement list and current house-count list (to give table booking in restaurant, arrange movie tickets and so forth assistance) |
| Reservation | - Prepare movement list to give to various relevant departments and subdivisions, perform reservation amendments and pre-registration of guest-with-reservation. | - Information about repeat guest (in order to offer a personalized service), list of newly black listed guests, current room status position, revised room rate structure and discount policy. |

| Guest relation | - Inspect rooms for VIPs and perform their registration in their rooms, thereafter submit VIP's GRC to front desk. | - Movement list (in order to offer hospitality services to VIP, regular and group guest). |
|---|---|---|
| Night auditor | - Balance all section's & guests' accounts, reconcile room discrepancy, thereafter prepare various reports, summary sheets and submit to management. | - Receive all accounts along with supportive documents (to verify their correctness, reconcile & balance them and to prepare various reports for management review). |
| Telephone | - Prepare *traffic sheet* and provide guest call charges information to cashier (so that cashier can transfer charges into respective guest account). | - Current house-count, checked-in and checked-out guest list (in order to switch *on* or *off* telephone lines, handle incoming calls and to charge correct guest room). |
| Travel desk | - Provide vehicle charges (those arranged for sightseeing of guest) to cashier, so he can timely transfer to respective guest account. | - Current house-count list (to plan itinerary and to arrange vehicle for sightseeing, zoo visit and so forth). |
| Mail & message | - Provide received mail information to front desk (so that when guest comes back, it can intimate or send mail at guest's given address). | - Movement list (to sort out incoming mails and obtain message for correct guest). |

## Activity

Fill in the right side column with the correct hotel department(s)/subdivisions/sections/ staffs involved so as to understand the role of coordination in smooth running of hotel organization.

| S. No. | Questions | Department/Sub-Division/Section/ Staff Involved |
|---|---|---|
| 1 | Mr. Water Melon calls at front desk and requests to place wake-up call early in the morning. Front desk instructs......... to place a wake-up call. | |

| 2 | Guest intimates to front desk that he wants an early check-out. Now front desk needs to inform................so that all charges can be timely posted into respective guest account. | |
| 3 | Night auditor prepares a room discrepancy report and hands over to FOM. Now front office needs to check with................so that loss of future room revenue due to wrong room status can be prevented. | |
| 4 | Front office provides house-count and room-count on the basis of which................departments/sections prepare their routine task/mise en place. | |
| 5 | As room boy enters a room, he finds guest lying unconscious on his bed. Room boy should immediately inform to........................ | |

## CONCLUSION

Coordination is the base of good productivity as well as sound control system. With the help of coordination hotel can deliver its products and services in more satisfactory way. Coordination is not important among individual departments but also among subdivisions of these departments. A lack of coordination may prove as a cause of revenue loss and guest dissatisfaction. For example, if housekeeping department does not provide timely information that room has been cleaned then front office will not be able to sell it. Similarly, if front office does not intimate to housekeeping and other relevant departments and subdivisions about the guest check-in then how will all these subdivisions offer their supplies/amenities/expendables/meals/beverage and other services to the guest? Consequently, the situation will lead towards guest dissatisfaction which would reflect loss of customer and revenue.

| Terms (with Chapter Exercise) | |
|---|---|
| Warning | Verbal or written message that a behaviour or environment is unsafe, bad or wrong. |
| ? | Harmonization of efforts for the achievement of an organizational goal/objective. |
| Deferred payment | It refers to diversion of payment transaction towards the third party by the guest. |

| Supplies | It refers to all those necessary items that guest requires as a part of the hotel stay like toilet tissues, hangers and so forth. For hotels, it can be classified as room supplies and bathroom supplies. |
|---|---|
| Expendables | ? |
| Amenities | It refers to the desirable or useful feature or facility of a hotel or department. |
| Traffic sheet | ? |
| Itinerary | Travel document that records a route of journey with expected time of arrival & departure. |
| Occupancy forecast | Reservation status of any specific date or duration. Simply, it is room sale forecasting. |
| Guest Index | ? |
| Indent sheet | It is an internal control document used to receive the commodity/article from the storeroom. |
| Requisition sheet | It is a control document prepared by any of the departments in an attempt to purchase/acquire anything from outside the organization. |
| House count | Number of guests staying in the hotel. |
| ? | Number of rooms occupied during a given time/date. |
| VPO | Situation whereby hotel makes payment on behalf of resident guest or management for rendering any external services. In turn, to charge that guest or settle this transaction, hotel raises VPO voucher. |

# Appendices

## Express Check-Out Form

Date 21/01/18

Dear Sir/Madam,

Tomorrow is your scheduled check-out date. If you have changed your stay plan, please do contact with our Front Desk. Enclosed is your bill as of today. Tonight's room charge and any other late charges are not included in it. Please review and call the Front desk, if you have any question and if you wish to split the invoice, please let us know now to save your time at check-out.

### Our Check-Out Time is 12 Noon, Late Charges May Apply

Thank you for staying in our hotel. We look forward to serving you again, for reservation or information please contact us www.hotelsabc.com

Note. - Please do not enclose cash with this ECO form.

**Front Side of Express-Check Form**

## Express Check-Out Form

Departure Date/Day 21/01/18 (Monday)           Room No. 104

Name of Guest     Mr. Water Melon

Signature

> If you would like to have your bill mailed to a different location, please give the mail forwarding address below.
> Name      Mr. Water Melon
> Address   IHM-Campus, Airport Road, Gwalior
> Contact No.   09415151515

**Back Side of Express-Check Form**

## CURRENCY DECLARATION FORM (CDF)
(Foreign Exchange Regulation Act, 1973)

**Instructions for passengers:**

1) This form need not be completed in cases where the aggregate value of the foreign exchange brought in by the passenge in the form of currency notes, bank notes, or travellers cheques does not exceed U.S.$ 10,000/- or its equivalent and/o the value of foreign currency notes does not exceed U.S.$ 5,000 or its equivalent.

2) Passengers are advised to produce this form to a bank authorized to deal in foreign exchange or money changer at th time of conversion of foreign exchange into Indian rupees or reconversion of rupees into foreign exchange.

3) Visitors to India may please note that in case they do not wish to encash all the foreign exchange declared above the should retain this form with them for production to the Customs at the time of their departure from India to enable ther to take with them the unutilized balance.

4) Details of travellers' cheques/currency notes need not be furnished.

5) Foreign tourists need not indicate their address.

### To be completed by passenger

I .......... Mr. Water Melon .......... hereby, declares that the following foreign exchange is in my possession at the time of m arrival.

(Aggregate value only)

| S. No. | Name of the currency | Currency Notes | Travelers Cheques | Total |
|--------|---------------------|----------------|-------------------|-------|
| 01 | American Dollar | 10 Dollar Each Note | - | 1000 Dollars |
| 02 | American Dollar | 100 Dollar Each Note | - | 100 Dollars |
| | | | | **Total - 1100 Dollars** |

Signature .......................... Passport No . .......... USBD 2345628XY ..........

Nationality .......... American .......... Address in India .......... Hotel Intercontinental, New Delhi

### To be completed by Customs Officer

This is to certify that the above named person has brought with him foreign exchange as indicated above.

Date .......... 21/01/18 (Monday) ..........

(Stamp and Signature of Customs Officer)

### Space for Endorsement

| Date | Distinctive Number of Encashment Certificate | Amount changed | Stamp and Signature of Bank or Money Changer |
|------|---------------------------------------------|----------------|---------------------------------------------|
| 21/01/18 | ENC-21345BDA | 500 Dollars | |
| | | | |

# Encashment Certificate

Serial No. .......... ENC-12345 ........              Date   21/01/18 (Monday)

We hereby certify that we have purchased today foreign currency from .......... Mr. Water Melon .......... holder of Passport No. ..... USBC12345678BVCG .....

Nationality ..... American .......... and paid rupee equivalent as per details given below:

A. Details of Foreign Currency Notes / Coins / Travellers Cheques Purchased

| Currency Purchased (indicating clearly notes/coins &travellers cheques separately) | Amount | Rate | Rupee Equivalent | Stamp & Sign. of authorized dealer/ full-fledged money changer |
|---|---|---|---|---|
| 1 | 2 | 3 | 4 | 5 |
| Notes | 1000 Dollars | Rs. 62/- | Rs. 62000/- | |
| Coins | | | | |
| Travellers Cheque | 500 Dollars | Rs. 62/- | Rs. 31,000/- | |
| Total* | 1500 Dollars | | Rs.91,000/- | |

B. Details of Payments made in Rupees towards hotel Bills/Cost of Goods/Services

Opening Balance Rs...........

| Sr. No | Name and full address of hotel/company/ firm | No. and date of Bill | Date of payment | Amount of Bill (in rupees) | Balance left after deducting *total amount. | Stamp and Signature of authorised money changer with date and No. of RBI Licence. |
|---|---|---|---|---|---|---|
| 01 | Hotel Intercontinental, New Delhi | Bill no. 1001, dated 21/01/18 | 21/01/18 (Monday) | Rs. 90,000/- | Rs. 1000/- | |
| | | | | | | |

**Note.** This certificate should be carefully preserved to facilitate reconversion of the rupee balance, if any, into foreign currency at the time of departure from India and/or for payment in rupees of hotel bills, passage fare or freight on personal baggage, if necessary

# Guest Weekly Bill

| | | | | |
|---|---|---|---|---|
| GRC No | HSBA12345GK | Bill No. | BN 2021 | |
| Guest Name | Mr. Water Melon | Departure Date | 23/01/18 | |
| Room No. | STD Room 104 | Departure Time | 14:25 | |
| Arrival Time | 13:15 | Meal Plan | CP | |
| Arrival Date | 21/01/18 (Monday) | Room Rate | Rs. 25,000/- p.n | |
| | | No. of Pax | 02 | |

| Particulars | Day 1 | Day 2 | Day 3 | Day 4 | Day 5 | Day 6 | Day 7 | Total |
|---|---|---|---|---|---|---|---|---|
| Opening Balance | - | 14,880 | 17,238 | 19,226 | | | | |
| Dr. Entries | | | | | | | | |
| Room Rate | 25,000 | 25,000 | 25,000 | 25,000 | | | | 1,00,000 |
| Extra bed charges | 7,500 | 7,500 | 7,500 | 7,500 | | | | 30,000 |
| Breakfast | 1,000 | | 1,200 | 1,250 | | | | 3,450 |
| Lunch | | 1,500 | 1,500 | 1,500 | | | | 4,500 |
| Dinner | 1,400 | 1,500 | | 1,200 | | | | 4,100 |
| Coffee/Tea | | | | | | | | |
| Room Service | 1,200 | | 1,800 | 1,000 | | | | 4,000 |
| Restaurant | | 1,700 | | 1,500 | | | | 3,200 |
| Coffee Shop | | 1,300 | | 1,100 | | | | 2,400 |
| Bar | 1,500 | | 1,600 | | | | | 3,100 |
| News Stand | | | | | | | | |
| Telephone | | 500 | 350 | | | | | 8,50 |
| Laundry | 1,200 | | | 1,250 | | | | 2,450 |
| VPOs | | 1,200 | | | | | | 1,200 |
| Others | 2,000 | 1,500 | 2,200 | 2,000 | | | | 7,700 |
| CGST (5%) | 2,040 | 2,829 | 2,919 | 3,126 | | | | 10,914 |
| SGST (5%) | 2,040 | 2,829 | 2,919 | 3,126 | | | | 10,914 |
| Total | 44,880 | 62,238 | 64,226 | 68,778 | | | | |
| | | | | | | | | |
| Cr. Entries | | | | | | | | |
| Advance Deposit | 30,000 | | | | | | | 30,000 |
| PIA | | | | | | | | |
| Allowance/Discount | | | | | | | | |
| Interim Payment | | 45,000 | 45,000 | 60,000 | | | | 1,50,000 |
| Total | 30,000 | 45,000 | 45,000 | 60,000 | | | | |
| | | | | | | | | |
| | | | | | | | | |
| Closing Balance | 14,880 | 17,238 | 19,226 | 8,778 | | | | |

# Visitor Tabular Ledger

| Particulars | 101 | 102 | 103 | 104 | 105 | 106 | 107 | Total |
|---|---|---|---|---|---|---|---|---|
| GRC No. | HBF10987 | HBF10967 | HBF10988 | HBF109654 | HBF109777 | | | |
| Arrival Date/Time | 01/01/18 (13:15) | 01/01/18 (14:00) | 01/01/18 (13:10) | 01/01/18 (15:00) | 25/12/17 (11:50) | | | |
| Expected Date/Time | 03/01/18 (11:45) | 03/01/18 (11:00) | 04/01/18 (13:45) | 03/01/18 (14:00) | 03/01/18 (15:00) | | | |
| No. of Pax | 02 | 02 | 03 | 03 | 03 | | | |
| Meal Plan | CP | CP | EP | MAP | BP | | | |
| | | | | | | | | |
| **Dr. Entries** | | | | | | | | |
| Opening Balance | - | - | - | - | 19,226 | | | |
| Room Rate | 25,000 | 25,000 | 20,000 | 35,000 | 25,000 | | | 130,000 |
| Extra bed charges | | | 7,500 | 8,500 | 7,500 | | | 23,500 |
| Breakfast | | | 1,200 | 1,250 | 1,250 | | | 3,700 |
| Lunch | | 1,500 | 1,500 | | 1,500 | | | 4,500 |
| Dinner | 1,400 | 1,500 | | 1,200 | 1,200 | | | 5,300 |
| Coffee/Tea | | | | | | | | |
| Room Service | 1,200 | | 1,800 | 1,000 | 1,000 | | | 5,000 |
| Restaurant | | 1,700 | | 1,500 | 1,500 | | | 4,700 |
| Coffee Shop | | 1,300 | | 1,100 | 1,100 | | | 3,500 |
| Bar | 1,500 | | 1,600 | | | | | 3,100 |
| News Stand | | | | | | | | |
| Telephone | | 500 | 350 | | | | | 8,50 |
| Laundry | 1,200 | | | 1,250 | 1,250 | | | 3,700 |
| VPOs | | 1,200 | | | | | | 1,200 |
| Others | 2,000 | 1,500 | 2,200 | 2,000 | 2,000 | | | 9,700 |
| SGST (5%) | 1,615 | 1,710 | 1,808 | 2,640 | 3,126 | | | 10,899 |
| CGST (5%) | 1,615 | 1,710 | 1,807 | 2,640 | 3,126 | | | 10,899 |
| | | | | | | | | |
| **TOTAL** | 35,530 | 37,620 | 39,765 | 58,080 | 68,778 | | | 239,773 |
| | | | | | | | | |
| **Cr. Entries** | | | | | | | | |
| Advance Deposit | 30,000 | | | | | | | 30,000 |
| PIA | | | | | | | | |
| Allowance/Discount | | | | | | | | |
| Interim Payment | | 45,000 | 45,000 | 60,000 | 60,000 | | | 1,50,000 |
| Total | 30,000 | 45,000 | 45,000 | 60,000 | 60,000 | | | 240, 000 |
| | | | | | | | | |
| **Closing Balance** | 5,530 | 7,380 | 5,235 | -1,920 | 8,778 | | | |

# High Balance Notification

Room No. 104                                                    Date 21/01/18

Dear Sir/Madam,

We would like to inform you that your expenses are Rs. 75000 till 21/01/18. This amount is excess of the level of hotel's credit limit.

Please contact our duty manager or front office manager to establish how you wish to settle your account.

Yours Truly,

Front Office Manager

---

# Guest Arrival Errand Card

Bellboy: Mr. Moon Walk                                Room No. 104
Name of Guest: Mr. Jonhy Walker                       Date 21/01/18

## Luggage Description

| Suitcase | Briefcase | Handbag | Packets | Other | Remark |
|----------|-----------|---------|---------|-------|--------|
| 01 | 01 | - | - | 01 | |
| | | | | | |

Sig. of Bellboy                                        Sig. of Bell Captain

---

# Luggage Tag

Room No. 104                                           Tag No. 009
Name of Guest – Mr. Jonhy Walker                       No. of Luggage- 04

Remarks-          03 bags and 01 suitcase transferred into taxi

Handled By

## Lobby Control Sheet

Date 21/01/18

Sr. No. 3123

Bell Captain: Mr. Blue Wale

Shift: Morning Shift

| Bell Boy No. | Room No. | Check-In | Check-Out | Outside Call | Room Change | Time | | Remarks |
|---|---|---|---|---|---|---|---|---|
| | | | | | | In | Out | |
| B.B.01 | 104 | ✓ | | | | 11:25 | 11:35 | |
| B.B.02 | 106 | | | | ✓ | 14:15 | 14:30 | |

## Visitors Paid Out

Serial No. 32001

Date 21/01/18

I acknowledge receipt of Rs. 2000/- (Rupees Two Thousand Only) for the payment made on my behalf towards Flower Bouquetagainst Bill No.2314 dated 21/01/18.

Guest Registration No.   HBF10987
Guest Room No.   101

Guest Signature

## Allowance / Discount Voucher

Serial No. 3145

Date- 21/01/18

Please credit the account of Mr. / Mrs. Water Melon
Guest Registration No. HBF10987 for Rs. 1000/- (Rupees One Thousand Only) towards allowance.

Explanation /Remarks Compensation in reference to In-Room Meal Service in Guest Room No. 104

Authorized Sig.

Cashier/F. O. Assist. Signature

# Account Ageing/Receivables

For the Month January 2018                                   Sr. No. 3123

| Name of Guest/Company | Balance Amount | Due Days | | | Remarks |
| --- | --- | --- | --- | --- | --- |
| | | Upto 30 Days | 60 Days | 90 Days | |
| Mr. Water Melon | Rs. 1,25,000 | Rs. 105,000 | Rs. 20,000 | - | |
| Comp. Blue Moon | Rs. 2,50,000 | Rs. 1,75,000 | Rs. 50,000 | Rs. 25,000 | |
| Ms. Pumpkin | Rs. 1,05,000 | Rs. 75,000 | Rs. 25,000 | Rs. 5,000 | |
| Mr. Blue Mango | Rs. 1,00,000 | | | *Rs. 100,000 | Bad Debt |
| **Total** | **Rs. 5,80,000** | **Rs. 3,55,000** | **Rs. 95,000** | **Rs. 1,30,000** | |

Note-* This amount has not received since last 90 days thus, hotel made it bad debt and going to split then write-off in upcoming five years.

# XYZ Travel Services

Serial No.      3147
Reference no.      CBA 241
Date-      21/01/18

Hotel Name: White Lemon Intercontinental Hotel
Accommodation Reserved: For Standard Suite Room           Centre- Gwalior 16 Pargana

Arriving on:     21/01/18 (Monday)     commencing with     Afternoon lunch
Departing on:     23/01/18 (Wednesday)     andterminating with     Afternoon Tea

Additional Facilities:           Free All Indoor Games and Mountain Trekking

Name of Guest:        Mr. Yellow Banana

# Guest Registration Card

Serial No.    AC- 3147
Regd. No.    CBA 241

Date    21/01/18 (Monday)

| | |
|---|---|
| Name of Guest    Mr. Yellow Banana | Room No.    104 |
| No. of Pax    03 Pax | Room Type    Standard Suite Room |
| Nationality    Indian | Room Rate    Rs. 25000 per day |
| Arriv. Date/Time 21/01/18 (Monday) | Payment details- |
| Exp. Date of Dep 23/01/18 (Wednesday) | Cash    ☐ |
| Company    Blueberry International | Credit Card    ▧ |
| Address    Airport Road, Gwalior | Others    ☐ |
| Remarks | If credit card then which Visa |

Sig. of Front Desk Agent                    Sig, of Guest

**Note**. The check-out time is 12 noon. The room rates are exclusive of all taxes and services charges. No early-departures are allowed. Late departures are subject to the approval of management.

# Movement List

For the Date/Day of- 21/01/2018 (Monday)        Sr. No. 3123

| Expected Arrivals | | | | | | Expected Departures | | | | | |
|---|---|---|---|---|---|---|---|---|---|---|---|
| Room No. | Type | Guest Name | No. of Pax | ETA | Stay Period | Room No. | Type | Guest Name | No. of Pax | ETD | Remark |
| 101 | STD | Mr. White | 02 | 11:20 | 02 days | 103 | STD | Ms. Pink | 02 | 12:00 | |
| 102 | STD | Mr. Blue | 01 | 11:00 | 03 days | 104 | STD | Mr. Purple | 02 | 11:50 | |
| 105 | STD | Mr. Lemon | 02 | 12:00 | 02 days | 202 | SUT | Mr. Black | 02 | 11:30 | |
| 201 | SUT | Mr. Orange | 02 | 13:00 | 03 days | | | | | | |
| | | | | | | | | | | | |

Room Count: 105 rooms
House Count: 375 guests

# Guest History Card

Sr. No. 3123

Name of Guest    Mr. Water Melon

Nationality        Indian

Address          Airport Road, Gwalior,

Date/Day of     21/01/2018 (Monday)

| Room No. | DOA | DOD | No. of Pax | Room Type | Room Rate | Total Bill | Mode of Payment | Remarks |
|---|---|---|---|---|---|---|---|---|
| 101 | 21/01/18 | 23/01/18 | 02 Pax | Suite Room | Rs. 25,000 per night | Rs. 1,05,000 | Credit Card | Visa Card |
| | | | | | | | | |

Guest Likes      .................................................................................................

Guest Dislikes   .................................................................................................

Sig. of Front Desk Agent

# Arrival Notification Slip

Sr. No. 3123

Date/Day of     21/01/2018 (Monday)

| Room No. | Guest Name | Time of Arrival | No. of Pax | Room Type | Room Rate | Expected date of Departure | Remarks |
|---|---|---|---|---|---|---|---|
| 101 | Mr. Water Melon | 14:30 | 02 Pax | Suite Room | Rs. 25,000 per night | 23/01/18 | HWC |
| | | | | | | | |

Sig. of Front Desk Agent

# Room Change Slip

Serial No.                AC- 3147

Date/Day        21/01/18 (Monday)
Time            14:30

Name of Guest        Mr. Yellow Banana
No. of Pax           03 Pax

From Room No.        104          From Room Type        Standard Room
To Room No.          108          To Room Type          Standard Room
Change in RR    No ☐    Yes ☐     No.. of Room Nights   02 nights

Remarks

Sig. of Front Desk Agent

# Wake Up Call Sheet

Sr. No. 3123

Date/Day of        21/01/2018 (Monday)

| Room No. | Guest Name | Time of Call | Morning Tea | Newspaper | Door Knob Menu | Remarks |
|----------|------------|--------------|-------------|-----------|----------------|---------|
| 101 | Mr. Water Melon | 05:30 a.m. | Tea for 02 Pax | Times of India | Yes (Continental Breakfast) | HWC |
| 108 | Mr. White Orange | 04:30 a.m. | Thé Complete | Times of India | No | - |
| 201 | Ms. Pink Lady | 04:00 a.m. | - | Hindustan Times | No | - |

Sig. of Agent

# Reservation Form

Sr. No. 3123

Date/Day     21/01/2018 (Monday)

| | | | |
|---|---|---|---|
| Guest Name | Mr. Water Melon | Contact No. | 9415151515 |
| Arrival Date | 21/01/18 (Monday) | E. Time of Arrival | 11:45 |
| Departure Date | 23/01/18 (Wednesday) | E. Time of Departure | 14:25 |
| No. of Pax | 02 Pax | No. of Rooms/Type | 01 Suite Room |

**Room Rate**

☐ Rack Rate    ☐ Corporate Rate    ☑ Consortia Rate    ☐ Package Rate    ☐ Courtesy Rate

**Transportation**

☐ Airport to Hotel      ☐ Hotel to Airport      ☑ Both      ☐ None

**Billing Instruction**

☑ Guest account     ☐ Room on Company     ☐ All Expenses on Company     ☐ Others

**Guaranteed By**

Company

Advance Cash

Credit Card     Maestro Card, Rs. 20,000 advance deposit

Confirmation     ☑ Yes     ☐ No     Remarks

Approved By     Mr. Red Lion     Taken By     White Elephant

Date     21/01/18 (Monday)

Sig. of Reservation Agen

# Reservation Confirmation Slip

Sr. No. 3123

Name of Guest  Mr. Water Melon                    R. Confirmation No.    RSVN 122012234

Address        IHM-Gwalior, Airport Road, Gwalior

| Arrival Date and Time | Departure Date | Room Type | Room Rate | No. of Pax | Room Preference | Special Instruction |
|---|---|---|---|---|---|---|
| 21/01/18 (Monday) at 12:30 noon | 23/01/18 (Wednesday) | Standard Suite | Rs. 25, 000 per night | 02 | Room on Ground floor | Cotton Pillows in my room |

Remarks ......................................................................................................

Method of Payment    ☐ Company Billing    ☐ Cash    ☐ Credit Card    ☐ Others

Sig. of Reservation Agent

# Reservation Cancellation/Amendment Form

Serial No.        AC- 3147

Amendment       ☐                  Confirmation No. ........... RSVN No. 2345678
Cancellation    ☐                  Cancellation No. ........... CANCNo. 3456124

Name of Guest ........... Mr. Yellow Banana
No. of Pax    ........... 03 Pax
EDA          ........... 21/01/18 (Monday)          EDD       ........... 23/01/18 (Wednesday)
Room Required ........... 02 Standard Rooms          Room Rate ........... Rs. 25,000/- per night

Remarks ..............................................................................................

Date ........... 20/12/17 (Monday)

Sig. of Front Desk Agent

# Night Report

Sr. No. 3123

| | | | | | |
|---|---|---|---|---|---|
| Date & Day | 21/01/18 (Monday) | | | | |
| Total Lettable Rooms | 125 Rooms | Room Count | 65 Rooms | Indians | 75 Pax |
| Total Revenue | Rs. 15, 75, 250/- | House Count | 125 Guests | Foreigners | 50 Pax |

| Room Description | | | Revenue Break-Up | |
|---|---|---|---|---|
| Single Rooms  35 Rooms | Reserved Guest | 60 Guests | Room Revenue | Rs. 12,25, 500/- |
| Twin Rooms  25 Rooms | Walk-Ins | 05 Guests | Revenue from Other Outlets- | |
| Double Rooms  05 Rooms | Walking Guest | Nil | - Restaurant | Rs. 1,75,550/- |
| | No. Shows | 03 Guests | - Bar | Rs. 1,25,500/- |
| Room Count  65 Rooms | Cancellation | 02 Rooms | - Laundry | Rs. 42, 500/- |
| | | | - Others | Rs. 06,200/- |
| Rooms Sold  65 Rooms | | | Revenue Details | |
| Rooms Vacant  50 Rooms | House Count | 125 Guests | - Cash | Rs. 02,50,000/- |
| OOO Rooms  05 Rooms | | | - Credit Card | Rs. 12,50,000/- |
| Compl. Rooms  05 Rooms | | | - Travl. Check | Rs. 50,000/- |
| | | | - Others | Rs. 25,250/- |

Sig. of Agent

# Room Discrepancy Report

Sr. No. 3123

Date & Day    21/01/18 (Monday)

| Room No. | FO Status | HK Status | Investigation Remarks |
|---|---|---|---|
| 101 | Occupied | Vacant | Room is Occupied But Guest Sleep Out |
| 205 | Vacant | Occupied | Room is Vacant, HK's Status is By-Mistake |

Remarks

Sig. of Night Auditor                                         Sig. of Assistant Manager

# Rooming List

Date 21/01/18 (Monday) ..........     Sr. No. 001 ......

Name of Leader Mr. Water Melon ........................     Source of Booking     E-mail

| Guest Name | Room Preference | Room Type | Designation | Nationality | Passport No. | Duration of Stay | Remarks |
|---|---|---|---|---|---|---|---|
| Mr. X | Non-smoking Room | Single | Liaison Officer | Chinese | G45889246 | 04 May to 10 May 2018 | |
| Mr. Y | Smoking Room | Double | Project Manager | Australian | E79568933 | 05 May to 09 May 2018 | HWC |
| Mr. Z. | Smoking Room | Suite | Vice President | American | 107225422 | 04 May to 08 May 2018 | VVIP |

Sig. of Group Leader

# Lost and Found Register

Sr. No. 3123 ....

| Date & Time | Room No. | Article Description | Name of Address of Guest | Finder's Detail | Acknowledgement |
|---|---|---|---|---|---|
| 21/01/18 (Monday) At 14:00 | 101 | A wristwatch of HMT company (brown coloured leather band) | Mr. Water Melon, IHM-Campus, Airport Road, Gwalior | Found by a Room Attendant (Mr. Black Jack) during afternoon shift. | |
| | | | | | |

# Uniform Register

Page No. 15 ......
Sr. No. 3123 ....

| Date | Laundry No. | Uniform Received at...... | Uniform Cleared at...... | Remarks |
|---|---|---|---|---|
| 21/01/18 (Monday) | LAU-2125 | 13:30 | 13:35 | Mending work needed at cuff link. |
| | | | | |

Uniform Room Supervisor

| | | | |
|---|---|---|---|
| • Foundation of Lodging Management | Hayes and Ninemeier | | Pearson Education |
| • Crisis Management and Communication | Dr. J. Smith and Ms. Robin Denney | | Journal |
| • Crisis Management | Gerry Nutt | | Middleton High School |
| • Mass Communication in India | Keval J Kumar | | Jaico Publishing House |
| • Communicating Today | Raymond Zeuschner | | Allyn and Bacon |
| • Communication | Barker/Gaut | | Allyn and Bacon |
| • Thinking Through Communication | Sarah Trenholm | Second | Allyn & Bacon |
| • Communication Mosaics | Julia T Wood | Second | Wadsworth |
| • Information Technology and Tourism | Werthner, Hannes, | 1999 | Wien, New York (Springer) |
| A Challenging Relationship | Klein, Stefan | | |
| • Hotel Lobby Design – Study of Parameters of Attractions | T. Dhiraj | | Thesis cum Project Work |
| • Study on Hotel Classification | The Joint WTO & IH&RA | | Report |
| • Front Office Management | S. K. Bhatnagar | 2003 | Frank Brothers |
| • Introduction to Hospitality Industry | Gerald W. Lattin | Fourth | AH&LA |
| • Tourism Planning & Development | J. K. Sharma | First | Kanishka Publication |
| • Cases in Hospitality Marketing & Management | Robert C. Lewis | Second | John Wiley & Sons. |
| • Hotel Front Office | Bruce Braham | Series | Virtue & Company Ltd. |
| • The Hotel Receptionist | Grace & Jane Paige | Second | Holt, Rinehart & Winston |
| • Managing Accounting for Hospitality & Tourism | Richard Kotas | Third | Thomson & Learning |

| | | | |
|---|---|---|---|
| • Managerial Accounting for Hotels | G. R. Kulkarni | 2003 | Ridhiraj Enterprise |
| • Leadership & Management in the Hospitality Industry | Robert H. Woods & Judy Z. King | Second | AH&LA |
| • The Hotel Book-Keeper Receptionist | M. Dunseath & J. G. Ransom | Second | Barrie & Rockliff (London) |
| • The Business of Hotels | S. Medlik | Third | Butterworth Heinemann |
| • Computer Systems in the Hotel & Catering Industry | Braham B. | 1988 | Cassell (London) |
| • Hotel, Hostel & Hospital Housekeeping | Branson J. C. & Lennox M. | Third | Hodder & Stoughton |
| • Professional Management of Housekeeping Operations | Thomas J.A. Jones | Fourth | John Wiley & Sons Inc. |
| • Managing Housekeeping Operations | A. A. Nitschek, and William D. | Third | AHLA |
| • Managerial Accounting in the Hotel & Catering Industry | Harris P & Hazzard P. | Third | Stanley Thornes |
| • Managing Hotels | Venison P. | Second | Oxford |
| • Hotel Reception | White P. B. & Beckley H. | Second | Hodder & Stoughton |
| • Hotels of the Future | Horwarth & Horwarth | 1988 | IHA (Paris) |
| • Accommodation Management | Hurst R. | | Butterworth Heinemann |
| • Front of House Operation | Hotel & Catering Comp. | | Macmillan (London) |
| • Worldwide Hotel Industry | Horwarth International | | HI – New York |
| • Travel, Tourism & Hospitality | Robert H. & Joy Howard | 2001 | Hospitality Press Pty. Australia |
| • Manpower Management in the Hotel & Catering Industry | Hornsey T. & Dann D. | 1984 | Batsford – London |

Cleaning methods 326
Cleaning process 311
Cleaning schedule 311
Cloakroom 274
Colour 410
Colour wheel 411
Combined account 110
Combined linen and uniform room 272
Commissionaire 48
Competitor based room rates 74
Complaint handling process 216
Concierge 45
Concierge desk 181
Condominium hotels 11
Corporeal attribute 22
Corridor, cleaning of 355
Cotton 387
Credit card payment 115
Credit limit, establishment of 223
Customer based rates 74

**D**
Decentralized kitchen system 18
Deodorants 325
Desk control room 270
Detergents 318
Dining room linen 290
Disinfectants 324
Door attendant 48
Doorman 48
Drunken guests 209
Dry cleaning 383
Duty manager 46

**E**
Early check-in 205
Ecotel hotels (nature-friendly hotel) 12
Effective selling tools 459
Electronic key card 449
Emulsion paints 413
Enamel paints 413

Engineering and maintenance department 19
Executive housekeeper's cabin/office 270
Exercise room, cleaning of 357
External threats 443
Extra charges 208

**F**
Fabric 386, 393
Fabric softener 375
Fibre 386
Fibre flooring 421
Fire threat 441
Five-star hotel, front office department of 41
Fixtures and fittings, cleaning 362
Floatel hotels 6
Floor decoration (rangoli) 405
Floor pantries 273
Floor polish 331
Flooring 417-418
Flower arrangement 403
Fluorescent lighting 407
Food and beverage department 18
Food production 18
Front desk agent 43
Front desk supervisor 42
Front office cashier 46
Front office department 17, 24, 33
    equipment 36
    functions of 34
    layout of 35
    organizational structure 33
    sections 37
Front office manager 42
Front office staff, duties and responsibilities of 42
Front office, coordination with interrelated departments 471
Full service hotel 13
Fully-automated check-out 123
Functional areas, cleaning of 357
Furniture polish 330

**G**

Game zone 185

Gardener 286

Garni hotels 7

Group guest check-in 92
    procedure 95

Group guest departure 173
    procedure 97, 121

Group guests reservation 146

Group guests registration 225

Group market segment 54

Guest 52

Guest accounts 108

Guest arrival, task force during 255

Guest check-in procedure 93

Guest check-in, preparation for 91

Guest check-out procedure 96, 118, 124, 171

Guest complaints 215

Guest cycle 240

Guest cycle activities 250

Guest departure, task force during 257

Guest from different targeted market
    segments, registration of 225

Guest luggage handling 168
    during check-out 169

Guest mail 191

Guest registration 220

Guest registration process, documents for 226
    methods of 227, 232

Guest relation desk 179
    functions of 179

Guest reservation, preparation for 145

Guest room amenities and supplies 266

Guest room cleaning 336

Guest room linen 290

Guest segmentation parameters 53, 54, 58

Guest status, identification of 222

Guest's credibility 224

Guests & guest rooms, security of 439

**H**

Hard flooring 419

Head gardener 286

Heavy equipment store 273

Heritage hotels 11

Highway hotels 6

Holistic cleaning 343

Horticulture section 285

Horticulture/flower room 275

Hostel 7

Hotel 3
    classification of 6
    departments of 28
    organizational structure of 16
    supportive departments of 29

Hotel assets, security of 440

Hotel brochure 459

Hotel cash, security of 440

Hotel entrance, cleaning of 353

Hotel lighting 406

Hotel lobby, cleaning of 354

Hotel mail 191

House account 110

Housekeeping attendants/runners 284

Housekeeping coordinator 284

Housekeeping department 18, 25
    functions of 264
    layout of 269
    organizational structure 278
    structural foundation of 269

Housekeeping equipment 264
    care and cleaning of 360

Housekeeping supervisors 283

Hubbart formula 78

**I**

Incandescent lighting 407

Incidental account 108

Incoming mails 192

Independent hotels 9

In-house areas 270

In-house laundry, advantages and
    disadvantages of 369

G

Game zone 288
Gardener 266
Garni hotels 7
Group guest check-in 92
  procedure 93
Group guest departure 23
  procedure 92, 127
Group guests reservation 149
Group guests registration 225
Group market segment 54
Guest 62
Guest accounts 198
Guest arrival, task force during 253
Guest check-in procedure 93
Guest check-in, preparation for 91
Guest check-out procedure 96, 118, 124, 161
Guest complaint 215
Guest cycle 240
Guest cycle activities 250
Guest departure, task force during 253
Guest from different targeted market
  segments, registration of 225
Guest luggage handling 168
  during check-out 168
Guest mail 191
Guest registration 220
Guest registration process, documents for 226
  methods of 227, 232
Guest relation desk 129
  functions of 129
Guest reservation, preparation for 145
Guest room amenities and supplies 266
Guest room cleaning 336
Guest room linen 290
Guest segmentation parameters 53, 54, 58
Guest status, identification of 222
Guest's credibility 224
Guests & guestrooms, security of 139

H

Hard flooring 419

Head quarters 288
Heavy equipment store 22
Heritage hotels 11
Highway hotels 6
Holistic cleaning 343
Horticulture section 283
Horticulture/flower room 276
Hostel 7
Hotel 5
  classification of 6
  departments of 25
  organizational structure of 16
  supportive departments of 29
Hotel assets, security of 140
Hotel brochure 459
Hotel cash, security of 140
Hotel premises, cleaning of 354
Hotel lighting 400
Hotel lobby, cleaning of 351
Hotel mail 191
House account 110
Housekeeping attendants, reports 284
Housekeeping coordinator 284
Housekeeping department 16, 25
  functions of 264
  layout of 265
  organizational structure 278
  structural foundation of 269
Housekeeping equipment 264
  care and cleaning of 304
Housekeeping supervisors 283
Hubbart formula 78

I

Incandescent lighting 407
Incidental account 108
Incoming mails 192
Independent hotels 9
In-house area 220
in-house laundry, advantages and
  disadvantages of 309

Scanty baggage 198
Security department 20
Semi-automated check-out 123
Semi-automatic system of reservation 154
Semi-residential hotels 9
Silk 388
Situation handling 195
Skippers 197
Small economy hotel, organizational
    chart for 279
Small hotel, organizational structure of 39
SMERF 55
Soap and emulsifier 324
Solid waste 433
SPAS 14, 184
Special rates 82
Staff uniform 307
Stain removal 396
    by chemical methods 397
Standard rooms 64
Standard weaves 394
Starch 374
Stayover 201
Store department 20
Suite rooms 65
Sundry things, care and cleaning of 364
Supplementary accommodation 15
Supply store 273
Support centre 177
Support departments, coordination
    with 71
Swimming pool 275, 439
    cleaning of 356
Synthetic fibre 389

**T**
Tailor's/sewing room 271
Telephone operator 45
Telephone protocol 135
Telephone section 131
    functions of 132

Textile 394
Texture paints 414
Timeshare hotels 10
Toilet and window cleanser 320
Tourism industry, role and importance of 4
    components of 5
Transient guest check-in 91
Transient guest check-in procedure 93
Transient guest departure procedure 96, 120
Transient guest reservation, preparation
    for 145
Transient hotel 7
Transient market segment 56
Travel agents voucher 116
Traveller's cheque/check 113
Triacetate rayon 393
Turndown service 347

**U**
Under-stay 206
Uniform room 271, 291
Uniformed service 178
Uniformed service department 26
Unusual situation handling 210

**V**
Vegetable fibre 387
VIP rooms, items and supplies in 348
Viscose rayon 392

**W**
Walking guests 199
Walking-out guest 200
Wall covering 417
Waste management 433
Whitener 374
Wood rot, control of 432
Wooden flooring 420
Wool 388
Woven fabric 393